Encyclopaedia of Mathematical Sciences

Volume 50

Editor-in-Chief: R. V. Gamkrelidze

Springer
Berlin
Heidelberg
New York
Barcelona
Budapest
Hong Kong
London
Milan
Paris
Santa Clara
Singapore
Tokyo

A.V. Arhangel'skii (Ed.)

General Topology II

Compactness,
Homologies of General Spaces

Springer

Consulting Editors of the Series:
A. A. Agrachev, A. A. Gonchar, E. F. Mishchenko,
N. M. Ostianu, V. P. Sakharova, A. B. Zhishchenko

Title of the Russian edition:
Itogi nauki i tekhniki, Sovremennye problemy matematiki,
Fundamental'nye napravleniya, Vol. 50
Obshchaya Topologiya-2
Publisher VINITI, Moscow 1989

Cataloging-in-Publication Data applied for

Die Deutsche Bibliothek - CIP-Einheitsaufnahme

General topology. - Berlin ; Heidelberg ; New York ; London ;
Paris ; Tokyo ; Hong Kong ; Barcelona : Springer.
  Einheitssacht.: Obščaja topologija <engl.>
  Literaturangaben
NE: EST
  2. Compactness, homologies of general spaces / A. V.
  Arhangel'skii (ed.). - 1995
    (Encyclopaedia of mathematical sciences ; Vol. 50)
    ISBN 3-540-54695-2 (Berlin ...)
    ISBN 0-387-54695-2 (New York ...)
NE: Archangel'skij, Aleksandr V.; GT

Mathematics Subject Classification (1991):54DXX, 55NXX

ISBN 3-540-54695-2 Springer-Verlag Berlin Heidelberg New York
ISBN 0-387-54695-2 Springer-Verlag New York Berlin Heidelberg

Typesetting: Camera-ready copy from the translator using a Springer $\mathrm{T_E X}$ macro package
SPIN: 10012902   41/3142 - 5 4 3 2 1 0 - Printed on acid-free paper

List of Editors, Authors and Translators

Editor-in-Chief

R.V. Gamkrelidze, Russian Academy of Sciences, Steklov Mathematical Institute, ul. Vavilova 42, 117966 Moscow, Institute for Scientific Information (VINITI), ul. Usievicha 20a, 125219 Moscow, Russia, e-mail: gam@ips.eps.msk.su

Consulting Editor

A.V. Arhangel'skii, Department of Mathematics, University of Moscow, Leninskie Gory, 119899 Moscow, Russia, and Department of Mathematics, Ohio University, 321 Morton Hall, Athens, Ohio 45701-2979, USA, e-mail: aarhange@oucsace.cs.ohiou.edu

Authors

A.V. Arhangel'skii, Department of Mathematics, University of Moscow, Leninskie Gory, 119899 Moscow, Russia, and Department of Mathematics, Ohio University, 321 Morton Hall, Athens, Ohio 45701-2979, USA, e-mail: aarhange@oucsace.cs.ohiou.edu
E. G. Sklyarenko, Department of Mathematics, University of Moscow, Leninskie Gory, 119899 Moscow, Russia

Translator

J. M. Lysko, College of Arts and Science, Science Division, Widener University, One University Place, Chester, PA 19013, USA, e-mail: lysko@cs.widener.edu

Contents

I. Compactness

A.V. Arhangel'skii

Translated from the Russian
by Janusz M. Lysko

Contents

 A.V. Arhangel'skii

Introduction

Compactness is related to a number of fundamental concepts of mathematics. Particularly important are compact Hausdorff spaces or compacta. Compactness appeared in mathematics for the first time as one of the main topological properties of an interval, a square, a sphere and any closed, bounded subset of a finite dimensional Euclidean space. Once it was realized that precisely this property was responsible for a series of fundamental facts related to those sets such as boundedness and uniform continuity of continuous functions defined on them, compactness was given an abstract definition in the language of general topology reaching far beyond the class of metric spaces. This immensely extended the realm of application of this concept (including in particular, function spaces of quite general nature). The fact, that general topology provided an adequate language for a description of the concept of compactness and secured a natural medium for its harmonious development is a major credit to this area of mathematics. The final formulation of a general definition of compactness and the creation of the foundations of the theory of compact topological spaces are due to P.S. Aleksandrov and Urysohn (see Aleksandrov and Urysohn (1971)).

In this work we will consider the main topological aspects of the theory of compact spaces, as well as certain fundamental results and principles of this theory associated with applications, and we will indicate briefly some of these applications. In particular we will discuss such results associated with compactness, as the theorems of Stone-Weierstrass, Krein-Milman, Alaoglu, Gel'fand-Kolmogorov. These theorems play a key role in the theory of function spaces.

We will also shed some light on the fundamental role of compactness in topological algebra. We will consider in this survey the Stone spaces of Boolean algebras, character theory and Pontryagin duality theory for compact groups as well as questions of the topological structure of compact groups (such as dyadicity).

The notation and terminology in this work are essentially the same as in Engelking (1977) and Juhász (1980). By a space we mean a topological space; a priori no separation axioms are assumed.

§1. Compactness and Its Different Forms: Separation Axioms

1.1. Different Definitions of Compactness. A space X is called *compact* if every open cover of X contains a finite subcover. A compact Hausdorff space will be called a *compactum*. We shall see that the Hausdorff separation axiom has a great impact on the properties of compact spaces.

A set F contained in X is called *compact* if F as a subspace of X is compact. A finite subset of any space is compact, while a discrete space is compact if and only if it is finite. The first, very narrow (but very important) class of compact spaces is described by the following result.

Theorem 1. *A subset F of the Euclidean space $\mathbb{R}^n$ is compact if and only if it is closed and bounded in $\mathbb{R}^n$.*

An intuitive view of compactness is the following: compact spaces are those in which infinite sets are "crowded" and are forced to "accumulate". Certain subtle points arise in the mathematical formulation of this concept: notice that the definition stated above would appear to have nothing to do with this intuitive view. We call a point x in a space X a point of *complete accumulation of a set $A \subset X$* if the intersection of every neighborhood O_x of x with A has the same cardinality as the set A.

Theorem 2 (see Arhangel'skii and Ponomarev (1974)). *A space X is compact if and only if every infinite set $A \subset X$ has a complete accumulation point in X.*

The proof of this theorem is based on the theory of ordered sets. The "easy" part is based on the following fact: the cardinality of a union of a finite number of infinite sets is equal to the cardinality of the "largest" set.

The following is a convenient, dual characterization of compactness: a space X is compact if and only if every family of closed subsets of X with the finite intersection property has a non-empty intersection. The proof is obvious. The intuitive idea of compact spaces as "absolutely" complete spaces, that is, spaces containing all the points that are "potentially" close to them, is associated with this characterization. This idea is reflected in the following results. If a space X is Hausdorff and if a point $x \in X$ is not isolated, then the system of the closures of neighborhoods of this point in the space $Y = X - \{x\}$ determines a family of closed subsets of Y with the finite intersection property that has the empty intersection. Thus we have proved the following.

Theorem 3. *A compact subset of a Hausdorff space is a closed subset of this space.*

Example 1. Let X be an infinite set and let x^* be a point in X. Let us define the open subsets of X to be the empty set, the set X itself and all subsets of X that contain x^* and have a finite complement in X. In the resulting (compact) space $(X, \mathcal{T}_A)$ the compact set $\{x^*\}$ is not only non-closed, but in addition its closure is the whole X. We observe, however, that X is not a T_1-space, see Engelking (1977).

Example 2. On an infinite set X we define the topology $\mathcal{T}_1$ consisting of the empty set and the complements of all finite subsets of X. Then $(X, \mathcal{T}_1)$ is a compact T_1-space, each of whose infinite proper subsets, is compact and non-closed because the closure of every infinite set in $(X, \mathcal{T}_1)$ coincides with X.

A Hausdorff space that is closed in every Hausdorff space containing it is called *H-closed.* Not every *H*-closed space is compact, but we have the following theorem of P.S. Aleksandrov and Urysohn.

Theorem 4 (Aleksandrov and Urysohn (1971)). *A regular T_1-space is H-closed if and only if it is compact.*

The above results allow us to interpret compactness as absolute closedness. A proof of Theorem 4 could be obtained from the following result (see Aleksandrov and Urysohn (1971)).

Theorem 5. *A Hausdorff space X is H-closed if and only if every open cover of X contains a finite subfamily of open sets $U_1, \ldots, U_k$ whose closures cover X.*

An important feature of compactness is its preservation under many operations. In particular, any closed subspace of a compact space is compact (Theorem 3 shows that the requirement of being closed is essential). By contrast, the property of being *H*-closed is not hereditary with respect to closed subspaces. We have the following theorem of M. Stone and M.G. Katetov, see Engelking (1977), Katetov (1940).

Theorem 6. *A Hausdorff space X is compact if and only if each closed subspace of X is H-closed.*

1.2. Relative Compactness. One of the important properties associated with compactness is relative compactness. Frequently it happens that infinite subsets of a subspace Y of a space X accumulate at points of X not necessarily belonging to Y. In such situations it is natural to introduce the concept of relative compactness of Y in X. The simplest solution would be to define relative compactness of Y in X as compactness of the closure $\overline{Y}$. However, the presence in $\overline{Y}$ of limit points of infinite subsets of Y is not sufficient to claim that such points will be present for infinite sets contained in the "remainder" $\overline{Y} \setminus Y$. We proceed therefore in the following way.

A subset Y of the space X is called *(relatively) compact* in X if for each family of subsets of Y with the finite intersection property the intersection of the closures in X of its elements is nonempty. The following is an equivalent condition: every open cover of X contains a finite subfamily covering Y. There exists a Hausdorff space X and a subset Y of X such that Y is compact in X but Y is not compact.

Example 3. Take any compactum $(X, \mathcal{T})$ containing an infinite closed subset B. Put $Y = X \setminus B$ and denote by $\mathcal{E}_B$ the family of all subsets of the form $\{x\} \cup (Y \cap U)$, where $U \in \mathcal{T}$ and $x \in U$. The family $\mathcal{B} = \mathcal{E}_B \cup \mathcal{T}$ is a base of some topology $\mathcal{T}_B$ on X that is stronger than $\mathcal{T}$. The space $(X, \mathcal{T}_B)$ is Hausdorff and each subset of Y has the same closure in $(X, \mathcal{T}_B)$ as in $(X, \mathcal{T})$. Therefore Y is compact in $(X, \mathcal{T}_B)$. However B is an infinite closed discrete subspace of the space $(X, \mathcal{T}_B)$ and therefore $(X, \mathcal{T}_B)$ is not compact.

For a broad class of spaces the question of relative compactness is solved by the following result (see Arhangel'skii and Ponomarev (1974)).

Theorem 7. *If X is a regular T_1-space and Y is a subset of X that is compact in X, then $\overline{Y}$ is compact.*

Theorem 7 implies that the space $(X, \mathcal{T}_B)$ in Example 3 is not regular.

1.3. Countable Compactness. Countable compactness is a form of compactness which, at a first glance, seems more elementary than compactness. This impression is associated with the fact that countable compactness is closer to intuition; its definition involves only countable sets and countable families of sets and does not depend on more delicate concepts such as a point of complete accumulation or a family of subsets with the finite intersection property. Because of this, countable compactness appeared before compactness. This concept, however, proved to be less successful. But as one of the facets of topology and as one of the elements of topological methodology, countable compactness is of considerable importance. The question of the mutual relationship (equivalence) between countable compactness and compactness is of paramount importance in many situations and constructions (Grothendieck's theorem on closures of countably compact sets in function spaces may serve as an example; see §7 for more details).

A space X is called *countably compact* if for every infinite set $A \subset X$ there exists a point $x \in X$ every neighborhood of which contains infinitely many points of A (such points are called *limit points* or $\aleph_0$-*accumulation points* of A). The following is an equivalent condition: every countable cover of X contains a finite subcover. Accordingly, a space is compact if and only if it is both countably compact and Lindelöf (see Engelking (1977)). Recall that a space is called *Lindelöf* if each open cover of this space contains a countable subcover. The last condition is responsible for the fact that previously compactness was called *bicompactness*.

It is also possible to define countable compactness by the following condition: the intersection of any descending sequence $\{F_n : n \in \mathbb{N}^+\}$ of nonempty closed subsets is nonempty.

One should not regard countable compactness merely as an underdeveloped form of compactness. In mathematical considerations in which constructions are essentially of a countable nature (as is often the case in the application of general topology, for example, to functional analysis), arguments involving ordinary converging sequences and, accordingly, the concept of countable compactness are preferred. The theory of Eberlein compact spaces (and other spaces closely related to them: the Corson, Gul'ko, Talagrand and Radon-Nikodým compact spaces) serves as an example.

Example 4. The space $T(\omega_1)$ of ordinals less than the first uncountable ordinal ω_1 equipped with the order topology is not compact since it is impossible to select a finite subcover from the open cover of $T(\omega_1)$ by the sets $T_\alpha = \{\beta < \omega_1 : \beta \leq \alpha\}$, where $\alpha \in T(\omega_1)$. Since every countable set A contained in $T(\omega_1)$ has the least upper bound in $T(\omega_1)$ and all spaces T_α are compact, the space $T(\omega_1)$ is countably compact.

For metrizable spaces countable compactness is equivalent to compactness. The reason is that in metrizable spaces the topology is determined by the behavior of their countable subsets. There exist first-countable, countably compact but non-compact spaces. The space $T(\omega_1)$ serves as an example of such space. The equivalence of countable compactness and compactness is guaranteed by the presence of certain, even quite weak, properties related to paracompactness. One of the first results of this type is: every metacompact countably compact space is compact (see Engelking (1977)). The assumption of metacompactness in this statement can be substantially weakened by replacing it, for example, by weak $\delta\theta$-refinability (see Bula and Turzański (1986)). We will discuss one of the most general results in this direction.

A countable collection $\mathcal{E} = \{\gamma_n : n \in \mathbb{N}^+\}$ of families of subsets of a space X is called an *interlacing* on X if together these families cover X and for every $n \in \mathbb{N}^+$, every $V \in \gamma_n$ is open in $\bigcup \gamma_n$. A space X is called *pure* (see Arhangel'skii (1980b)) if for every maximal family ξ of subsets of X with the finite intersection property that has the empty intersection, there exists an interlacing $\mathcal{E} = \{\gamma_n : n \in \mathbb{N}^+\}$ on X that is δ-separated from ξ in the following sense: for every $n \in \mathbb{N}^+$ and for every $x \in \bigcup \gamma_n$ there is a countable subfamily $\eta \subset \xi$ such that $\mathrm{St}_{\gamma_n}(x) \cap (\bigcap \eta) = \emptyset$ (where $\mathrm{St}_{\gamma_n}(x) = \bigcup \{U \in \gamma_n : x \in U\}$). We have the following result (Arhangel'skii (1980b)).

Theorem 8. *Every countably compact pure space is compact.*

The class of pure spaces contains metaLindelöf spaces, subparacompact spaces, spaces complete in the sense of Hewitt, spaces with a G_δ-diagonal and weakly $\delta\theta$-refinable spaces (see Arhangel'skii (1980b), Vaughan (1984)). Theorem 8 allows us to conclude that in the presence of any of these properties countable compactness is equivalent to compactness.

We will consider in detail two of the most important types of countably compact spaces. A space X is called ω-*bounded* if the closure of every countable subset of X is compact; an example is the space $T(\omega_1)$. A space X is called *sequentially compact* if every sequence $\{x_n : n \in \mathbb{N}^+\}$ in X contains a convergent subsequence. All countably compact subspaces of ordered spaces are sequentially compact. Both ω-bounded and sequentially compact spaces are countably compact. But not every compactum is sequentially compact! Every countably compact sequential (for definition see Engelking (1977)) space is sequentially compact. Later we will encounter situations in which these types of countable compactness behave quite differently.

In the class of first-countable (and even sequential) spaces countable compactness is equivalent to sequential compactness. A sufficient condition for a countably compact Tikhonov space X to be sequentially compact is that every nonempty closed subspace Y of X contains a point y at which Y is first-countable.

Let us mention that countable compactness, ω-boundedness and sequential compactness are hereditary with respect to closed subspaces.

1.4. Relative Countable Compactness. It is natural to say that a subset Y of a space X is (relatively) *countably compact* in X if for every infinite set $A \subset Y$ there exists a point in X each of whose neighborhoods contains infinitely many points of A. This property behaves more capriciously than relative compactness. Even in a Tikhonov space the closure of a relatively countably compact set does not have to be countably compact (compare with Theorem 7).

Example 5. A family of sets is called *quasi-disjoint* if the intersection of any two different elements of it is finite. Consider a maximal, quasi-disjoint family $\{S_\alpha : \alpha \in A\}$ of infinite subsets of the set $\mathbb{N}$ of natural numbers. We may assume that $A \cap \mathbb{N} = \emptyset$. Define a topology on the set $X = A \bigcup \mathbb{N}$ in the following way: all points of $\mathbb{N}$ are isolated in X, and an arbitrary basic neighborhood of any point $\alpha \in A$ consists of the point α and all but a finite number of points of S_α. The space X is Tikhonov, the set $\mathbb{N}$ is both countably compact and dense in X, but the space X itself is not countably compact since A is an infinite closed discrete subspace of it. In addition the space X is a locally compact locally metrizable Moore space. The question concerning the conditions under which the closure of a countably compact subset of X is a countably compact subspace of X is of fundamental importance, in particular, in functional analysis (see Burke and Davis (1981), Lindenstrauss (1972)) (Grothendieck's theorem, Eberlein compact spaces, etc.).

In connection with Example 5 the notion of a *countably pracompact* (or *weakly countably compact*) space arises naturally as a space X containing a dense subspace that is countably compact in X. Most of the results concerning countably compact spaces are also true in the class of countably pracompact spaces (see, in particular, Arhangel'skii (1984a)). The space constructed in Example 5 is countably pracompact but not countably compact. Since $\mathbb{N}$ is countable, every infinite subset of $\mathbb{N}$ has a complete accumulation point in X. Nevertheless, by Theorem 7 in §2, the set $\mathbb{N}$ is not compact in X. Therefore Theorem 2 in §1 cannot be generalized to the case of relative compactness.

1.5. Pseudocompact Spaces. One of the most important properties (particularly from the point of view of applications) of compact and countably compact spaces is the following: every continuous, real-valued function defined on such a space is bounded and attains its maximum and minimum. Completely regular spaces with this property are called *pseudocompact*. The next theorem provides a simple internal characterization of pseudocompactness (see Engelking (1977)).

Theorem 9. *A Tikhonov space X is pseudocompact if and only if every discrete family of nonempty open subsets of X is finite.*

As with countable compactness, pseudocompactness is equivalent to compactness in the presence of several properties related to paracompactness and metrizability. The following theorem is due to Scott and Watson (see Scott (1979), Watson (1985)).

Theorem 10. *Every metacompact pseudocompact space is compact.*

The following two facts play a crucial role in the proof of Theorem 10.

Theorem 11. *Every pseudocompact space X has the Baire property, namely the intersection of a countable family of dense open subsets of X is dense in X.*

Proposition 1. *For every open, point-finite cover of a pseudocompact space X the set of points at which this cover is locally finite is a dense and open subset of X.*

Pseudocompact metaLindelöf spaces (in contrast to countably compact, metaLindelöf spaces) do not have to be compact (D.B. Shakhmatov, Watson, see Shakhmatov (1984)). If a pseudocompact space admits a one-to-one continuous map onto a metrizable space then it is also metrizable and, by Theorem 10, compact (N.V. Velichko, see Arhangel'skii (1984a)).

Pseudocompactness and countable compactness are perceived initially as very similar properties. For example, for normal T_1-spaces pseudocompactness and countable compactness are equivalent. It is true, however, that pseudocompactness is by far a more general concept in the class of Tikhonov spaces and that in a number of constructions it behaves quite differently from compactness and countable compactness. The key difference between these concepts is the fact that pseudocompactness is not hereditary with respect to closed subspaces. The following two statements show, to what extent the principle of "heredity" is violated (see Engelking (1977)).

Proposition 2. *A Tikhonov space is countably compact if and only if each of its closed subspaces is pseudocompact.*

Theorem 12 (Noble (1969)). *Every Tikhonov space can be embedded as a closed subspace in a pseudocompact Tikhonov space.*

It would not be an overstatement to say that Theorem 12 highlights the fundamental difference between countable compactness and pseudocompactness. This theorem leads to a class of problems that can be given the following general formulation: for which classes $\mathcal{P}$ of topological spaces is it true that every space from class $\mathcal{P}$ can be embedded as a closed subspace of some pseudocompact space belonging to $\mathcal{P}$? Some results in this direction will be presented later. See also papers of Bell (1980), Matveev (1984), Shakhmatov (1984).

Still another property of pseudocompactness that sharply distinguishes it from compactness and countable compactness is illustrated by

Proposition 3. *If a subspace Y of a Tikhonov space X is pseudocompact, then every subspace Z such that $Y \subset Z \subset \overline{Y}$ is also pseudocompact.*

The connection between pseudocompactness and relative compactness is expressed in the following result.

Proposition 4. *If a subspace Y of a Tikhonov space X is countably compact in X, then the space $\overline{Y}$ is pseudocompact.*

Moreover, a space is pseudocompact if it contains a dense, countably compact (in itself) subspace.

Corollary 1. *Every countably pracompact Tikhonov space is pseudocompact.*

In particular, the space X from Example 5 is pseudocompact. In addition, X is a locally compact Moore space. Consequently, not every pseudocompact Moore space is metrizable; therefore, not every pseudocompact space with a G_δ diagonal is metrizable (see Arhangel'skii and Ponomarev (1974)).

Not every pseudocompact space is countably precompact (see Shakhmatov (1984)).

1.6. Separation Axioms and Properties Related to Compactness. In the presence of compactness the relationship between the separation axioms simplifies significantly. We have the following result (see Engelking (1977)).

Theorem 13. *Every compactum is normal.*

Moreover, every compactum is a Tikhonov space. This fact is particularly important since being Tikhonov, in contrast to being normal, is hereditary with respect to arbitrary subspaces. This implies that every subspace of a compactum is a Tikhonov space.

Theorem 13 does not extend to countably compact Hausdorff spaces. Such spaces do not have to be normal and they do not even have to be regular.

Theorem 14 (Vaughan (1984)). *Every countably compact, first-countable Hausdorff space is regular.*

It is impossible to strengthen the above result to complete regularity. The following theorem is quite useful.

Theorem 15. *Every locally compact Hausdorff space is Tikhonov.*

There exists a locally compact countably compact Hausdorff space that is not normal (see Engelking (1977)). However, if a locally compact Hausdorff space is normal, then according to the most recent results (see Balogh (1987)), it is consistent with ZFC to assume that such space is collectionwise normal. On the other hand collectionwise normality of a locally compact Hausdorff space does not imply either paracompactness or metacompactness (for exam ple consider the space $T(\omega_1)$ of all countable ordinals). However, we have the following result.

Theorem 16. *Every locally compact, paracompact Hausdorff space is strongly paracompact.*

◁ The space X can be represented as a direct sum of σ-compact (and therefore, Lindelöf) spaces (see Engelking (1977)). ▷

1.7. Star Characterizations of Countable Compactness and Pseudocompactness. Let γ be a family of sets in a space X and let $A \subset X$. Recall that the set $\mathrm{St}_\gamma(A) = \bigcup\{U \in \gamma : U \cap A \neq \emptyset\}$ is called the *star* of A with respect to γ (see Engelking (1977)). By recursion with respect to $n \in \mathbb{N}^+$ we define the $(n+1)$st star of A with respect to γ as $\mathrm{St}_\gamma^{n+1}(A) = \mathrm{St}_\gamma(\mathrm{St}_\gamma^n(A))$. In particular the sets $\mathrm{St}_\gamma^2(A)$ and $\mathrm{St}_\gamma^3(A)$ will be called respectively the *double* and the *triple* stars of A. In place of $\mathrm{St}_\gamma^n(\{x\})$ we will write $\mathrm{St}_\gamma^n(x)$.

Every cover γ of a space X generates coarser covers of this space, namely the families $\mathrm{St}^n(\gamma) = \{\mathrm{St}_\gamma^n(x) : x \in X\}$, where $n \in \mathbb{N}^+$. In addition, it is convenient to assume that $\mathrm{St}^0(\gamma) = \gamma$. If γ is an open cover, then so are all the $\mathrm{St}^n(\gamma)$.

A space X is called *n-pseudocompact*, where $n \in \mathbb{N}^+$, if for every open cover γ the cover $\mathrm{St}^n(\gamma)$ of X contains a finite subcover. If for every open cover γ of X there is an $n \in \mathbb{N}^+$ such that $\mathrm{St}^n(\gamma)$ contains a finite subcover of X, then X is called *∞-pseudocompact*.

It is clear that 0-pseudocompactness coincides with compactness. The following theorem was proved by Houston (see Engelking (1977)).

Theorem 17. *A regular T_1-space is countably compact if and only if it is 1-pseudocompact.*

The theory of n-pseudocompact spaces was developed by Matveev (see Matveev (1984)). He obtained the following, unexpected result.

Theorem 18. *For every Tikhonov space X the following conditions are equivalent: (a) X is pseudocompact, (b) X is 3-pseudocompact, (c) X is n-pseudocompact for some n, (d) X is ∞-pseudocompact.*

In accordance with the above theorem, there are only four types of n-pseudocompactness: 0-pseudocompactness (compactness), 1-pseudocompactness (countable compactness), 2-pseudocompactness and 3-pseudocompactness (pseudocompactness). Matveev obtained the following interesting result about the relationship between countable pracompactness and the above system of concepts: every countably pracompact Tikhonov space is 2-pseudocompact. He also constructed a locally compact, 2-pseudocompact Hausdorff space that is not countably pracompact and also gave an example of a pseudocompact but not 2-pseudocompact space. We note that Theorem 12 can be strengthened in the following way: every Tikhonov space Y can be embedded as a closed subspace of a Tikhonov space X containing a dense countably compact subspace. The space X in addition can be taken to be countably precompact and 2-pseudocompact.

§2. Compactness and Products

The operation of the topological product of topological spaces exhibits more clearly than anything else the superiority of "true" compactness over countable compactness, pseudocompactness and other properties of compactness type. Compactness is preserved under formation of the product without any restrictions on the cardinality of the family of factors and their separation properties, while other properties of compactness type can be destroyed already by the operation of the product of two spaces or even by the operation of topological squaring.

2.1. Tikhonov's Theorem on Compactness of the Product. Possibly the most important theorem of general topology and one of the most fundamental theorems of all mathematics is the following theorem of Tikhonov (see Engelking (1977)).

Theorem 1. *The product of any family of compact spaces is compact.*

The Hausdorff separation axiom is also preserved under the operation of the product and therefore the product of any family of compacta is a compactum.

By Theorem 1 the following spaces are compacta: the *generalized Cantor discontinua D^τ*, the products of an arbitrary number of copies of the two point space $D = \{0, 1\}$, and the *Tikhonov cubes I^τ*, the products of an arbitrary number of copies of the interval $I = [0, 1]$. However, Theorem 1 enables one to not only to construct interesting and important examples of compacta but it also provides a complete "constructive" description of the class of all compacta.

Theorem 2 (see Engelking (1977)). *A space is a compactum if and only if it is homeomorphic to a closed subspace of a Tikhonov cube I^τ.*

Of no less importance is the following theorem of Tikhonov that provides a complete description of Tikhonov spaces.

Theorem 3 (see Engelking (1977)). *A space X is Tikhonov if and only if it is homeomorphic to a subspace of some Tikhonov cube I^τ.*

Theorem 3 is of fundamental importance in the theory of Hausdorff compact extensions of Tikhonov spaces (see §6).

By applying Theorem 1 we can also obtain some interesting examples of noncompact topological spaces. This way, by deleting an arbitrary point x from the cube I^τ with $\tau > \aleph_0$, we obtain the "punctured" Tikhonov cube $I^\tau(-)$, which is a locally compact, pseudocompact space that is not countably compact and consequently not normal (in I^τ every point is a limit of a sequence in its complement). A slightly more general example is constructed in the following way.

Example 1. Let A be an uncountable set, $I_\alpha = I = [0,1]$, for all $\alpha \in A$ and let $I^A = \prod\{I_\alpha : \alpha \in A\}$ be the Tikhonov cube. For each $B \subset A$ let us denote by $I^B(0)$ the subspace $\{x \in I^A : x_\alpha = 0$ for all $\alpha \in A \setminus B\}$ of the cube I^A and let us call it a B-face of this cube. Consider a set $B \subset A$ of cardinality smaller than that of A and remove from I^A the face $I^B(0)$. The resulting space Y_B is open and dense in I^A. Moreover, for every $B' \subset A$ of cardinality less than that of A the set Y_B is mapped onto the whole space $I^{B'}$ under the natural projection $\pi_{B'} : I^A \to I^{B'}$ (since $B \cup B' \neq A$). Consequently, the space Y_B is pseudocompact and locally compact.

Consider an arbitrary Tikhonov space X. Using Theorem 3 it is possible to select sets A and B in the space from Example 1 in such way that X is a subspace of the face $I^B(0)$. Put $X_B = Y_B \cup X$. Since $I^B(0)$ is closed in I^A the set X is closed in the space X_B. Clearly $Y_B \subset X_B \subset I^A$ and Y_B is dense in X_B. Consequently, the space X_B is also pseudocompact. This provides the proof of Theorem 12 of §1: every Tikhonov space can be embedded as a closed subspace in a pseudocompact space.

2.2. Products of Countably Compact Spaces. In this and the following sections we assume that all spaces under consideration are Tikhonov, even though this is not always essential.

It is easy to verify that the product of a compact space and a countably compact space is countably compact. However, already the product of two countably compact spaces does not have to be countably compact or even pseudocompact (Novak, Terasaka, see Engelking (1977) and Vaughan (1984)). At present the question whether, within ZFC, it is possible to construct a countably compact topological group, whose square is not countably compact, remains open. Malykhin proved (see Malykhin (1987a)), that the existence of such group is consistent with ZFC. In connection with the above, of particular interest is the question: under what conditions is countable compactness preserved under the formation of a product?

Theorem 5 (see Vaughan (1984)). *The product of any family of ω-bounded spaces is ω-bounded and therefore countably compact.*

Another result of the same nature is the following.

Theorem 6 (see Engelking (1977)). *The product of any countable family of sequentially compact spaces is sequentially compact and therefore countably compact.*

The product of a countably compact space and a countably compact k-space (in particular, a countably compact sequential space) is countably compact (see Engelking (1977)).

In addition, the product of a countably compact space and a sequentially compact space is countably compact (Mrówka, see Engelking (1977)). Frolik constructed an example of a countably compact space X, all finite powers of which are countably compact, while its countable power $X^{\aleph_0}$ is not countably compact and not even pseudocompact (see Vaughan (1984)). The following subtle result was obtained by Scarborough and A. Stone (see Engelking (1977)).

Theorem 7. *The product of a family of cardinality not exceeding $\aleph_1$ of sequentially compact spaces is countably compact.*

Let us point out that not every compactum is countably compact. For example, the Tikhonov cube I^c, where $c = 2^{\aleph_0}$ is not sequentially compact (it contains a topological copy of the *Stone-Čech compactification* βN of natural numbers). The answer to the question of sequential compactness of the spaces $D^{\aleph_1}$ and $I^{\aleph_1}$ is independent of the axioms of set theory (Booth, see Engelking (1977)).

2.3. Products of Pseudocompact Spaces. From the results of the previous section it is clear that the product of two pseudocompact spaces does not have to be pseudocompact. However, the following is true.

Theorem 8 (Tamano, see Engelking (1977)). *The product of a pseudocompact space and a pseudocompact k-space is pseudocompact.*

Corollary 1. *The product of a pseudocompact space and a compactum is pseudocompact.*

Corollary 2. *The product of a pseudocompact space and a pseudocompact sequential space is pseudocompact.*

Theorem 9 (see Engelking (1977)). *The product of a pseudocompact space and a sequentially compact space is pseudocompact.*

The results quoted above are analogous to the results concerning countable compactness. In contrast, the following brilliant and surprising result of Comfort and Ross is specific in nature.

Theorem 10 (see Comfort and Ross (1966)). *The product of an arbitrary family of pseudocompact topological groups is a pseudocompact topological group.*

Every pseudocompact group is homeomorphic to a dense subspace of a Dugundji compactum (see Sect. 7.8). Therefore the space of a pseudocompact topological group is κ-metrizable (see Shchepin (1973)). It is now possible to derive Theorem 10 from the following result of Chigogidze (see Chigogidze (1982)).

Theorem 11. *The product of an arbitrary family of pseudocompact κ-metrizable spaces is pseudocompact.*

Unfortunately, there is no good internal characterization of those countably compact spaces X whose product with any countably compact space is countably compact (see Vaughan (1984)).

2.4. Total Countable Compactness and Total Pseudocompactness. In this section we consider only Hausdorff spaces. Sequentially compact spaces and ω-bounded spaces have certain general properties which we are going to identify now. A space X is called *totally countably compact* if every infinite set $A \subset X$ contains an infinite set B whose closure in X is compact (see Vaughan (1984)). Let us notice that total countable compactness, in contrast to countable compactness and ω-boundedness, does not have a convenient description in the language of open covers. One of the advantages of total countable compactness over sequential compactness is included in the fact that every compact space is totally countably compact. An example of a countably compact Tikhonov space that is not totally countably compact was given in 2.2.

Recall that the *k-leader of the space* $(X, \mathcal{T})$ is defined as the set X equipped with the strongest of all topologies $\mathcal{T}'$ on X inducing on each compact subset of the space $(X, \mathcal{T})$ the same topology as $\mathcal{T}$. With this definition the k-leader of an arbitrary space is a k-space. We have the following result.

Theorem 11 (see Noble (1969)). *A space $(X, \mathcal{T})$ is totally countably compact if and only if its k-leader is countably compact. In particular, every countably compact k-space is countably compact.*

Theorem 12 (see Noble (1969)). *(a) If X is totally countably compact and Y is countably compact then the product $X \times Y$ is countably compact.*

(b) The product of a countable family of totally countably compact spaces is totally countably compact.

(c) The product of a family of cardinality $\leq \aleph_1$ of totally countably compact spaces is countably compact.

The product of a sufficiently large family (for example, of cardinality $> 2^{\aleph_0}$) of totally countably compact spaces does not have to be countably compact (see Vaughan (1984)). It is consistent with ZFC that there exists a family of sequentially compact spaces whose product is not sequentially compact. So far the attempts to construct such family of spaces within the ZFC system have been unsuccessful.

The analogy with total countable compactness allows us to identify an interesting subfamily in the class of pseudocompact spaces. A Tikhonov space X is called *totally pseudocompact* if for each infinite family $\xi = \{U_\alpha : \alpha \in A\}$ of nonempty open subsets of X there exists a compactum $\Phi \subset X$ intersecting an infinite number of elements of ξ (see Noble (1969)):$|\{\alpha \in A : \Phi \cap U_\alpha \neq \emptyset\}| \geq \aleph_0$. The following result was obtained by Noble (see Noble (1969)).

Theorem 13. *(a) The product of a pseudocompact space and a totally pseudocompact space is pseudocompact.*

(b) The product of any family of totally pseudocompact spaces is totally pseudocompact.

The class of all pseudocompact spaces whose product with every pseudocompact space is pseudocompact is also closed with respect to the operation of the product.

2.5. Compactness with Respect to a Fixed Ultrafilter (ξ-Compactness). The importance of the generalization of countable compactness considered in this section is associated with the fact that the Tikhonov's theorem on the compactness of the product fully extends to such spaces.

Denote by ω^* the family of all free ultrafilters on the set ω of all natural numbers. Thus if $\xi \in \omega^*$ then ξ is a maximal family with the finite intersection property of subsets of ω, and $\bigcap \xi = \emptyset$.

Let us fix $\xi \in \omega^*$. A point x of the space X will be called a ξ-*limit* of a sequence $f : \omega \to X$ of points of X if for every neighborhood U of x the set $\{n \in \omega : f(n) \in U\}$ belongs to ξ. A sequence $f : \omega \to X$ is called ξ-*convergent* (in X) if it has a ξ-limit in X.

No two different points of a Hausdorff space can be ξ-limits of the same sequence of points of this space, since an ultrafilter cannot contain disjoint sets. A point of X is a limit point of a sequence $f : \omega \to X$ if and only if it is its ξ-limit for some $\xi \in \omega^*$.

Let $\xi \in \omega^*$. A space X is called ξ-*compact* if every sequence $f : \omega \to X$ is ξ-convergent in X (see Vaughan (1984)). The following is one of the most important results concerning ξ-compactness (see Vaughan (1984)).

Theorem 14. *For every $\xi \in \omega^*$ the product of every family of ξ-compact spaces is ξ-compact.*

Theorem 14 implies that not every countably compact space is ξ-compact for some $\xi \in \omega^*$. As an example we can take any countably compact space Y whose square is not countably compact. The theorem below describes a relationship between ξ-compactness and ω-boundedness.

Theorem 15 (see Vaughan (1984)). *A Tikhonov space X is ω-bounded if and only if it is ξ-compact for every $\xi \in \omega^*$.*

Even though ξ-compactness is preserved by topological products, it is possible to select, for every $\xi \in \omega^*$, a countably compact Tikhonov space X and a ξ-compact Tikhonov space Y such that the product $X \times Y$ is not countably compact (not even pseudocompact) (see Vaughan (1984)).

We now describe a totally countably compact space that is not ξ-compact for any $\xi \in \omega^*$. As a result of this construction we will obtain several interesting conclusions.

Example 1. Let us take the Stone-Čech compactification $\beta\mathbb{N}$ of the set $\mathbb{N}$ of natural numbers (this remarkable compactum is considered in detail in §6). For every $x \in \beta\mathbb{N} \setminus \mathbb{N}$ denote by B_x the subspace $\beta\mathbb{N} \setminus \{x\}$ of $\beta\mathbb{N}$. For every $x \in \mathbb{N}^* = \beta\mathbb{N} \setminus \mathbb{N}$ the subspace B_x is totally countably compact, but not ω-bounded. This follows from the fact that there are no non-trivial convergent sequences in $\beta\mathbb{N}$ (see §6). Since $\bigcap\{B_x : x \in \beta\mathbb{N} \setminus \mathbb{N}\} = \mathbb{N}$, the discrete space $\mathbb{N}$ is homeomorphic to a closed subspace of the product $B = \prod\{B_x : x \in \beta\mathbb{N} \setminus \mathbb{N}\}$, specifically with its "diagonal" consisting of all the points of the form $\{n, n, \ldots, n, \ldots\}$, where $n \in \mathbb{N}$. Consequently, the space B is not countably compact. Consider the direct sum $\sum_{\oplus}\{B_x : x \in \beta\mathbb{N} \setminus \mathbb{N}\}$ of the spaces B_x and attach to it one additional point z^*. On the set $Z = \sum_{\oplus}\{B_x : x \in \beta\mathbb{N}\setminus\mathbb{N}\}\cup\{z^*\}$ introduce a topology in the following way: every B_x is both closed and open in Z, and the base at z^* is formed by complements of finite sums of sets B_x. We obtain a totally countably compact space Z, that is not ξ-compact for any $\xi \in \omega^*$. Indeed, the space $B_\xi = \beta\mathbb{N} \setminus \{\xi\}$ is not ξ-compact (the sequence $\mathbb{N}$ does not have a ξ-limit in B_ξ), but it is a closed sub space of the space Z.

2.6. $\sum$-Products of Compact Spaces. Let $X = \prod\{X_\alpha : \alpha \in A\}$ be the topological product of spaces X_α, τ - an infinite cardinal number, and let $x^* = \{x_\alpha : \alpha \in A\}$ be a fixed point in X. Then by the $\sum_\tau$ *-product* of spaces X_τ *with the basic point* x^* we mean the subspace of the product X, consisting of all the points $x = \{x_\alpha : \alpha \in A\}$, for which the cardinality of the set $\{\alpha \in A : x_\alpha \neq x_\alpha^*\}$ of those coordinates at which the points x and x^* are different, does not exceed τ. This space is denoted by $\sum_\tau \prod\{X_\alpha : \alpha \in A\}$ and the point x^* is specified separately. With $\tau = \aleph_0$ the $\sum_\tau$-product is called the $\sum$-*product* and is denoted by $\sum\prod\{X_\alpha : \alpha \in A\}$.

The operation of the $\sum$-product, by contrast with the topological product, does not preserve compactness. Nevertheless, when applied to compact spaces, this operation produces countably compact (even ω-bounded) spaces. In this way we obtain a broad class of natural examples of countably compact spaces. Among them is the space $\sum\prod I^\tau$ which is the $\sum$-product of τ copies of the interval $I = [0, 1]$ with the basic point $0 = \{0, 0, \ldots, 0, \ldots\}$.

The interest associated with $\sum$-products is due to the fact that, although not preserving compactness, they can preserve other important properties of topological spaces that are lost under the operation of the product. A typical example of a result of such nature is the following theorem of Noble (see Arhangel'skii (1978)).

Theorem 16. *The $\sum$-product of an arbitrary family of first-countable spaces is a Fréchet-Urysohn space.*

In particular, for every $\tau \geq \aleph_0$ the space $\sum\prod I^\tau$ is a countably compact Fréchet-Urysohn space, while already the cube $I^{\aleph_1}$ is not sequential (it even has uncountable tightness). Every subspace of the space $\sum\prod I^\tau$ is also a

Fréchet-Urysohn space and, a fortiori, a k-space. At the same time every Tikhonov space can be embedded in some I^τ for a suitably selected τ.

§3. Continuous Mappings of Compact Spaces

3.1. Theorem on Compactness of the Image and Its Consequences. One of the most fundamental properties of compactness is its invariance under continuous mappings.

Theorem 1. *If a space Y is the continuous image of a compact space X, then Y is compact.*

A similar statement is true for countable compactness, pseudocompactness and for practically all properties of compactness type. In any attempt to describe in an axiomatic way which topological properties should be called properties of compactness type, the axiom of preservation under continuous mappings should be included.

Theorem 1 gains particular power in the class of Hausdorff spaces. If X is a compact space and Y is Hausdorff, then for a continuous mapping $f : X \to Y$ the image of any compact set $\Phi \subset X$ is a compact subset of Y. According to Theorem 3 in §1, the set $f(\Phi)$ is closed in Y. Since every closed subset of X is compact, we arrive at the following conclusion.

Theorem 2. *Every continuous mapping of a compact space X into a Hausdorff space Y is closed, that is, the image $f(F)$ of every closed subset F of X is closed in Y.*

Example 1. Theorem 2 is truly "Hausdorff" in nature: the identity mapping *id* of the interval $[0, 1]$ with the usual topology onto the same interval with the canonical T_1-topology (in which the only closed subsets are finite sets and the whole set $[0, 1]$) is continuous, one-to-one, but not closed: thus the inverse of *id* is not continuous.

Theorems 1 and 2, together with Theorem 1, §1 and Tikhonov's theorem on the compactness of the product are related to a number of fundamental principles associated with compactness. To a great extent precisely these theorems are responsible for the prominent place occupied by compactness in general topology and its applications.

Let us point out several important corollaries to Theorems 2 and 1.

Corollary 1. *Every one-to-one continuous mapping of a compact space onto a Hausdorff space is a homeomorphism.*

◁ Indeed, every continuous closed one-to-one mapping of one space onto another is a homeomorphism. ▷

The following two results are associated with the characterization of compact subsets of finite dimensional Euclidean spaces as bounded closed subsets of E^n (see §1).

Corollary 2. *The image of a compact space under a continuous mapping into the Euclidean space E^n is closed and bounded in E^n.*

Corollary 3. *Every continuous real-valued function on a compact space X is bounded and attains its minimum and maximum.*

The last statement is one of the fundamental principles of mathematical analysis.

The following is another classical result in the theory of compact spaces: every continuous function on a compact space is uniformly continuous. If the topology of a compactum X is generated by a metric ρ then uniform continuity is understood in the traditional way. In the general case it is necessary to define uniform continuity in the sense of uniform topology.

The family $C(X)$ of all real-valued continuous functions on a compactum has several important properties. There exist a number of natural topologies on the ring $C(X)$, among them the topology of uniform convergence and the topology of point-wise convergence (for more information on this subject see §7).

3.2. Continuous Images of "Standard" Compacta. There exists a number of deep results on the representation of compacta as continuous images of compacta with simpler structure or of a single, standard compactum. Theorems of such nature are known in other areas of mathematics and their role in the classification of mathematical objects is clear.

Let us begin with the following classical result (Aleksandrov, see Engelking (1977)).

Theorem 3. *Every non-empty metrizable compactum is a continuous image of the Cantor set (see Engelking (1977)).*

It is known that the Cantor set is homeomorphic to the countable product $D^{\aleph_0}$ of $\aleph_0$ copies of the discrete space $D = \{0,1\}$. The following natural question (P.S. Aleksandrov) arises naturally in connection with Theorem 3: is every compactum of weight $\leq \tau$ (that is, having a base of cardinality $\leq \tau$) a continuous image of the generalized Cantor discontinuum D^τ? Marczewski gave the negative answer to this question (see Engelking (1977)), which raised the problem of describing those compacta that are continuous images of some D^τ. Such spaces are called *dyadic* (see §6).

The Cantor discontinuum is a zero-dimensional compactum of countable weight. The negative answer to the question of P.S Aleksandrov has lead to more general formulations of the problem and during their study the following theorems were obtained.

Theorem 4 (see Aleksandrov and Pasynkov (1977)). *Every compactum of weight $\leq \tau$ is the continuous image of a zero-dimensional compactum of weight $\leq \tau$ (included in D^τ).*

Theorem 5 (see Parovichenko (1963)). *If* $2^{\aleph_0} = \aleph_1$, *then there exists a zero-dimensional compactum Φ of weight $2^{\aleph_0}$ that can be mapped continuously onto any compactum of weight $\leq 2^{\aleph_0}$.*

It is not known whether the assumption $2^{\aleph_0} = \aleph_1$ can be eliminated from the above theorem.

Continuous images of the interval were described by Hahn and Mazurkiewicz (see Arhangel'skii and Ponomarev (1974), Engelking (1977)).

Theorem 6. *In the class of Hausdorff spaces the non-empty connected locally connected metrizable compacta, and only such spaces, are continuous images of the unit interval.*

In particular the interval I can be mapped continuously onto the square $I \times I$ (Peano curve), onto the cube I^3 and onto the Hilbert cube $I^{\aleph_0}$.

3.3. Open Mappings of Compacta and Dimension. Clearly, the mappings appearing in the statement of Theorems 3, 4, 5 are not open since the image of a zero-dimensional space under a continuous mapping that is both open and closed is a zero-dimensional space. In connection with the above, the following theorem of Pasynkov becomes of interest (see Aleksandrov and Pasynkov (1977)).

Theorem 7. *Every non-zero-dimensional compactum can be represented as a continuous image of a one-dimensional compactum under a continuous mapping that is both closed and open and has zero-dimensional inverse images of points.*

The behavior of dimension under continuous open mappings between compact spaces has been studied very thoroughly. First of all, the conditions under which the dimension can increase were clarified. Let us mention that dimension-raising continuous open mappings of compacta may be of interest in coding theory.

Keldysh constructed a continuous, open mapping of the three-dimensional cube onto an arbitrary cube I^n, where $n > 3$, under which inverse images of all points are connected (see Aleksandrov and Pasynkov (1977)). Another delicate example constructed by Keldysh, namely a continuous open mapping of a one-dimensional compactum onto the square with zero-dimensional inverse images of points (see Keldysh (1959)), plays a fundamental role in the proof of Theorem 7 (see Aleksandrov and Pasynkov (1977)).

3.4. Mardešić's Factorization Theorem.

Theorem 8 (see Aleksandrov and Pasynkov (1977)). *Let $f : X \to Y$ be a continuous mapping of a finite-dimensional (in the sense of $\dim$) compactum X onto a compactum Y. Then there exist a compactum Z and continuous mappings $g : X \to Z$ and $h : Z \to Y$ such that $\dim Z \leq \dim X$ and the weight of Z does not exceed the weight of Y.*

With the help of Theorem 8 one can prove the theorem of A.V. Zarelua and Pasynkov (see Aleksandrov and Pasynkov (1977)) which states that for every natural number n and every cardinal number $\tau \geq \aleph_0$ there exists a compactum of weight τ and dimension n that is universal with respect to embeddings, that is, a compactum X for which $\dim X = n$, $w(X) = \tau$ and such that every compactum of weight $\leq \tau$ and dimension not exceeding n is homeomorphic to a compact Hausdorff subspace of X.

3.5. Continuous Images of Ordered Compacta. Both the interval and the Cantor discontinuum are examples of *ordered* compacta, that is, of spaces whose topology is generated by a linear order. Some ordered compacta are not metrizable. As an example we can take an arbitrary sufficiently large interval of the ordinals. Every non-empty linearly ordered compactum is zero- or one-dimensional. The following natural question can be raised in connection with the results of Sect. 4.2 (see in particular Theorems 3 and 7): which compacta can be represented as a continuous image of an ordered compactum? The following theorem shows that not all compacta have this property.

Theorem 9 (see Engelking (1977)). *If a compactum Y is the continuous image of a linearly ordered compactum then every subspace of Y is normal (and even more, collectionwise normal).*

In the last statement it is impossible to replace the condition of normality by paracompactness (for example consider the space $T(\omega_1 + 1)$).

The standard proof of Theorem 9 is based on the fact that all ordered compacta are hereditarily normal and this last property is preserved under continuous closed mappings.

Recall that a T_1-space is called *monotonically normal* (see Gruenhage (1984)) if it is possible to assign to each point $x \in X$ and to each open neighborhood U of x an open set $G(x, U)$ in such way that the following conditions are satisfied: (1) $x \in G(x, U) \subset U$; (2) if $x \in U \subset V$, where U and V are open in X, then $G(x, U) \subset G(x, V)$, and (3) if $x, y \in X$ and $x \neq y$, then $G(x, X \setminus y) \cap G(y, X \setminus x) = \emptyset$.

Every monotonically normal space is hereditarily collectionwise normal and every compactum that is the continuous image of an ordered compactum is monotonically normal. The following interesting question remains open: is every monotonically normal compactum the continuous image of an ordered compactum? (see Nikiel (1986)).

The characterization of continuous images of ordered compacta given by Bula and Turzański (see Bula and Turzański (1986)) could be useful in a solution of the last problem. The surveys related to this subject include Gruenhage (1984), Nikiel (1986). Let us mention here the following result of Treybig (see Gruenhage (1984)).

Theorem 10. *If the product of infinite compacta X and Y is the continuous image of an ordered compactum, then X and Y are metrizable.*

◁ This follows from the fact that such a product can be monotonically normal only if the compacta X and Y are metrizable. ▷

The theme of continuous mappings of compacta is one of the main subjects in general topology and it appears in all areas of this discipline. We shall return to it on numerous occasions.

3.6. Pseudocompactness and Continuous Mappings. Already the definition describes pseudocompactness in terms of continuous functions on a space. A slightly more general view of this property is provided by the following result.

Theorem 11. *A Tikhonov space is pseudocompact if and only if its image under any continuous mapping into a metrizable space is compact.*

◁ In the class of Tikhonov spaces pseudocompactness is preserved under continuous mappings. This property is equivalent to compactness in the class of metrizable spaces. ▷

Theorem 11 is nicely complemented by the following result of Veličko (see Arhangel'skii (1984a)).

Theorem 12. *Every continuous, one-to-one mapping of a pseudocompact space onto a metrizable space is a homeomorphism.*

Theorem 12 can be generalized in the following way (A.V. Arhangel'skii, (1986)): every continuous mapping of a pseudocompact space onto a metrizable space is an $\mathbb{R}$-*quotient* (that is, a *real quotient*, see Arhangel'skii (1987a)).

Theorem 11 indicates a possibility of a natural strengthening of the concept of pseudocompactness in which the cardinal number $\aleph_0$ ceases to play a privileged role. A Tikhonov space X is called τ-*pseudocompact* (where τ is an infinite cardinal number) if for every continuous mapping f of X into a Tikhonov space of weight $\leq \tau$ (into $\mathbb{R}^\tau$) the image $f(X)$ is compact Hausdorff.

Example 2. Let $\lambda > \tau \geq \aleph_0$. By deleting a single point from the Tikhonov cube I^λ we obtain a locally compact τ-pseudocompact κ-metrizable space that is not compact and that contains a closed discrete subspace of cardinality λ (see Engelking (1977)). A space with similar properties will be obtained if one deletes from the cube I^λ an arbitrary compact subset of weight $\leq \tau$. From this it follows easily (see §2) that for every $\tau \geq \aleph_0$, each Tikhonov space is homeomorphic to a closed subspace of a τ-pseudocompact space. Consequently, τ-pseudocompactness cannot imply any property of Tikhonov spaces that is hereditary with respect to closed subspaces. In particular, τ-pseudocompactness is not stronger than countable compactness for any cardinal number τ. Of course, a τ-pseudocompact space of weight $\leq \tau$ is compact.

Proposition 1. *If a Tikhonov space X is τ-bounded (that is, the closure of each set $A \subset X$ of cardinality $\leq \tau$ is compact), then X is τ-pseudocompact.*

3.7. Continuous Mappings and Extremally Disconnected Compacta. An important feature of compactness is the possibility of reduction of continuous mappings.

A mapping f of a topological space X onto a topological space Y is called *irreducible* if no proper closed subset F of X is mapped by f onto Y.

Proposition 2. *Let a mapping $f : X \to Y$ of a space X onto a space Y satisfy the following condition: for each point $y \in Y$ its inverse image $f^{-1}(y)$ is a compact subspace of X. Then there exists a closed subspace X_1 of X, such that $f(X_1) = Y$ and the restriction $f_1 = f|X_1 : X_1 \to Y$ of f to X_1 is irreducible.*

◁ We order the family $\mathcal{F}$ of all closed subspaces of the space X that are mapped by f onto Y by the relation opposite to inclusion: $Z_1 \le Z_2$ if and only if $Z_2 \subset Z_1$. Since the space $f^{-1}(y)$ is compact for each $y \in Y$, the ordered family $(\mathcal{F}, \le)$ is inductive (see Arhangel'skii and Ponomarev (1974), Engelking (1977)). Consequently, by Zorn's lemma, it has a maximal element Z^*. The space Z^* has the desired property. ▷

From Proposition 2 we obtain the following.

Theorem 13. *For each continuous mapping f of a compact space X onto a T_1-space Y there exists a closed subspace X_1 of X such that $f(X_1) = Y$ and the restriction $f_1 = f|X_1 : X_1 \to Y$ is irreducible.*

If a mapping $f : X \to Y$ is such that $f(X) = Y$ and the set of all points $x \in X$ for which $f^{-1}f(x) = \{x\}$ is dense in X, then f is irreducible.

However, there exists an irreducible continuous mapping of a compactum onto a compactum for which the inverse image of every point contains more than one point (see Engelking (1977)). In connection with the above the following theorem is of particular interest.

Theorem 14. *For each irreducible continuous mapping f of a metrizable compactum X onto a compactum Y, the set $\{x \in X : f^{-1}f(x) = \{x\}\}$ is dense in X.*

Irreducibility is a particularly effective property when combined with the closedness of a mapping and therefore irreducible mappings of compacta are of special interest. In the class of Hausdorff spaces a mapping is simultaneously open and irreducible if and only if it is a homeomorphism. Compacta for which there exists an irreducible mapping of one onto another are much closer than arbitrary compacta X and Y, one of which can be mapped continuously onto another. The following proposition provides a reason why the inverse image under an irreducible continuous mapping of a compactum shares a variety of properties of the image.

Proposition 3. *If $f : X \to Y$ is an irreducible mapping that is both continuous and closed, then the inverse image under f of a dense subset of Y is dense in X.*

Theorem 15. *If a compactum X can be mapped continuously and irreducibly onto a compactum Y, then the spaces X and Y have the same density and the same π-weight.*

One quite simple but fundamentally important corollary to Theorem 15 is the following: the cardinality of each compactum admitting an irreducible continuous mapping onto a compactum X is bounded by the cardinal number $2^{2^{|X|}}$. From this it follows that for an arbitrary compactum X it is possible to talk about the set of all compacta, determined up to a homeomorphism, admitting an irreducible continuous mapping onto X. Associated with this is one of the most interesting categorical properties of irreducible continuous mappings of compacta that was discovered by Gleason (see Gleason (1958)).

Theorem 16. *For every compactum X there exist a compactum $X^{\bullet}$ and an irreducible continuous mapping p of $X^{\bullet}$ onto X, such that the following condition is satisfied:*

(*) for every continuous $\phi : Z \to X$, where Z is compact Hausdorff and $\phi(Z) = X$, there exists a continuous mapping $g : X^{\bullet} \to Z$, for which $\phi \circ g = p$.

Using Theorem 16 it is easy to verify that this $X^{\bullet}$ is determined uniquely, up to a homeomorphism, and the mapping p is determined uniquely up to a certain natural equivalence relation (see Arhangel'skii and Ponomarev (1974)). This allows us to define the *absolute* of an arbitrary compactum X as the space $X^{\bullet}$ appearing in Theorem 16. The same theorem allows us to view the absolute of a compactum X as the largest compactum that can be irreducibly and continuously mapped onto X. It is true that the absolute of the absolute of a compactum X coincides with the absolute of X. In the language of category theory the absolutes of compacta are characterized as projective objects in the category of compacta and their irreducible continuous mappings. What properties of a space are also shared by its absolute? When do compacta X and Y have the same absolute (that is, when are they *co-absolute*)? These and similar questions were considered in great detail by V.I. Ponomarev who extended the theory of absolutes far beyond the class of compacta (see Arhangel'skii and Ponomarev (1974), Ponomarev and Shapiro (1976)).

According to Theorem 15 the absolute of a separable space is separable. A compactum X is co-absolute with a metrizable compactum if and only if the π-weight of X is countable (see Ponomarev and Shapiro (1976)). However, the absolute of an infinite compactum never has a countable base. This is a result of the following internal characterization of absolutes of compacta given by Gleason (see Gleason (1958)).

Theorem 17. *The following statements about a compactum X are equivalent:*

(a) there exists a compactum Y, whose absolute is X

(b) X is its own absolute: $X^{\bullet} = X$.

(c) the compactum X is extremally disconnected.

Recall that a space X is called *extremally disconnected* if the closure of every open subset of X is open in X. Every extremally disconnected Tikhonov space is zero-dimensional since it has a base consisting of sets that are both closed and open. But not every zero-dimensional compactum is extremally disconnected. We have the following simple result.

Theorem 18 (see Engelking (1977)). *No extremally disconnected Tikhonov space X contains a nontrivial converging sequence.*

◁ Let the sequence $\{x_n : n \in \mathbb{N}^+\}$ be convergent to a point x^* in X and assume that $x_{n'} \neq x_{n''}$ for $n' \neq n''$. Using induction we construct mutually disjoint neighborhoods O_{x_n} of the points x_n. The sets $G_0 = \bigcup \{O_{x_n} : n \text{ even}\}$ and $G_1 = \bigcup \{O_{x_n} : n \text{ odd}\}$ are open and disjoint, but their closures intersect since $x^* \in \overline{G}_0 \cap \overline{G}_1$. Consequently, at least one of the sets $\overline{G}_0, \overline{G}_1$ is not open which gives a contradiction. ▷

Theorem 18 implies that the absolute even of the simplest infinite compactum, a convergent sequence, is not metrizable. An extremally disconnected compactum is metrizable if and only if it is finite.

Theorem 17 can be strengthened in the following way: the closure of every discrete subspace of cardinality $\aleph_0$ of an extremally disconnected compactum is homeomorphic to $\beta\mathbb{N}$, the Stone-Čech compactification of natural numbers. This shows that infinite sets in extremally disconnected compacta have the tendency to "disperse".

The inverse image of an isolated point under an irreducible continuous mapping of a T_1-space consists of precisely one isolated point. The image of an isolated point under an irreducible continuous mapping of a T_0-space is also an isolated point. Consequently, the absolute of a compactum X has isolated points if and only if X has isolated points. These remarks lead to the following.

Example 3. There exist non-co-absolute metrizable compacta. Such are, for example, the Cantor discontinuum and a convergent sequence (with the limit). In the absolute of the first one there are no isolated points, while in the absolute of the second, such points exist. Note that both absolutes are separable but not metrizable.

It is true, however, that all metrizable compacta without isolated points are co-absolute with the Cantor set (and therefore with each other), see Malykhin and Ponomarev (1975), Ponomarev and Shapiro (1976).

Example 4. The Stone-Čech compactification $\beta\mathbb{N}$ of the set $\mathbb{N}$ of natural numbers is the absolute of the convergent sequence $S = \{0\} \cup \{\frac{1}{n} : n \in \mathbb{N}^+\}$. An irreducible mapping of $\beta\mathbb{N}$ onto S is obtained by assigning to each $n \in \mathbb{N}$ the point $\frac{1}{n+1} \in S$ and to every point of the remainder $\beta\mathbb{N} \setminus \mathbb{N}$ the point 0. The compactum $\beta\mathbb{N}$ is extremally disconnected while its closed subspace $\beta\mathbb{N} \setminus \mathbb{N}$ does not have this property. Therefore the property of being extremally disconnected is not hereditary with respect to closed subspaces. It is true, however, that every dense subspace of an extremally disconnected space is extremally disconnected.

The theory of extremally disconnected compacta is canonically associated with the theory of Boolean algebras. Extremally disconnected compacta and only such spaces are the Stone spaces of complete Boolean algebras (see §8 and Sikorski (1960)).

Extremally disconnected compacta have a high degree of saturation. The following result of Balcar and Franek serves as evidence: every extremally disconnected compactum of weight τ can be mapped continuously onto the Cantor discontinuum D^τ of weight τ (see Uspenskij (1983)). An infinite extremally disconnected compactum X is never *topologically homogeneous* (Frolik, Kunen, see Arhangel'skii (1978), Frolik (1987)), that is, there are always points $x, y \in X$ for which there exists no homeomorphism of X onto itself which carries x to y. This result can be proved with the help of the following theorem of Frolik on the nature of homeomorphisms of extremally disconnected spaces (see Arhangel'skii and Ponomarev (1974)).

Theorem 19. *Let f be a homeomorphism of an extremally disconnected compactum X into itself. Then X can be represented in the form $X = B_0 \cup B_1 \cup B_2 \cup B_3$, where B_0, B_1, B_2, B_3 are mutually disjoint open-and-closed subsets of X, $f(B_i) \cap B_i = \emptyset$ for $i = 1, 2, 3$, and $f(x) = x$ for all $x \in B_0$.*

Theorem 19 implies that the set of fixed points of such a homeomorphism f is both closed and open in X. In particular, if $f(X)$ is nowhere dense in X then $f(x) \neq x$ for all $x \in X$, in other words, f is fixed point free.

3.8. Scattered Compacta and Their Images. A space is called *scattered* if every nonempty subspace Y of X contains an isolated point (in Y).

The image of a scattered compactum under a continuous mapping (onto a Hausdorff space) is a scattered compactum. Therefore no scattered compactum can be mapped continuously onto the interval $I = [0, 1]$. It turns out that this property characterizes such spaces.

Theorem 20 (see Choban and Dodon (1979), Semadeni (1954)). *A compactum is scattered if and only if it cannot be mapped continuously onto the interval $[0, 1]$.*

In §5 we give a result of Shapirovskij characterizing the compacta that do not admit a continuous mapping onto the Tikhonov cube $I^{\aleph_1}$ (onto the cube I^τ). Scattered compacta are of interest in many situations. There exists a variety of remarkable relationships between the cardinal invariants of such spaces. For example, the cardinality of a scattered compactum does not exceed its hereditary Lindelöf number. If a scattered compactum is first countable then it has to be countable (see Choban and Dodon (1979)). Scattered compacta have subtle applications to constructions in mathematical logic. Scattered spaces form a fairly narrow class of spaces. On the other hand, as demonstrated by the next two (quite unexpected) results, the spaces in which every compact subspace is scattered form an exceptionally broad class.

Theorem 21 (see Arhangel'skii (1978)). *Every Tikhonov space contains a dense subspace every compact subset of which is scattered.*

Theorem 22 (Bregman, Shostak, Shapirovskij, see Bregman and Shapirovskij (1984)). *The following statement is consistent with the ZFC axioms of set theory: every Tikhonov space can be represented as a union of two subspaces every compact subset of which is scattered.*

§4. Metrizability Conditions for Compact, Countably Compact and Pseudocompact Spaces

Metrizability criteria for compact spaces are both simpler and more original than in the general case. It is worth pointing out that there exists a variety of very diverse metrizability theorems related to compact spaces. It is possible to consider the question of metrizability of compact spaces as a small part of the theory of cardinal invariants of such spaces (see the following section). This part, however, is so special and important that it deserves a separate treatment.

When one considers metrizability criteria it is very important to investigate what are the limits of their application, which of them are true only for compact spaces and which can be generalized to countably compact or even pseudocompact spaces. This provides us with a better understanding of the relationship between different types of compactness.

4.1. Classical Results and the Theorem of Chaber. The following is a well known classical result.

Theorem 1 (Urysohn, see Engelking (1977)). *A compactum is metrizable if and only if it has a countable base.*

Recall that a family $\mathcal{P}$ of subsets of the space X is called a *network* in X (see Arhangel'skii and Ponomarev (1974)) if for every $x \in X$ and every neighborhood O_x of x there exists a $P \in \mathcal{P}$, such that $x \in P \in O_x$.

Theorem 2 (see Arhangel'skii and Ponomarev (1974)). *Every pseudocompact space X with a countable network is metrizable and compact.*

◁ A space with a countable network is Lindelöf, and pseudocompact Lindelöf spaces are compact. In the product $X \times X$ there exists a countable network of closed sets. It follows that the diagonal $\Delta = \{(x,x) : x \in X\}$ is a G_δ-set in $X \times X$. It is enough to apply the metrizability theorem of Shneider (see Engelking (1977)). ▷

Proposition 1. *If X is a compactum with a G_δ diagonal (that is if Δ is a G_δ set in $X \times X$), then X is metrizable.*

The following two facts are useful in the proof of Theorem 2.

Proposition 2 (see Arhangel'skii and Ponomarev (1974)). *Every Tikhonov space with a countable network can be mapped in a one-to-one, continuous way onto a Tikhonov space with a countable base.*

Proposition 3 (Velichko, see Arhangel'skii (1984a)). *A one-to-one continuous mapping of a pseudocompact space X onto a metrizable space Y is a homeomorphism (consequently, X is a compactum with a countable base (see Theorem 12)).*

Proposition 1 cannot be generalized to pseudocompact spaces. In particular, there exists a non-metrizable, pseudocompact (separable) Moore space X (see Engelking (1977)). The square $X \times X$ of this space is also a Moore space and thus all closed subsets of $X \times X$ (in particular the diagonal Δ) are of the type G_δ in $X \times X$. However, Chaber obtained the following subtle result (see Engelking (1977)).

Theorem 3. *If X is a countably compact Hausdorff space and the diagonal Δ is of the type G_δ in $X \times X$, then X is a metrizable compactum.*

Example 1. The unit ball V of the Hilbert space is compact in the weak topology (see Bourbaki (1969a)). In the strong topology the ball V has a countable base $\mathcal{B}$. The base $\mathcal{B}$ is a countable network of V in the weak topology. Consequently, the ball V is metrizable and has a countable base.

Let a space X be the union of two subspaces X_1 and X_2, both of which have a countable base. Then X has a countable network and therefore if X is pseudocompact, then it is metrizable. This leads to the following.

Corollary 1. *If a pseudocompact space is a union of two subspaces each having a countable base, then it is metrizable, has a countable base and is compact.*

Let $f : X \to Y$ be a continuous mapping of a space X with a countable base onto a space Y. Then $\{f(U) : U \in \mathcal{B}\}$ is a countable network in Y. Thus from Theorem 2 we obtain the following.

Corollary 2. *If a pseudocompact space X is a continuous image of a space with a countable base then X is metrizable, separable and compact.*

4.2. Theorems of Dow and Tkachenko. The following peculiar metrizability theorem was obtained by Dow.

Theorem 4. *If every subspace of cardinality $\leq \aleph_1$ of a compactum X is metrizable then X is metrizable.*

The proof of Theorem 4 is based on the following result (Tkachenko, Juhász, see Juhász (1980)): if every subspace of cardinality $\leq \aleph_1$ of a Hausdorff space X has a countable base then the space X has a countable base.

4.3. Point-countable and σ-Point-finite Bases. The following classical result was obtained by A.S. Mishchenko (see Mishchenko (1962)).

Theorem 5. *If a base $\mathcal{B}$ of a countably compact T_1-space X is point-countable then it is countable.*

This theorem implies that a compact space X that is the image of a metrizable space under a continuous, open mapping with separable inverse images of points is metrizable (see Arhangel'skii and Ponomarev (1974)). Fillipov proved (see Fillipov (1969)) that the last statement remains true if the word "open" is replaced with "quotient".

Mishchenko's theorem was generalized in several different ways (see in particular, §5 in Arhangel'skii and Ponomarev (1974)). Let us recall that a *pseudobase* of a space X is a family $\mathcal{B}$ of open subsets of X such that every point in X is the intersection of some subfamily of $\mathcal{B}$. Arhangel'skii and Proizvolov proved that if X is a compactum with a point-countable pseudobase $\mathcal{B}$ then $\mathcal{B}$ is countable and X is metrizable (see Arhangel'skii and Ponomarev (1974)).

Every metrizable space has a point-countable base. Indeed, every σ-point-finite base is a point-countable base. How much larger than the class of all metrizable spaces is the class of all Tikhonov spaces with a point-countable base? The *Michael's line* obtained by strengthening the usual topology of the real line by making all irrational numbers isolated points is an example of a nonmetrizable paracompact space with a point-countable base (even a σ-disjoint base). There exist also nonmetrizable Lindelöf spaces with a σ-disjoint base (see Engelking (1977)). On the other hand, the existence of a point-countable base implies the following property related to paracompactness: every open cover of a space with a point-countable base has a point-countable open refinement (such spaces are called *metaLindelöf*, see Burke (1984)).

Using Theorem 5 one can show that every locally compact Hausdorff space with a point-countable base is metrizable. On the other hand not every Moore space is metaLindelöf and moreover not every Moore space has a point-countable base (an example is the *Nemytskij plane*).

Theorem 5 cannot be generalized to pseudocompact spaces (see Shakhmatov (1984), Watson (1985)): there exists a nonmetrizable pseudocompact space with a point-countable base. However, every pseudocompact space with a σ-point-finite base is metrizable (Uspenskij (1984)).

Let us recall that a base $\mathcal{B}$ of a space X is called *uniform* if for each $x \in X$ every infinite family of distinct elements of $\mathcal{B}$ containing x constitutes a base at x.

Corollary 3 (see Scott (1979), Watson (1981)). *Every pseudocompact space with a uniform base is metrizable (and thus is a compactum).*

4.4. Quasi-developments and $\delta\theta$-Bases. The concept of a quasi-development was introduced by Bennett (see Burke (1984)). It generalizes both the concept of a development and the concept of a σ-disjoint base.

A *quasi-development* of a space X is defined as a sequence $\{\gamma_n : n \in \mathbb{N}^+\}$ of families γ_n of open subsets of X such that for every point $x \in X$ and every neighborhood O_x of x there exists $k \in \mathbb{N}^+$, such that $x \in \bigcup\{U \in$

$\gamma_k : x \in U\} \subset O_x$. A space is called *quasi-developable* if it is regular and has a quasi-development. Clearly *Moore spaces* are characterized as quasi-developable spaces in which closed sets are of the type G_δ. On the other hand we have the following, subtle *Bennett's theorem* (see Burke (1984)): a quasi-developable compactum is metrizable.

It is true that all regular spaces with a σ-point finite base are quasi-developable (Aull, see Burke (1984)). In connection with the above, the concept of a $\delta\theta$-base is of interest since it generalizes both the concept of a development and the concept of a point-countable base (see Burke (1984)). A sequence $\{\gamma_n : n \in \mathbb{N}^+\}$ of families of open subsets of X is called a *$\delta\theta$-base* (*θ-base*) of this space if for every point $x \in X$ and every neighborhood O_x of the point x there exists $k \in \mathbb{N}^+$ and $U \in \gamma_k$ such that $x \in U \subset O_x$ and the family of elements of γ_k containing x is countable (finite).

Bennett and Lutzer proved (see Burke (1984)) that a regular space has a θ-base if and only if it has a quasi-development. The following is a far-reaching generalization of Theorem 5 (in the class of Tikhonov spaces).

Theorem 6 (see Burke (1984)). *Every countably compact regular space with a $\delta\theta$-base is metrizable.*

4.5. Strongly $\aleph_0$-Noetherian Bases. In this section we will present a natural generalization of the concept of a point-countable base. A base $\mathcal{B}$ of a space X is called *strongly $\aleph_0$-noetherian* if for every (nonempty) element V of $\mathcal{B}$ the family of elements of $\mathcal{B}$ containing V is countable. It is clear that a point-countable base is strongly $\aleph_0$-noetherian. The Tikhonov cube I^τ has a strongly $\aleph_0$-noetherian base (see Peregudov (1976)). Thus a compactum with a strongly $\aleph_0$-noetherian base does not have to be metrizable. However, we have the following.

Theorem 7 (Peregudov (1976)). *If each closed subspace of a compactum X has a strongly $\aleph_0$-noetherian base, then X is metrizable.*

Theorem 8 (Peregudov (1976)). *Every hereditarily normal compactum with a strongly $\aleph_0$-noetherian base is metrizable.*

Theorem 9 (Peregudov (1976)). *If $2^{\aleph_0} < 2^{\aleph_1}$, then every compactum of cardinality $\leq 2^{\aleph_0}$ with a strongly $\aleph_0$-noetherian base is metrizable.*

4.6. Rank of a Base and Metrizability Conditions for Compacta. The rank of a family $\mathcal{F}$ of sets is *finite* if there exists a natural number n such that if $U_1, U_2, \ldots, U_{n+1} \in \mathcal{F}$ and the sets U_i, U_j are not comparable by inclusion for $i \neq j, 1 \leq i, j < n+1$, then $\bigcap_{i=1}^{n+1} U_i = \emptyset$. The smallest $n \in \mathbb{N}^+$ with this property is called the *rank of the family* $\mathcal{F}$ and is denoted by $r(\mathcal{F})$ (see Gruenhage and Nyikos (1978), Nyikos (1977)). A family $\mathcal{F}$ has a *subinfinite rank* if there exists no infinite subfamily $\gamma \subset \mathcal{F}$ with $\bigcap \gamma \neq \emptyset$ such that no two different elements of γ are comparable by inclusion. Clearly the rank of a family $\mathcal{F}$ does not exceed 1 if for every two elements U, V of $\mathcal{F}$ one of the following conditions is satisfied: $U \subset V, V \subset U, U \cap V = \emptyset$. The concept of the rank of a base is essentially connected with the concept of dimension: in particular bases of rank 1 can easily be constructed in zero-dimensional spaces with a countable base. Bases of this type were already mentioned in the characterization of the space of irrational numbers given by P.S. Aleksandrov and Urysohn (see Arhangel'skii and Ponomarev (1974)). Every T_1-space with a base of rank 1 is normal and zero-dimensional. The following result was proved by Nagata and Arhangel'skii (see Gruenhage and Nyikos (1978)).

Theorem 10. *Let X be metrizable. Then $\dim X < n$ if and only if the space X has a base of rank $\leq n$.*

The concept of the rank of a base is also directly related to some properties of paracompactness type. This is illustrated by the following.

Theorem 11 (see Gruenhage and Nyikos (1978)). *Every space with a base of a subinfinite rank is metacompact.*

Theorem 12. *Every space with a base of rank 1 is hereditarily paracompact.*

Arhangel'skii proved that every compactum with a base of rank 1 is metrizable. This theorem was substantially generalized in the work of Gruenhage and Nyikos (see Gruenhage and Nyikos (1978)).

For every T_1-space with a base of a finite rank the weight is equal to density (see Gruenhage and Nyikos (1978)). If X is a regular space with countable Suslin number and with a base of finite rank, then X is Lindelöf (see Gruenhage and Nyikos (1978)). These results cannot be generalized to spaces with a base of subinfinite rank since such a base exists in the "arrow" space as well as in the Pixley-Roy space $\mathrm{Fin}(\mathbb{R})$ (see Example 3). A regular space with a base of finite rank and countable Suslin number has a point-countable base (see Gruenhage and Nyikos (1978)). Every space with a base of finite rank has a base of rank 1 at every point and moreover is radial (see Arhangel'skii (1980a)). For every space with a base of subinfinite rank the weight is equal to the network weight (see Gruenhage and Nyikos (1978)). A space with a base of subinfinite rank has countable tightness if and only if it is a Fréchet-Urysohn space.

Theorem 13 (see Gruenhage and Nyikos (1978)). *If a compactum has a base of finite rank then it is metrizable.*

Theorem 14 (Gruenhage and Nyikos (1978)). *A compactum with a noetherian base of a subinfinite rank is metrizable.*

Let us recall that a family of sets is called *noetherian* if every increasing chain of elements of it is finite.

Theorem 15 (Gruenhage and Nyikos (1978)). *Every compactum with a base of subinfinite rank is first countable.*

Example 2. The square $I \times I$ with the topology induced by the lexicographical order (see Engelking (1977)) is a nonmetrizable compactum with a base of subinfinite rank. Moreover, the Suslin number of this space is uncountable.

The *rank of a family* $\mathcal{F}$ of sets is *countable* if every subfamily of $\mathcal{F}$ consisting of mutually noncomparable elements and having a nonempty intersection is countable. Biriukov proved the following.

Theorem 16 (Biriukov (1981)). *If a compactum has a noetherian base of countable rank then it is first countable.*

Theorem 17 (Biriukov (1981)). *A dyadic compactum with a π-base of subinfinite rank is metrizable.*

Example 3. The *Aronszajn continuum* is a nonmetrizable compactum with a noetherian base of countable rank (see Gruenhage and Nyikos (1978)).

Let us point out that the property of having a noetherian base of subinfinite rank is preserved under the operation of a finite product (see Gruenhage and Nyikos (1978)).

4.7. Symmetrics and Metrizability of Compacta. Recall that a *symmetric* on a set X is a nonnegative, real-valued function d on the set $X \times X$ such that $d(x,y) = d(y,x)$, for all $x, y \in X$, and $d(x,y) = 0$ if and only if $x = y$. A symmetric d generates the topology $\mathcal{T}_d$ according to the rule: a set $U \subset X$ is open if for every $x \in U$ there exists an $\epsilon > 0$ such that $Q_\epsilon(x) = \{y \in X : d(x,y) < \epsilon\} \subset U$. The sets $Q_\epsilon(x)$ are not necessarily open and they can have empty interior (see Arhangel'skii and Ponomarev (1974)). This accounts for the fact that the space $(X, \mathcal{T}_d)$ does not have to be first countable. In addition the topology $\mathcal{T}_d$ does not have to satisfy the Hausdorff separation axiom (see Nedev (1971)).

We can ask the following natural question: what restrictions on a symmetric d guarantee metrizability of the space $(X, \mathcal{T}_d)$?

Theorem 18 (see Arhangel'skii and Ponomarev (1974)). *Every symmetrizable compactum is metrizable.*

◁ If a compactum is first countable in the topology generated by a symmetric then it is metrizable (Nemytskij, see Arhangel'skii and Ponomarev (1974)). Arhangel'skii and Stojanovskij proved (see Arhangel'skii and Ponomarev (1974)) that the topology of a symmetrizable compactum satisfies the first axiom of countability (see below for a more general result). ▷

Theorem 19 (see Nedev (1971)). *If a Lindelöf space X is symmetrizable then every closed subset of X is a G_δ-set in X.*

In a space X with the properties described in Theorem 19 every point is also a G_δ-set. In compacta the last condition is equivalent to the first axiom of countability. Therefore Theorem 19 implies that every symmetrizable compactum is first countable. It is impossible to generalize Theorem 18 to all Lindelöf spaces. It is also not true in the class of pseudocompact spaces. There exists a nonmetrizable, pseudocompact Moore space (the *Isbell-Mrówka space*, see Gillman and Jerison (1960)) and it is known that every Moore space is symmetrizable. However, every countably compact symmetrizable regular space is metrizable (see Nedev (1971)).

The following section contains some metrizability theorems for compacta presented as special cases of general theorems about cardinal invariants. Certain new metrizability theorems for compacta are also included in this section.

§5. Cardinal Invariants in the Class of Compacta

A *cardinal invariant* is a function defined on the class of all topological spaces or on any of its subclasses whose values are infinite cardinal numbers and has the property that for homeomorphic spaces the function assumes the same value. In general topology the theme of cardinal invariants plays a crucial role. We will provide some reasons why this is the case.

First, general topology has a clearly expressed set-theoretic nature that is manifested primarily by the fact that the objects of its study are mostly infinite sets. Thus the fundamental concept of the Cantor's set theory, namely the concept of a cardinal number, becomes one of main tools of general topology. It became clear that with the help of cardinal numbers it is possible to describe many phenomena that occur in topological spaces: their "size", peculiarities of their local structure, the nature of the operation of the closure and even the ability of a space to change (through the behavior of cardinal invariants under continuous mappings). For this reason cardinal invariants arise naturally and are useful in practically all areas of general topology.

Secondly, the cardinal numbers form quite a sophisticated and highly organized system. It is possible to perform on them the operations of addition, multiplication and power, and these operations are subject to harmonious and nontrivial laws of cardinal arithmetic. In addition the class of cardinal numbers is well ordered according to their size. Because of this cardinal numbers, and for the same reasons the cardinal invariants, are not only a convenient instrument of representing information about the structure of topological spaces, but more importantly they are an effective tool in the processing of this information. The main result of this process are the various relationships (primarily inequalities) between several cardinal invariants. The fact that the class of cardinal numbers is well ordered enables one to prove many relationships between cardinal invariants based on the principle of transfinite induction.

Finally the theory of cardinal invariants is very sensitive to the selection of the class $\mathcal{P}$ of topological spaces on which it is considered. The relationship between cardinal invariants varies greatly, depending on the selection of $\mathcal{P}$. For example, in the class of metrizable spaces five types of cardinal invariants coincide while they behave very differently in the class of Tikhonov spaces.

These invariants are: density, Suslin number, Lindelöf number, extent and spread (for definitions see below). The last fact allows us to prove nonmetrizability of some specific spaces or to decide whether a given space belongs to a class $\mathcal{P}$. The peculiarity mentioned above leads to separate theories of cardinal invariants depending on the class $\mathcal{P}$ and this makes this theory one of the most efficient tools of general topology.

Of a particularly distinctive nature is the theory of cardinal invariants on the class of compacta, to which this section is devoted. The definitions of cardinal invariants are given in the case of arbitrary T_1-spaces. Until the end of this paragraph the word "space" will mean a T_1-space. The symbol τ will be used only to denote infinite cardinal numbers.

5.1. Network Weight, Diagonal Number and Weight of Compacta. A family S of subsets of a space X is called a *network* of X if for every point $x \in X$ and every neighborhood O_x of x there exists $P \in S$ such that $x \in P \subset O_x$ (A.V. Arhangel'skii, see Arhangel'skii and Ponomarev (1974)). Every base is a network and also the family of all one point subsets is a network. The *network weight (weight)* of a space X is defined as the smallest cardinal number τ such that X has a network (base) of cardinality $\leq \tau$ and is denoted by $nw(X)$ ($w(X)$ respectively). The cardinality of an infinite space X (that is the cardinality of the set of its points) is denoted by $|X|$. If X is finite then we put $|X| = \aleph_0$. It is clear that for every infinite space the following inequalities are satisfied:

$$nw(X) \leq w(X) \quad \text{and} \quad nw(X) \leq |X|.$$

There exists a countable Tikhonov space X of uncountable weight. For such a space $nw(X) = |X| = \aleph_0 < w(X)$ (see Arhangel'skii and Ponomarev (1974)). However, for compacta we have the following theorem (Arhangel'skii, see Arhangel'skii and Ponomarev (1974), Engelking (1977)).

Theorem 1. *For every compactum X, $nw(X) = w(X)$.*

The simplest proof of this theorem makes use of the concept of the *diagonal number* of X. It is defined as the smallest cardinal number τ such that there exists a family γ of cardinality not exceeding τ of open sets in the space $X \times X$ whose intersection is the diagonal $\Delta_X = \{(x,x) : x \in X\}$ of $X \times X$ (this number is denoted by $\Delta(X)$).

Proposition 1. *If X is a Hausdorff space then the diagonal Δ_X is closed in $X \times X$.*

Proposition 2. *If X is regular and S is a network in X, then the family $\{\overline{P} : P \in S\}$ of the closures of elements of S is also a network in X.*

If $nw(X) \leq \tau$, then $nw(X \times X) = nw(X) \leq \tau$ (network weight does not increase under countable products). If, in addition, X is regular then by Proposition 2 the space $X \times X$ has a network S of cardinality $\leq \tau$ consisting of closed sets. But the diagonal Δ_X is a closed subset of $X \times X$ (by Proposition

1). Therefore $\Delta_X = \bigcap \gamma$, where $\gamma = \{(X \times X) \setminus P : P \in S \text{ and } P \cap \Delta_X = \emptyset\}$ is a family of open subsets of X and $|\gamma| \le |S| \le \tau$. Consequently, $\Delta(X) \le |\gamma| \le \tau$. This proves the following.

Proposition 3. *If X is regular then $\Delta(X) \le nw(X)$.*

Theorem 1 is now a consequence of the following result (see Arhangel'skii and Ponomarev (1974)).

Theorem 2. *For every compactum X, $\Delta(X) = w(X)$.*

The proof of Theorem 2 is based on the ideas of the next section. That is where we will give its outline.

Corollary 1. *For every compactum X, $w(X) \le |X|$.*

Corollary 2. (see Arhangel'skii and Ponomarev (1974)). *If a compactum Y is the continuous image under a mapping f of an arbitrary space X, then $w(Y) \le w(X)$.*

◁ Indeed, if S is a network in X, then the family $S' = \{f(P) : P \in S\}$ is a network in Y and $|S'| \le |S|$. ▷

In particular every countable compactum has a countable base, and consequently is metrizable. Moreover, every compactum with a countable network is metrizable. From Proposition 2 we obtain the following.

Corollary 3. *If a compactum is a continuous image of a space with a countable base then it is metrizable.*

The above statement is true also for pseudocompact spaces, since every pseudocompact space with a countable network is a compactum.

Corollary 4 (see Arhangel'skii and Ponomarev (1974)). *Let a compactum X be represented as the union of its subspaces, $X = \bigcup\{X_\alpha : \alpha \in A\}$, where $|A| \le \tau$, and let $w(X_\alpha) \le \tau$ for every $\alpha \in A$. Then $w(X) \le \tau$ (provided that $\tau \ge \aleph_0$).*

◁ In each X_α consider a base $\mathcal{B}_\alpha$ of cardinality $\le \tau$. Then $S = \bigcup\{\mathcal{B}_\alpha : \alpha \in A\}$ is a network of the compactum X and $|S| \le \tau$ (since $\tau \cdot \tau = \tau$). ▷

In the case when $\tau = \aleph_0$ Corollary 4 was proved by Smirnov (see Engelking (1977)).

Corollary 5. *If a compactum X is a union of a countable family of separable and metrizable subspaces, then it is metrizable (and separable).*

If a locally compact Hausdorff space X has a network S of cardinality $\le \tau$, then it has a base of cardinality $\le \tau$. Indeed, it is possible to add a single point to X in such a way that $\alpha(X) = X \cup \{a\}$ is a compactum. The family $\tilde{S}$ consisting of the elements of S and the one point set $\{a\}$ is a network of cardinality $\le \tau$ of the compactum $\alpha(X)$. Consequently, $w(X) \le w(\alpha(X)) \le \tau$. Actually, network weight coincides with weight for all p-spaces (see Arhangel'skii (1965a)).

Example 1. A compactum that is a union of two metrizable subspaces does not have to be metrizable. Let A be an uncountable set of cardinality τ and let a_τ be any element not belonging to A. Define a topology on the set $A_\tau = A \cup \{a_\tau\}$ in the following way: a set $U \subset A_\tau$ is open if and only if it is contained in A or its complement $A_\tau \setminus U$ is finite. The resulting space A_τ is called the *supersequence of Aleksandrov* (of length τ). It is an example of a nonmetrizable compactum (the first axiom of countability fails at the point a_τ). However A_τ is a Fréchet-Urysohn space. It is clear that $A_\tau = A \cup \{a_\tau\}$ is a union of two discrete (thus metrizable) subspaces, one of which is open and the other closed.

In Theorems 1 and 2 and in the Corollaries 1, 2 and 3 both assumptions, compactness of X and the presence of Hausdorff axiom, are essential.

Example 2. In the space $\mathbb{Q}$ of rational numbers let us "glue together" the set of all integers. The quotient space Y obtained in this process is a countable normal space of uncountable weight (see Arhangel'skii and Ponomarev (1974)) (at the point $\{\mathbb{N}\} \in Y$ the first axiom of countability fails; this fact can be verified by the standard Cantor diagonal argument). The space Y is a continuous image of a countable, discrete space. It can be represented as a union of a countable family of spaces with a countable base, namely, one-point spaces. Thus $|Y| = \Delta(Y) = nw(Y) < w(Y)$ and Corollaries 2 and 3 can not be generalized to the space Y.

Example 3. The set $\mathbb{R}$ of all real numbers equipped with the canonical T_1-topology (the only open sets are the empty set and the complements of finite sets) is a compact T_1-space of weight $2^{\aleph_0}$ with countable network weight. A countable base of the standard topology on $\mathbb{R}$ is a network of this space.

In connection with the concept of the diagonal number Hušek introduced the interesting concept of a *space with a small diagonal* (see Hušek (1977)). This is a space X with the property that for every uncountable set $A \subset (X \times X) \setminus \Delta_X$ there exists a neighborhood U of the diagonal Δ_X in $X \times X$ whose complement contains uncountably many points of $A : |A \setminus U| > \aleph_0$. It is clear that if $\Delta(X) = \aleph_0$ then X is a space with a small diagonal. For Tikhonov spaces the converse is not true, however Zhou proved (see Zhou (1982)) that the claim that compacta with a small diagonal are metrizable is consistent with the ZFC (Juhász and Szentmiklóssy have recently proved that this claim holds under CH). It is not known whether it is possible, within ZFC, to prove metrizability of compacta with a small diagonal. Important partial results were obtained by Hušek and van Douwen. Using the Continuum Hypothesis Hušek proved that a compactum with a small diagonal is metrizable in each of the following cases: a) the tightness of X is countable (see Hušek (1977)); b) X is separable; c) the weight of X does not exceed $2^{\aleph_0}$. Metrizability of every linearly ordered compactum with a small diagonal was established by van Douwen, while Lutzer generalized the last result to Lindelöf spaces.

5.2. Pseudocharacter and Character in the Class of Compacta. The *pseudocharacter* $\psi(A, X)$ *of the set* $A \subset X$ in the space X is defined as the smallest infinite cardinal number τ such that X contains a family γ of open subsets for which $|\gamma| \leq \tau$ and $\bigcap \gamma = A$. Accordingly, $\Delta(X) = \psi(\Delta_X, X \times X)$.

The *character* $\chi(A, X)$ *of the set* $A \subset X$ in the space X is defined as the smallest infinite cardinal number τ such that A has a base of cardinality $\leq \tau$ of neighborhoods in X (a family γ is called a *base of neighborhoods of a set* A in X if it consists of open subsets of X, $A \subset \bigcap \gamma$, and for every open set U containing A there exists $V \in \gamma$ such that $V \subset U$). If $x \in X$, we put $\psi(x, X) = \psi(\{x\}, X)$ and $\chi(x, X) = \chi(\{x\}, X)$.

It is always true that $\psi(A, X) \leq \chi(A, X)$ and in particular that $\psi(x, X) \leq \chi(x, X)$. In addition, $\psi(A, X) \leq |X|$ for every $A \subset X$ since $A = \bigcap\{X \setminus \{x\} : x \in X \setminus A\}$ (we consider only T_1-spaces). On the other hand the relation $\chi(y, Y) \leq |Y|$ is not always true.

Example 4. For the closed subspace $\mathbb{N}$ of the space $\mathbb{Q}$ of rational numbers (with the standard topology) we have: $\chi(\mathbb{N}, \mathbb{Q}) > \aleph_0 = \psi(\mathbb{N}, \mathbb{Q})$. Let us mention that $\mathbb{Q}$ is a countable space with a countable base! For the space Y in Example 2 and the point $y = \{\mathbb{N}\} \in Y$ it is true that $\chi(y, Y) = \chi(\mathbb{N}, \mathbb{Q}) > \aleph_0$.

One of the most amazing and most frequently used properties of compacta is the equality of character and pseudocharacter for closed subsets (see Arhangel'skii and Ponomarev (1974)).

Theorem 3. *For any compactum* X *and every closed subset* A *of* X *we have* $\psi(A, X) = \chi(A, X)$.

This theorem cannot be generalized to subsets that are not closed.

Corollary 6. *A compactum* X *is first countable if and only if every point of* X *is a* G_δ-*set in* X.

For Tikhonov spaces the analogous statement is not true (see Example 2). Similarly, Corollary 6 cannot be generalized to the class of compact T_1-spaces (see Arhangel'skii and Ponomarev (1974)).

◁ *Sketch of proof of Theorem 2.* Let X be a compactum and let $\Delta(X) \leq \tau$. The last condition implies that $\psi(\Delta_X, X \times X) \leq \tau$. But then, according to Theorem 3, $\chi(\Delta_X, X \times X) \leq \tau$. Consider a base $\mathcal{B}$ of neighborhoods of the set Δ_x in $X \times X$ such that $|\mathcal{B}| \leq \tau$. For any $W \in \mathcal{B}$ we denote by γ_W the family of all open subsets U of X whose square $U \times U$ is contained in W. Then we have $\bigcup \gamma_W = X$ and since X is compact, we can select from γ_W a finite subcover μ_W. The family $\mathcal{E} = \bigcup\{\mu_W : W \in \mathcal{B}\}$ is a base of X. In addition, it is true that $|\mathcal{E}| \leq |\mathcal{B}| \leq \tau$. ▷

The theorem on the equality of character and pseudocharacter for compacta can be partially generalized to countably compact and pseudocompact spaces. We have the following.

Proposition 4. *Every pseudocompact spaces satisfies the first axiom of countability at every point that is a* G_δ-*set in this space.*

◁ Let X be pseudocompact, $x \in X$, and $\psi(x, X) \leq \aleph_0$. Then X is regular and there exists a countable collection $\xi = \{U_n : n \in \mathbb{N}^+\}$ of open subsets of X such that $\overline{U}_{n+1} \subset U_n$ and $\{x\} = \bigcap\{U_n : n \in \mathbb{N}^+\}$. We claim that ξ is a base of X at the point x. Assume the contrary. Then there exists an open set V such that $x \in V$ and $U_n \setminus \overline{V} \neq \emptyset$ for all $n \in \mathbb{N}^+$. If we put $W_n = U_n \setminus \overline{V}$ then the family $\{W_n : n \in \mathbb{N}^+\}$ is discrete in the space X, is infinite and consists of nonempty, open subsets, which leads to a contradiction with pseudocompactness of X. ▷

At the same time there exists a countably compact, regular space X of cardinality $\tau = 2^{\aleph_0}$ whose character at each point is equal to $2^\tau > \tau = |X| \geq \psi(X)$ (see §2).

5.3. First Countable Compacta. A compactum satisfying the first axiom of countability does not have to be metrizable.

Example 5 (*"Double arrow"* compactum). On the set $X = [0,1] \times [0,1]$ consider the lexicographical order $<$ defined in the following way $(x', y') < (x'', y'')$ if $x' < x''$, or $x' = x''$ and $y' < y''$. The space X equipped with the topology induced by this order is a compactum and the subspaces $X_0 = [0,1] \times \{0\}$ and $X_1 = [0,1] \times \{1\}$ of the space X are homeomorphic to the *"arrow"* space (see Arhangel'skii and Ponomarev (1974)). Consequently, every subspace of the space X is separable, that is X is hereditarily separable. However X does not have a countable base (or a countable network) since they do not exist in the "arrow" space. Therefore the compactum X is not metrizable. Since X is separable and linearly ordered, it satisfies the first axiom of countability (see §5).

The class of compacta satisfying the first axiom of countability is different from the class of metrizable compacta. However we have the following.

Theorem 4 (Arhangel'skii, see Arhangel'skii and Ponomarev (1974)). *The cardinality of every compactum satisfying the first axiom of countability does not exceed $2^{\aleph_0}$.*

The following general inequality is true (see Arhangel'skii and Ponomarev (1974)): if X is a compactum and if $\chi(X) \leq \tau$ (that is, if $\chi(x, X) \leq \tau$ for all $x \in X$), then $|X| \leq 2^\tau$.

This theorem can be applied in may situations (see Hödel (1984)). It can be made more precise: the cardinality of every uncountable compactum satisfying the first axiom of countability is equal to $2^{\aleph_0}$ (see Arhangel'skii and Ponomarev (1974)). Theorem 4 and the last remark cannot be generalized to the Fréchet-Urysohn compacta. As a counterexample we may take the Aleksandrov supersequence A_τ for $\tau > 2^{\aleph_0}$ (see Example 1).

Example 6. Consider an arbitrary regular cardinal number $\tau > \aleph_0$ and take any well ordered set $(Z, <)$ of cardinality τ together with the topology induced by the order $<$. Consider the subspace Y of Z consisting of all isolated points of Z and all $z \in Z$ for which there exists a countable subset $B \subset Z \setminus \{z\}$ such that $z \in \overline{B}$. The space Y is countably compact, normal and satisfies the first axiom of countability. It has the same cardinality as Z, that is $|Y| = \tau$.

According to the above example there exist countably compact Tikhonov spaces of arbitrarily high cardinality that satisfy the first axiom of countability.

Theorem 4, however, can be generalized to Lindelöf spaces.

Theorem 5 (see Arhangel'skii and Ponomarev (1974)). *Every Lindelöf (Hausdorff) space satisfying the first axiom of countability has cardinality not exceeding $2^{\aleph_0}$.*

Gryzlov generalized Theorem 4 to the case of compact T_1-spaces satisfying the first axiom of countability (see Gryzlov (1980)). However, the following question remains open: is it true that the cardinality of each Lindelöf T_1-space satisfying the first axiom of countability does not exceed $2^{\aleph_0}$? The paper of Mrówka (see Engelking (1977)) on this subject contains an error. As mentioned above, in compacta the first axiom of countability is equivalent to the statement that all points are G_δ-sets in a space. In the class of Lindelöf spaces the second condition is substantially weaker than the first one. It is not known whether there exist Lindelöf (Hausdorff, regular) spaces of arbitrarily high cardinality in which all points are G_δ-sets. The cardinality of such a space can exceed $2^{2^{\aleph_0}}$? Is it possible to construct such spaces of cardinality greater than $2^{\aleph_0}$ within the framework of ZFC? Shelah proved the last statement (see Juhász (1980)) using additional assumptions. Within the framework of ZFC only the following estimate is known (see Arhangel'skii and Ponomarev (1974)): the cardinality of every Lindelöf space of countable pseudocharacter is less than the first cardinal number that is measurable in the sense of Ulam (if such a cardinal number exists). Let us recall that all Lindelöf spaces, by our definition, are Tikhonov, but in Theorem 5 it is enough to assume the Hausdorff separation axiom.

Let us mention that the class of compacta satisfying the first axiom of countability is closed with respect to countable products and with respect to closed subspaces but is not preserved by continuous maps.

The first axiom of countability has a nontrivial global generalization: the existence of a point-countable base (that is a base for which the intersection of any uncountable subfamily is empty). Mishchenko proved (see Arhangel'skii and Ponomarev (1974), Mishchenko (1962)) that every compactum with a point-countable base is metrizable. The same statement is no longer true for pseudocompact spaces (Shakhmatov, Watson, see Shakhmatov (1984)).

The idea of the *weak first axiom of countability* was brought to general topology from the theory of symmetrizable spaces. We will recall here the necessary definitions. Assume that with every point x of the space X is associated a family $\{P_n(x) : n \in \mathbb{N}^+\}$ of subsets of X such that the following conditions are satisfied: (a) $x \in P_{n+1}(x) \subset P_n(x)$ for all $n \in \mathbb{N}^+$ and all $x \in X$, (b) a set $U \subset X$ is open if and only if for every $x \in U$ there exists $n \in \mathbb{N}^+$ such that $P_n(x) \subset U$. We say then that the space X satisfies the weak first axiom of countability. The importance of this concept is clear from the following examples. The intersection of any two metrizable topologies always satisfies the weak first axiom of countability but does not even have to have the Fréchet-Urysohn property. Every symmetrizable space satisfies the weak first axiom of countability, while it may fail to satisfy the first axiom of countability. Even for countable Tikhonov spaces the weak first axiom of countability is not equivalent to the first axiom of countability. Relevant examples can easily be constructed (see Arhangel'skii and Ponomarev (1974)). However, it is quite difficult to see the difference between the two axioms in the class of compacta. Only by using the Continuum Hypothesis was Yakovlev able to construct a compactum with the weak first axiom of countability that does not satisfy the first axiom of countability (see Yakovlev (1976)). In spite of the similarity of both axioms for compacta it is impossible to generalize Theorem 4 to the class of spaces satisfying the weak first axiom of countability. Malykhin used additional set theoretic assumptions to construct a compactum satisfying the weak first axiom of countability and of cardinality exceeding $2^{\aleph_0}$ (see Malykhin (1982)).

The class of compacta satisfying the weak first axiom of countability is contained in the class of sequential compacta (see Arhangel'skii and Ponomarev (1974)). However, the intersection of the class of all compacta that satisfy the weak first axiom of countability with the class of all Fréchet-Urysohn compacta consists precisely of all compacta satisfying the first axiom of countability.

5.4. Perfectly Normal Compacta. Between the classes of all metrizable compacta and of all first countable compacta there lies a remarkable class of perfectly normal compacta. A normal space in which every closed subset is a G_δ-set is called *perfectly normal*. All metrizable compacta are perfectly normal. The "double arrow" compactum (see Example 4) is perfectly normal. Accordingly, perfect normality is not equivalent to metrizability in the class of compacta. However, in this class of spaces this property is much stronger than the first axiom of countability. This follows easily from the following result of Katetov (see Engelking (1977)).

Theorem 6. *A compactum X is metrizable if and only if its square its perfectly normal.*

◁ The diagonal Δ_X in the square of a Hausdorff space X is closed and, by the hypothesis, is of type G_δ. It remains to apply Theorem 2. ▷

Theorem 6 reveals also a fundamental deficiency of the class of perfectly normal spaces, namely the fact that it is not closed under products. In particular, the compactum "double arrow" is perfectly normal, while its square is not, by Theorem 6, since the "double arrow" compactum is not metrizable. Thus the square of the "double arrow" compactum is an example of a space that is first countable but not perfectly normal. It is not clear whether within ZFC it is possible to construct a perfectly normal locally connected nonmetrizable compactum.

It is also possible to look at perfectly normal compacta from a different point of view reflected by the following result.

Theorem 7 (Smirnov, see Engelking (1977)). *A compactum X is perfectly normal if and only if every subspace of X is Lindelöf.*

A compactum is first countable if and only if the complement of every point is Lindelöf.

A remarkable feature of the class of perfectly normal compacta is its preservation under continuous maps. With respect to this property the class of perfectly normal compacta is similar to the class of metrizable compacta.

Theorem 8. *Let X be a compactum. Then the following conditions are equivalent:*

a) every Hausdorff space that is a continuous image of X satisfies the first axiom of countability,

b) every Hausdorff space that is a continuous image of X is a perfectly normal compactum.

Theorem 8 implies that the class of first countable compacta is not closed under continuous maps. If a compactum X is not perfectly normal but satisfies the first axiom of countability, then among its continuous images there is always a compactum that is not first countable.

Weiss and Watson considered the class of continuous images of the "double arrow" compactum and proved that not all separable perfectly normal compacta belong to it.

There is an interesting connection between perfect normality of compacta and their hereditary separability. The *Suslin continuum* (that is a nonmetrizable connected perfectly normal ordered compactum), if it exists, can be used as an example of a perfectly normal, nonseparable compactum. However, under Martin's axiom and the negation of the Continuum Hypothesis (notation: MA+¬CH) it is possible to prove (see Juhász (1980)) that every perfectly normal compactum is separable. Shapirovskij proved, using ZFC, that the density of every perfectly normal compactum does not exceed $\aleph_1$ (see Arhangel'skii (1978)). In particular under MA+¬CH the Suslin continuum does not exist (see Juhász (1980)). Szentmiklóssy proved, using MA+¬CH, that every hereditarily separable compactum is perfectly normal (see Juhász (1980)). This fact combined with Theorem 6 implies that the following characterization of metrizable compacta is consistent with ZFC: a compactum X is metrizable if and only if its square $X \times X$ is hereditarily separable.

The following question was raised in connection with Theorem 6. Let the product of the spaces X and Y be a perfectly normal compactum. Is it true then that at least one of the compacta X and Y is metrizable? Let us mention that the product of a metrizable compactum and a perfectly normal compactum is always a perfectly normal compactum. Rudin proved that it is consistent with ZFC that there exist nonmetrizable compacta whose products are perfectly normal (see Juhász (1984)). At present, it is not known whether such compacta can be constructed within ZFC.

5.5. Continuous Images of First Countable Compacta. Even the simplest continuous maps do not have to preserve the property of being first countable in the class of compacta.

Example 6. Let X be a first countable but not perfectly normal compactum. Consider a closed subset F of X that cannot be represented as a countable intersection of open subsets and "glue together" all points of F. The resulting quotient space is a compactum that fails to satisfy the first axiom of countability at the point F.

It is true, however, that the image of a first countable space under a continuous closed map is always a Fréchet-Urysohn space (see Arhangel'skii and Ponomarev (1974)). Consequently we have the following.

Theorem 9. *In the class of Hausdorff spaces a continuous image of a first countable compactum is a Fréchet-Urysohn compactum.*

Certain important constructions appearing in functional analysis result in spaces that are Fréchet-Urysohn compacta. It is true that Eberlein compact spaces belong to this class (see Arhangel'skii (1984a): Eberlein compact spaces do not have to satisfy the first axiom of countability).

However, it is not true that every Fréchet-Urysohn compactum is a continuous image of a first countable compactum. Indeed, according to Theorem 4, the cardinality of an image of a first countable compactum does not exceed $2^{\aleph_0}$. At the same time there exist Fréchet-Urysohn compacta of arbitrarily large cardinality (see Example 1). But even among Fréchet-Urysohn compacta of cardinality not exceeding $2^{\aleph_0}$ not all are continuous images of first countable compacta. Simon constructed a Fréchet-Urysohn compactum whose square is not Fréchet-Urysohn (see Simon (1980)). Clearly, the class of continuous images of first countable compacta is closed under countable products. This leads to the problem of finding an "internal" characterization of continuous images of first countable compacta. This problem does not have a complete solution but the following approach deserves attention.

We will say that a sequence $\xi = \{A_n : n \in \mathbb{N}^+\}$ of subsets of a space X converges to a point x of this space if for every neighborhood O_x of x there exists a number $m \in \mathbb{N}^+$ such that $A_n \subset O_x$ for all $n \geq m$. The families ξ and η are called *synchronized* if $A \cap B \neq \emptyset$ for all $A \in \xi$ and $B \in \eta$. A space is called *bisequential* if for every family η of subsets of X with the finite intersection property and for every point $x \in \bigcap\{\overline{P} : P \in \eta\}$ there exists a sequence $\{A_n : n \in \mathbb{N}^+\}$ of subsets of X converging to x and synchronized with η. Every bisequential space is Fréchet-Urysohn but the converse is not true. This follows easily, for example, from the fact that the product of a countable family of bisequential spaces is bisequential. Let us mention that the property of being bisequential is hereditary with respect to arbitrary subspace. We have the following strengthening of Theorem 9.

Theorem 10. *A compactum that is a continuous image of a first countable compactum is bisequential.*

All Aleksandrov supersequences A_τ belong to the class of bisequential compacta provided that the cardinal number τ is less than the first Ulam measurable cardinal. At present, it is not known whether every bisequential compactum of cardinality not exceeding $2^{\aleph_0}$ is a continuous image of a first countable compactum. It is convenient to introduce a special name for such compacta: they will be called *supersequential*.

Not every bisequential compactum is first countable (see Example 6), but every bisequential Tikhonov space satisfies the following condition that is very close to the first axiom of countability: for every point $x \in X$ there exists a sequence $\{U_n : n \in \mathbb{N}^+\}$ of open subsets of X that converges to x. At present we do not know answers to the following questions. Is it true that every nonempty bisequential compactum satisfies the first axiom of countability at least one point? Does every supersequential compactum satisfy the first axiom of countability at least one point? It is conjectured that the answer to the last question should be positive in ZFC.

5.6. Sequential Compacta and the First Axiom of Countability Almost Everywhere. A compactum X is called *sequential* (see Engelking (1977)) if every subset A of X that is not closed in X contains a sequence $\{x_n : n \in \mathbb{N}^+\}$ of points converging to a point in the complement $X \setminus A$ of A. Not every sequential compactum is a Fréchet-Urysohn space (see Arhangel'skii and Ponomarev (1974)). Associated with this is the fact that in general the property of being sequential is not hereditary (it is hereditary with respect to closed subspaces). In particular, Fréchet-Urysohn spaces are characterized as those sequential spaces each of whose subspaces is sequential.

The class of sequential compacta is closed with respect to countable products (see Arhangel'skii and Ponomarev (1974)) and in this respect it is a more natural extension of the class of first countable compacta than the class of Fréchet-Urysohn compacta. In addition, the product of a sequential space and a sequential compactum is sequential. A continuous image of a sequential compactum (under a map onto a Hausdorff space) is a sequential compactum.

All known examples of (nonempty) sequential compacta constructed within the ordinary (Cantor) set theory have one peculiarity: they all contain a point at which they satisfy the first axiom of countability. Moreover, as a rule this happens at most points. For example, in the space A_τ such points form a dense open subspace. We have the following theorem (see Arhangel'skii (1978, 1979a)).[1]

---

[1] Recently P. Kozmider has shown that the answer to this question and to the preceding two is consistently negative.

Theorem 11 (A.V. Arhangel'skii). *If $2^{\aleph_0} < 2^{\aleph_1}$, then in each sequential compactum the set of all points at which the compactum satisfies the first axiom of countability is dense.*

The strength of this theorem is demonstrated, for example, by the following.

Corollary 7 (see Arhangel'skii (1987b)). *Let $2^{\aleph_0} < 2^{\aleph_1}$. Then every topologically homogeneous sequential compactum satisfies the first axiom of countability (at all points), and consequently, its cardinality does not exceed $2^{\aleph_0}$.*

Let us recall that sequential compacta can have arbitrarily high cardinality (Example 1). Later we will indicate other corrolaries of Theorem 11.

$\triangleleft$ *A sketch of a proof of Theorem 11.*

Proposition 5 (see Arhangel'skii and Ponomarev (1974)). *In a sequential Hausdorff space the cardinality of the closure of a countable set does not exceed $2^{\aleph_0}$.*

The Continuum Hypothesis is used directly in considerations involving the following classical result of Čech and Pospišil (see Engelking (1977)).

Theorem 12. *If X is a nonempty compactum whose cardinality is strictly less than $2^{\aleph_1}$, then X contains points at which the space satisfies the first axiom of countability.*

Since for metrizability of topological groups it is enough that the first axiom of countability be satisfied at one point, Theorem 12 implies the following.

Corollary 8 (see Comfort (1984)). *If $2^{\aleph_0} < 2^{\aleph_1}$ and G is a compact topological group of cardinality $\leq 2^{\aleph_0}$, then G is metrizable.*

The following proposition plays a crucial role in the proof of Theorem 11. It involves the class of compacta of countable tightness that is slightly larger than the class of sequential compacta.

Lemma 1 (see Arhangel'skii (1978)). *In every nonempty compactum of countable tightness there exists a countable set A and a nonempty (closed) subset Φ of the type G_δ, such that $\Phi \subset \overline{A}$..*

Let X be a sequential compactum and suppose that we have a countable subset A and a closed subset Φ be as indicated in Proposition 1. According to Proposition 5 the cardinality of Φ does not exceed $2^{\aleph_0}$. Assuming that $2^{\aleph_0} < 2^{\aleph_1}$ and making use of Theorem 12 we conclude that the compactum Φ satisfies the first axiom of countability at some point y. Since Φ is a G_δ-subset of X, the point y is the intersection of a countable family of open subsets of X. According to Theorem 3 (or Corollary 6), compactum X satisfies the first axiom of countability at the point y. $\triangleright$

It is impossible to prove Theorem 11 without making use of additional assumptions. Malykhin proved, using the method of forcing, that the existence of a nonempty Fréchet-Urysohn compactum that does not satisfy the first axiom of countability at any point is consistent with ZFC. However, the compactum that he constructed is neither bisequential nor topologically homogeneous (see Malykhin (1987b)).

Without the Continuum Hypothesis one can obtain the next result by an argument following the lines of the proof of Theorem 11.

Theorem 13 (see Arhangel'skii (1978)). *For each sequential compactum X the set of all points at which the character of X does not exceed $2^{\aleph_0}$ is dense in X.*

The property of being a sequential compactum is related to several theorems about the unions of compacta. The example of the compacta A_τ (Example 1) indicates that not every compactum that is the union of two metrizable spaces satisfies the first axiom of countability. We have, however, the following.

Theorem 14 (see Arhangel'skii (1987b), Pytkeev and Yakovlev (1980)). *If a compactum X is the union of a countable family of metrizable subspaces, then it is sequential and satisfies the first axiom of countability at all points of some dense subset of X. If the Suslin number of this compactum is countable, then the compactum is separable.*

In the case of the union of two metrizable spaces this result can be made significantly more precise. Rudin proved that every compactum that can be represented as a union of two metrizable subspaces is an Eberlein compactum. However, a compactum that is a union of three metrizable subspaces does not have to be a Fréchet-Urysohn compactum (see Arhangel'skii and Ponomarev (1974)).

It is known that the countability of the Suslin number of a compactum does not restrict its cardinality. For each cardinal number τ the compacta D^τ and I^τ have a countable Suslin number (see Arhangel'skii and Ponomarev (1974)). Sequential compacta, as mentioned earlier, can also have arbitrarily large cardinality. It is true, however, that the two properties mentioned above, when combined together, seriously restrict the cardinality of a compactum.

Theorem 15 (see Arhangel'skii (1978)). *If the Suslin number of a sequential compactum X is countable, then the cardinality of X does not exceed $2^{\aleph_0}$.*

The proof of Theorem 15 is also based on Proposition 1 (and on the combinatorial techniques introduced by Hajnal and Juhász, see Juhász (1980)). It is worth mentioning that Theorem 15 is obtained using ZFC only. By adding Martin's Axiom combined with the negation of the Continuum Hypothesis the conclusion of this theorem can be strengthened.

Theorem 16 (see Arhangel'skii (1978)). *Under $\mathrm{MA} + \neg\mathrm{CH}$ every sequential compactum with countable Suslin number is separable.*

This result was generalized to the class of compacta of countable tightness by Shapirovskij (see Arhangel'skii (1978) and the previous paragraph).

The role of compactness in Theorems 11-15 is extremely important. None of these theorems can be generalized to the class of countably compact spaces, or to σ-compact (and consequently, to Lindelöf) spaces.

Example 7. Consider a cardinal number $\tau \geq 2^{\aleph_0}$ and let Y and Z be the $\sum$-product and the σ-product, respectively, of τ copies of the interval $I = [0,1]$ (or the discrete two point space $D = \{0,1\}$) with the origin as the base point (see Arhangel'skii and Ponomarev (1974)). Then Y is a countably compact Fréchet-Urysohn space, Z is a σ-compact Fréchet-Urysohn space and both Y and Z are dense in I^τ. Consequently the Suslin number of both Y and Z is countable. The cardinalities of the spaces Y and Z are equal to $\tau^{\aleph_0}$ and τ, respectively, and can therefore be arbitrarily large. Since $\tau > 2^{\aleph_0}$ the spaces Y and Z do not satisfy the first axiom of countability at any point (they even have uncountable π-character at every point) and none of them is separable. Finally, both Y and Z are topological groups.

5.7. Corson Compact Spaces and $\aleph_0$-Monolithicity. Compacta contained in the $\sum$-product of some family of intervals are called *Corson compact spaces*. Many compacta that are considered in functional analysis (see Arhangel'skii (1976, 1984a)) are of this type. All Corson compact spaces are also Fréchet-Urysohn spaces which implies that they are sequential. In addition, such spaces have one other important property, namely $\aleph_0$-monolithicity.

A space X is called $\aleph_0$-*monolithic* if the closure in X of every countable subset of X is a space with a countable network. In $\aleph_0$-monolithic compacta the closure of every countable subset is a compactum with a countable base since the network weight of compacta coincides with their weight (by Theorem 1). The property of $\aleph_0$-monolithicity is responsible for important features of the structure of topological spaces. In particular, every separable $\aleph_0$-monolithic space has a countable network and every separable $\aleph_0$-monolithic compactum is metrizable. Taking into account the last statement and Theorem 16 we obtain the following result.

Corollary 9 (see Arhangel'skii (1976)). *Under* MA$+\neg$CH *every Corson compact space with countable Suslin number is metrizable.*

A "naive" proof of this statement is not possible (see R. Pol (1982)). It is true, however, that for some subclasses of the class of Corson compact spaces this result can be obtained within ZFC. For example, Eberlein compact spaces with countable Suslin number are metrizable (see Rosenthal (1974)).

The following property of Corson compact spaces is also based on $\aleph_0$-monolithicity.

Theorem 17. *The set of points of a Corson compact space at which it satisfies the first axiom of countability is dense in this compactum.*

$\lhd$ Indeed, let X be a Corson compact space and let $A, \Phi \subset X$ be the sets described in Proposition 1. Then $\overline{A}$ is a metrizable compactum. Consequently, the set $\Phi \subset \overline{A}$ is also metrizable and all of its points are of the type G_δ in X. ▷

For Eberlein compact spaces (and even for Gul'ko compact spaces) a stronger statement is true. Such spaces contain dense subspaces metrizable by a complete metric (for more details see Arhangel'skii (1987a)). It is not always possible to find a dense metrizable subspace of a Corson compact space. This result was obtained by Todorčevič (see Arhangel'skii (1987a)).

5.8. Compacta of Countable Tightness. The concept of a space of countable tightness seems to be more natural than the concept of a sequential space if we look at it from the "internal" point of view of general topology. We consider a cardinal-valued function, tightness, defined on the class of all topological spaces, and we simply impose a bound (countability) on its values. The *tightness $t(X)$ of a space X* is defined as the smallest infinite cardinal τ, such that for every point $x \in X$ and every $A \subset X$ with $x \in \overline{A}$, there exists a set $B \subset A$ for which $x \in \overline{B}$ and $|B| \leq \tau$. If $t(X) \leq \aleph_0$, then we say that tightness of X is countable. Clearly, all first countable spaces, and consequently, all metrizable spaces have countable tightness. Little work is required to prove that every sequential space has countable tightness. It is not true, however, that every space of countable tightness is sequential. As a matter of fact, the tightness of a space never exceeds it cardinality. However, even a countable Tikhonov space does not have to be sequential.

Example 8. The subspace $Y = \mathbb{N} \cup \{p\}$ of the Čech-Stone compactification $\beta\mathbb{N}$ of the natural numbers $\mathbb{N}$ obtained by adding to $\mathbb{N}$ an arbitrary point p of $\beta\mathbb{N} \setminus \mathbb{N}$ has the indicated property.

Accordingly, the class of spaces of countable tightness is one of the most natural extensions of the class of first countable spaces. In studying sequential spaces we repeatedly deal with this class also because the property of being sequential is not generally preserved for subspaces, while every subspace of a sequential space has countable tightness. Let us point out that not every space of countable tightness can be obtained in this way. The space $Y = \mathbb{N} \cup \{p\}$ from Example 8 is not a subspace (even non-closed) of any Hausdorff, sequential space (see Arhangel'skii and Ponomarev (1974)). It is useful to keep in mind the following.

Proposition 6. *If every subspace of a space X is separable (that is, if X is hereditarily separable), then the tightness of X is countable.*

The class of all hereditarily separable spaces contains, in particular, all spaces with a countable network. The following formula is true.

$$t(X) \leq nw(X) \leq |X|. \tag{1}$$

Example 9. Consider the space $C_p(I)$ of all continuous real-valued functions on the interval $I = [0,1]$ in the topology of pointwise convergence. It has a countable network and, consequently, the tightness of $C_p(I)$ is countable. However, the space $C_p(I)$ is not sequential (see Arhangel'skii (1984a)).

For every compactum X the space $C_p(X)$ of all continuous, real-valued functions with the topology of pointwise convergence has countable tightness, but is sequential only in the case when X is scattered (see Arhangel'skii (1976, 1984a)). This indicates the importance of the concept of countable tightness to functional analysis. The last example demonstrates also to what extent the concept of countable tightness is different from that of sequentiality.

The notion of tightness plays a fundamental role in the study of the structure of compacta. Recall that a *set of type* G_τ in a space X is defined as a subset A of X that can be represented as the intersection of a family of cardinality $\leq \tau$ of open subsets of X. For every $A \subset X$ let us put $[A]_\tau = \bigcup\{\overline{B} : B \subset A, |B| \leq \tau\}$ and let us denote by $[A]^\tau$ the set of all points $x \in X$ such that every set P of type G_τ in X that contains x intersects A.

Proposition 7 (see Arhangel'skii (1978)). *For every compactum X and every $A \subset X$ the following formula is true:*

$$\overline{A} = [[A]_\tau]^\tau. \tag{2}$$

Proposition 7 leads to the following general version of Proposition 1 that plays a key role in the study of compacta by means of tightness.

Proposition 8 (Arhangel'skii (1978), Arhangel'skii and Ponomarev (1974)). *Let X be a nonempty compactum with $t(X) \leq \tau$. Then there exists a set $A \subset X$ and a nonempty (closed) set Φ of type G_τ in X, such that $|A| \leq \tau$ and $\Phi \subset \overline{A}$.*

As mentioned earlier, this statement cannot be generalized to countably compact spaces.

The proofs of Proposition 8 and of the next Theorem 18, as well as the proofs of many other theorems about cardinal invariants of compacta, make use of the following concept.

A transfinite sequence $\{x_\alpha : \alpha < \tau\}$ of points of a space X is called a *free sequence of length* τ in X if for every $\beta < \tau$

$$\overline{\{x_\alpha : \alpha < \beta\}} \cap \overline{\{x_\alpha : \beta \leq \alpha\}} = \emptyset. \tag{3}$$

We have the following.

Theorem 18 (Arhangel'skii (1978), Arhangel'skii and Ponomarev (1974)). *The tightness of an arbitrary compactum X is equal to the supremum of the lengths of free sequences in X.*

The concept of a free sequence played a fundamental role in the proof of the estimate of the cardinality of an arbitrary compactum satisfying the first axiom of countability (see Theorem 2). Proposition 8 implies the following result.

Proposition 9. *In every compactum of countable tightness the set of points at which character of the compactum does not exceed $2^{\aleph_0}$ is dense.*

If a compactum X of countable tightness is topologically homogeneous then, by Proposition 9, its character at every point does not exceed $2^{\aleph_0}$ and thus (see §5.3) $|X| \leq 2^{2^{\aleph_0}}$. There is no "naive" proof that every nonempty compactum of countable tightness satisfies the first countability axiom at some point (see §5.6). But it is also impossible to establish this fact by using the Continuum Hypothesis. Indeed, Fedorchuk constructed, with the help of the $\Diamond$ principle (see Fedorchuk (1976)), a hereditarily separable compactum of cardinality $2^{2^{\aleph_0}}$ that does not contain any nontrivial convergent sequence. Moreover, this compactum does not satisfy the first axiom of countability at any point. The following conclusion (due to Ostaszewski and Fedorchuk) is of particular importance: within ZFC it is impossible to prove that every compactum of countable tightness is sequential. However, in ZFC the last statement cannot be disproved either! First, Balogh deduced from the PFA (Balogh (1989)) that every compactum of countable tightness is sequential, and later this was improved so that to exclude large cardinals. However, it remains unknown whether it is possible to construct, in a "naive" way, a compactum of countable tightness that is not first countable at any point.

In connection with the previous results the next theorem of Shapirovskij relating to π-character of compacta of countable tightness is of particular interest. A family γ of nonempty open subsets of a space X is called a π-*base* at a point $x \in X$ if every neighborhood of x contains at least one element of γ. The minimum of the cardinalities of all π-bases of X at a point $x \in X$ is called the π-*character* at this point and is denoted by $\pi_\chi(x, X)$.

Theorem 19 (Shapirovskij, see Arhangel'skii (1978, 1979a)). *Let X be a compactum. Then $\pi_\chi(x, X) \leq t(X)$, for all $x \in X$.*

A strong restriction of topological properties of a compactum can be obtained if the assumption of countable tightness is combined with $\aleph_0$-monolithicity.

Theorem 20 (see Arhangel'skii (1976)). *Let X be an $\aleph_0$-monolithic compactum of countable tightness. Then X is a Fréchet-Urysohn space and the set of points at which X is first countable is dense in X.*

The proof of Theorem 20 is based on Proposition 1. Shapirovskij proved (see Shapirovskij (1980b)) that every compactum of countable tightness can be mapped by a perfect irreducible map onto some $\aleph_0$-monolithic compactum of countable tightness, more precisely, onto a Corson compact space contained in a $\sum$-product of intervals.

If we assume MA+$\neg$CH then every compactum of countable tightness whose Suslin number is countable is separable (see Arhangel'skii (1978)). Thus it is consistent with ZFC that every $\aleph_0$-monolithic compactum of countable tightness with countable Suslin number is metrizable.

Recall that the *spread of a space* X is defined as the supremum of the cardinalities of its discrete (in themselves) subspaces. The points of a free sequence always form a discrete subspace. Therefore, Theorem 18 implies the following.

Corollary 10 (Shapirovskij, Arhangel'skii, see Arhangel'skij and Ponomarev (1974), Arhangel'skii (1978)). *The tightness of a compactum does not exceed its spread.*

For topologically homogeneous compacta of countable tightness several estimates of cardinal invariants can be substantially strengthened.

Theorem 21 (Ismail, see Engelking (1977)). *If X is a topologically homogeneous compactum of countable tightness and with countable Suslin number, then $|X| \leq 2^{\aleph_0}$.*

In particular, the cardinality of every hereditarily separable topologically homogeneous compactum does not exceed $2^{\aleph_0}$. It is not clear whether it is possible to prove in ZFC that the cardinality of every homogeneous compactum of countable tightness does not exceed $2^{\aleph_0}$, or that every such compactum is first countable. However, it is known, that the cardinality of every topologically homogeneous sequential compactum does not exceed $2^{\aleph_0}$ (see Arhangel'skii (1978, 1987b)). Thus, according to the results of Balogh (see Balogh (1989)) it is consistent with ZFC to assume that the cardinality of every homogeneous compactum of countable tightness does not exceed $2^{\aleph_0}$. In the presence of Martin's axiom (MA) every homogeneous sequential compactum is a Fréchet-Urysohn space. From the above results, in conjunction with some results of Dow we obtain the following.

Theorem 22. *The statement that every homogeneous compactum of countable tightness is a Fréchet-Urysohn space is consistent with the ZFC system of axioms of set theory.*

The class of compacta of countable tightness is closed under countable products (Malykhin, see Arhangel'skii (1979b)). Without the assumption of compactness the product of a normal space of countable tightness with a countable Tikhonov space may no longer be of countable tightness (see Arhangel'skii (1979b)). A compactum that is a continuous image of a compactum of countable tightness also has countable tightness since under quotient maps tightness does not increase. Tightness is monotone with respect to arbitrary subspaces, however, not every countable space can be embedded in a compactum of countable tightness! (see Arhangel'skii (1979b), Arhangel'skii and Ponomarev (1974)).

5.9. Mappings of Compacta onto Tikhonov Cubes I^τ. Every uncountable metrizable compactum contains the Cantor perfect set (see Arhangel'skii and Ponomarev (1974), Engelking (1977)) and consequently can be mapped continuously onto the interval $I = [0, 1]$. In turn, the interval I can be mapped continuously onto the Hilbert cube $I^{\aleph_0}$. Thus every uncountable metrizable compactum can be mapped onto the Hilbert cube. The following question arises naturally: which compacta can be mapped continuously onto the Tikhonov cube $I^{\aleph_1}$ and onto the Tikhonov cubes I^τ, for $\tau > \aleph_0$?

The theory of cardinal invariants provides several obstructions to the existence of such maps. If the values of a cardinal invariant φ do not increase under continuous mappings of compacta onto Hausdorff spaces and in addition $\varphi(I^{\aleph_1}) > \aleph_0$, then no compactum X for which $\varphi(X) \leq \aleph_0$ can be mapped continuously onto $I^{\aleph_1}$ (and thus onto I^τ for $\tau > \aleph_0$).

Tightness belongs to the class of cardinal invariants for which the above requirements are satisfied (see §5.8). Consequently, we have the following.

Theorem 23. *If a compactum X has countable tightness then X cannot be mapped continuously onto $I^{\aleph_1}$.*

It is also impossible to map any (linearly) ordered compactum onto $I^{\aleph_1}$.

◁ Every subspace of an ordered compactum is normal (see Engelking (1977)). The same property is shared by all Hausdorff spaces that are continuous images of ordered compacta (this follows from the fact that continuous maps of compacta onto Hausdorff spaces are closed). However, $I^{\aleph_1}$ contains subspaces that are not normal (see Arhangel'skii and Ponomarev (1974)). ▷

We will identify one more class of compacta that does not include $I^{\aleph_1}$ and is preserved by continuous mappings onto Hausdorff spaces.

A space X is called *radial* if for every $A \subset X$ and $x \in \overline{A}$ there exists a chain ξ of subsets of X that is convergent to x and each element of which intersects A. Every ordered space is radial and the property of being radial is preserved under continuous open maps as well as under continuous closed maps (see Arhangel'skii (1980a)). In particular, every compactum that is a continuous image of an ordered compactum is radial. The cube $I^{\aleph_1}$, however, is not radial (see Arhangel'skii (1980a)). Consequently we have the following.

Theorem 24. *No radial compactum can be mapped continuously onto the Tikhonov cube $I^{\aleph_1}$.*

Shapirovskij solved, for any $\tau > \aleph_0$, the problem of a complete "internal" characterization of compacta that admit a continuous mapping onto the Tikhonov cube I^τ. Let us call a compactum τ-*saturated* if at each point its π-character is not less than τ. The following fundamental result was obtained by Shapirovskij.

Theorem 25. *A compactum X can be mapped continuously onto the Tikhonov cube I^τ (where $\tau > \aleph_0$) if and only if there exists a τ-saturated compactum Y included in X.*

Theorem 25 implies that if a finite product of compacta can be mapped continuously onto I^τ, then at least one of these compacta can be mapped continuously onto I^τ (see Shapirovskij (1980b)).

Corollary 11 (Shapirovskij (1980b)). *The Continuum Hypothesis $2^{\aleph_0} = \aleph_1$ is equivalent to the following statement: every nonempty compactum either contains a point at which its π-character is countable (that is, a point with a countable π-base), or it contains a subspace homeomorphic to $\beta\mathbb{N}$. Moreover, this compactum contains a topological copy of $\beta\mathbb{N}$ if and only if X contains some $2^{\aleph_0}$-saturated compactum.*

The following fact is essential in the proofs of the last statements: if a compactum X can be mapped continuously onto I^τ then X contains topological copies of all extremally disconnected (see §3.7) Tikhonov spaces of weight $\leq \tau$.
Shapirovskij also proved the next result (see Shapirovskij (1980b)).

Theorem 26. *Let X be a compactum with countable Suslin number and weight $\geq \tau$, where τ is the first cardinal number greater than $2^{\aleph_0}$. Then X can be mapped continuously onto the cube I^τ.*

For topologically homogeneous compacta with countable Suslin number the question concerning the existence of a continuous mapping onto the cube $I^{\aleph_1}$ has a particularly simple solution.

Theorem 27 (Arhangel'skii (1987b)). *Let $2^{\aleph_0} < 2^{\aleph_1}$ and let X be a topologically homogeneous compactum with countable Suslin number. Then X can be mapped continuously onto the Tikhonov cube $I^{\aleph_1}$ if and only if $|X| > 2^{\aleph_0}$.*

The condition $2^{\aleph_0} < 2^{\aleph_1}$ in the above theorem is necessary.
Finally, let us mention the following strong theorem of Balcar and Franek (see Uspenskij (1983) and also Arhangel'skii (1978)): every extremally disconnected compactum of weight τ can be mapped continuously onto the generalized Cantor discontinuum D^τ.

5.10. Dyadic Compacta. The *generalized Cantor discontinuum D^τ* of weight τ is defined as the topological product of τ copies of the two-point discrete space $D = \{0,1\}$. According to Tikhonov's theorem, all spaces D^τ are compacta. Unlike Tikhonov cubes I^τ, compacta D^τ are zero-dimensional since they have a base consisting of closed-and-open sets. Therefore not every Tikhonov space can be embedded into D^τ. In fact a Tikhonov space X is homeomorphic to a subspace of D^τ if and only if it is zero-dimensional and its weight does not exceed τ (see Engelking (1977)).
Let us consider the dual question: which compacta can be obtained as a continuous image of some compactum D^τ? P.S. Aleksandrov called such compacta *dyadic* (see Arhangel'skii and Ponomarev (1974)). Every metrizable compactum is a continuous image of the Cantor perfect set. Since the last space is homeomorphic to $D^{\aleph_0}$, all metrizable compacta are dyadic. The fact that not all compacta are dyadic can be established by using the theory of cardinal invariants. The following *theorem of Hewitt, Marczewski and Pondiczery* is one of the most important results of this theory.

Theorem 28. *The Suslin number of any product of separable spaces is countable.*

Theorem 28 implies that $c(D^\tau) = \aleph_0$ (and $c(I^\tau) = \aleph_0$) for every $\tau \geq \aleph_0$. The Suslin number does not increase under continuous maps. Thus we have the following.

Theorem 29. *The Suslin number of every dyadic compactum is countable.*

The fact that for any τ the Suslin number of D^τ is countable plays a crucial role in mathematical logic in a construction of the boolean-valued models of set theory that lead to the most important theorems about the independence of the Axiom of Choice, Continuum Hypothesis, etc.

Example 10. The space $T(\omega_1+1)$ of ordinal numbers not exceeding the first uncountable ordinal number ω_1 is a compactum whose set of isolated points is uncountable. It follows that the Suslin number of $T(\omega_1+1)$ is uncountable and thus this compactum is not dyadic.

The theory of dyadic compacta is the area of general topology that is rich with delicate and unexpected results (see below). It influences in a creative way many other areas of general topology. The development of this theory, in particular in the works of Shanin, Efimov, Engelking (as well as research associated with the Suslin problem), truly initiated the creation of the new large branch of general topology, the theory of cardinal invariants, which touches practically all other area of this discipline. This theory has an important direct connection with the theory of topological groups and functional analysis. In particular, Ivanovskij and Kuz'minov (independently) proved the following fundamental fact (see Ivanovskij (1958), Kuz'minov (1959)).

Theorem 30. *The space of every compact group is a dyadic compactum.*

Using the last theorem it is possible to obtain, by combining it with the theory of dyadic compacta, several delicate theorems concerning the structure of topological groups.

Theorem 30 is nicely complemented by the following result of Hewitt (see Comfort (1984)).

Theorem 31. *The space of an arbitrary zero-dimensional infinite compact group of weight τ is homeomorphic to D^τ (notice that D^τ is one of such groups).*

On the other hand, it was established that the classes of Milyutin and Dugundji compact spaces which arise in functional analysis in connection with questions concerning the existence of averaging operators and extensions of mappings are very close to the class of dyadic compacta (see Pełczyński (1968)).

In 1949 Esenin-Volpin obtained one of the main results of the theory of dyadic compacta: every first countable dyadic compactum is metrizable (see Arhangel'skii and Ponomarev (1974), Engelking (1977)). Later Efimov improved this result and provided one of its modern formulations (see Efimov (1965)).

Theorem 32. *If the set of points of countable π-character is dense in a dyadic compactum, then the compactum is metrizable.*

The last statement does not mean, however, that every separable dyadic compactum is metrizable. For example the generalized Cantor discontinuum D^c and the Tikhonov cube I^c, where $c = 2^{\aleph_0}$, are separable dyadic compacta. The following is another far reaching generalization of the theorem of Esenin-Volpin.

Theorem 33 (see Arhangel'skii and Ponomarev (1974), Engelking (1977)). *If the tighteness of a dyadic compactum is countable then the compactum is metrizable.*

The example of the spaces D^c and I^c shows that in the statement of Theorem 33 it is not enough to assume that a dyadic compactum contains a dense subspace of countable tightness. Theorem 33 implies, for example, that the intersection of the class of Corson compacta (Eberlein compacta), which are of great importance in functional analysis, with the class of dyadic compacta consists exactly of metrizable compacta.

Theorems 33 and 30 imply the following.

Corollary 13. *If the tightness of a compact topological group is countable then the group is metrizable.*

The following unexpected consequence of Theorem 32 can be obtained by using Lusin's conjecture (see Arhangel'skii (1978), Arhangel'skii and Ponomarev (1974)).

Corollary 14. *If $2^{\aleph_0} < 2^{\aleph_1}$, then every dyadic compactum X of cardinality not exceeding $2^{\aleph_0}$ is metrizable.*

$\triangleleft$ Under the above assumption the set of points of X at which it satisfies the first axiom of countability is dense in X. $\triangleright$

Combining this result with Theorem 30 we obtain the following.

Corollary 15. *If $2^{\aleph_0} < 2^{\aleph_1}$, then every compact group of cardinality not exceeding $2^{\aleph_0}$ is metrizable.*

It is impossible to remove the assumption $2^{\aleph_0} < 2^{\aleph_1}$ from the above corollaries. If $2^{\aleph_0} = 2^{\aleph_1}$, then $D^{\aleph_1}$ would be a nonmetrizable compact group (dyadic compactum) of cardinality $2^{\aleph_0}$.

Dyadic compacta are very rarely ordered spaces. This follows from the next result of A.V. Arhangel'skii.

Theorem 34 (see Arhangel'skii (1980a)). *A dyadic compactum is radial if and only if it is metrizable.*

Since every ordered space is radial we conclude that ordered dyadic compacta are metrizable.

The image of an ordered space under a continuous open map, as well as under a continuous closed map, is a radial space. Therefore Theorem 34 implies the following result.

Corollary 16. *No ordered space can be mapped by a continuous open (continuous closed) mapping onto any nonmetrizable dyadic compactum.*

From a technical point of view the following theorem of Efimov and Gerlits on embedding the generalized Cantor discontinua D^τ into dyadic compacta is very useful.

Theorem 35. *If X is a dyadic compactum of weight τ and if $cf(\tau) > \aleph_0$ (that is, the cardinal number τ is not representable as a countable sum of smaller cardinal numbers), then X contains a subspace homeomorphic to D^τ.*

Theorem 35 implies, for example, that a dyadic compactum X is metrizable in any of the following cases: (a) every discrete subspace of X is countable, (b) every subspace of X is normal, (c) X cannot be mapped continuously onto the cube $I^{\aleph_1}$ (that is, if the index of X is countable) (see Shapirovskij (1980b)).

An analogous result on embedding Tikhonov cubes I^τ into continuous images of Tikhonov cubes was obtained by Shchepin: if a nonmetrizable compactum X is a continuous image of a Tikhonov cube I^τ, then X contains a topological copy of I^τ (see Shchepin (1981)). This result implies the following theorem of Shchepin (see Shchepin (1981)).

Theorem 36. *If a finite dimensional compactum X is an absolute neighborhood retract in the class of compacta, then X is metrizable.*

A generalization of this theorem was obtained by Tsuda (see Tsuda (1986)).

5.11. Supercompacta and Extensions of the Class of Dyadic Compacta.
A family $\mathcal{P}$ of closed subsets of a space X is called a *closed prebase* of this space if every closed subset of X is an intersection of some subfamily of $\mathcal{P}$. A family of sets is called *linked* if every two elements of this family intersect. A closed subbase is called *binary* if each of its linked subfamilies has a nonempty intersection. A topological space is called *supercompact* (de Groot, see van Mill (1977)) if it has a binary, closed prebase. Supercompactness can be described also in a dual way: a space X is supercompact if and only if there exists a prebase $\mathcal{B}$ of the topology of X such that for any open cover of X by the elements of $\mathcal{B}$ it is possible to select a subcover consisting of at most two elements.

Every supercompact space is compact (*Alexander's lemma*, see Engelking (1977)). The product of any family of supercompact spaces is supercompact. Since every metrizable compactum is supercompact (Strok and Szymanski, see van Mill (1977)) it follows that all Tikhonov cubes and all Cantor cubes are supercompact. But not all compacta are supercompact (see Bell (1985)). Indeed, we have the following result of van Mill and Bell.

Theorem 37. *If the Stone-Čech compactification $\beta(X)$ of a space X is a continuous image of some supercompact Hausdorff space, then X is pseudocompact.*

In particular, the Stone-Čech compactification $\beta(\mathbb{N})$ of natural numbers is not supercompact. Now, since $\beta(\mathbb{N})$ is included in some Tikhonov cube, it is clear that supercompactness, in general, is not hereditary with respect to closed subspaces.

Supercompact Hausdorff spaces will be referred to later as *supercompacta*. It is not always easy to establish whether a given compactum is a supercompactum. The space $T(\omega_1 + 1)$ of all ordinal numbers not exceeding the first uncountable ordinal is a supercompactum. Van Mill constructed an example of a first countable, separable compactum that is not a continuous image of any supercompactum (see van Mill (1977)). A closed G_δ-subset of a supercompactum does not have to be a supercompactum (Bell, see Bell (1985)). For every lattice (see van Mill (1977)) there exists a natural interval topology associated with it. This topology is compact if and only if the lattice is complete. We have the following.

Theorem 38 (see van Mill (1977)). *Every compact lattice is a supercompactum.*

In particular every linearly ordered compactum is a supercompactum.

As was demonstrated by van Mill and Mills, a continuous image of a supercompactum can be a non supercompact Hausdorff space (see van Mill and Mills (1978) and also van Mill (1977)). It was recently proved (Bell) that not every dyadic compactum is a supercompactum. Bell calls a compactum *superadic* if it is a continuous image of a supercompactum. The class of superadic compacta is an interesting extension of the class of dyadic compacta. Not every Eberlein compact space is supercompact (Bell, see Bell (1985)), however the one-point compactification of every discrete space is always a supercompactum. It is not known whether every Eberlein compact space is superadic. Every infinite superadic compactum contains a nontrivial convergent sequence (van Douwen and van Mill, see van Mill (1977)). From this result it follows again that $\beta(\mathbb{N})$ is not supercompact.

If $\mathcal{P}$ is a family of sets then we will denote by $r(\mathcal{P})$ the smallest family of sets containing $\mathcal{P}$ that is closed with respect to the operations of the intersection and the union of finite number of elements. If a compactum X has a closed prebase $\mathcal{P}$ for which the elements of the family $r(\mathcal{P})$ are canonical closed sets (a closed set is called canonical if it is the closure of an open set), then X is called a *regular Wallman compactum* (see van Mill (1977)). If a compactum X has a binary closed prebase $\mathcal{P}$ for which $r(\mathcal{P})$ consists of canonical closed sets, then X is called *regularly supercompact* (see van Mill (1977)). Not all compacta are regular Wallman compacta. The concepts of a regular Wallman compactum and a regular supercompactum are applied in the theory of compact Hausdorff extensions. It is not clear whether every supercompactum is a regularly supercompact space, or whether there exists a supercompactum that is not a regular Wallman compactum (see van Mill (1977)).

A useful generalization of the concept of a dyadic compactum was introduced by Mrówka. He called a compactum Y *polyadic* if Y can be represented as a continuous image of some power of the Aleksandrov compactification of a discrete space (see Bell (1985)). All polyadic compacta are superdyadic, but not every polyadic compactum is a supercompactum (Bell, see Bell (1985)).

An interesting construction leading to the extension of the class of polyadic compacta was introduced by Bell (see Bell (1985)). Let S be an arbitrary infinite family of sets. Let us denote by $\mathrm{Cen}(S)$ the collection of all subfamilies of the family S (including the empty one) with the finite intersection property. For $P \in S$ let us put $P^+ = \{\xi \in \mathrm{Cen}(S) : P \in \xi\}$ and $P^- = \{\xi \in \mathrm{Cen}(S) : P \notin \xi\} = \mathrm{Cen}(S) \setminus P^+$. The family of all sets of the form P^+ and P^-, where $P \in S$, is considered as a closed-and-open prebase of the space $\mathrm{Cen}(S)$. Thus all sets P^+ and P^-, where $P \in S$, are closed-and-open in $\mathrm{Cen}(S)$ and their finite intersections form a base of the space $\mathrm{Cen}(S)$. The space $\mathrm{Cen}(S)$ is compact Hausdorff and zero-dimensional regardless of the original family S. If X is a a zero-dimensional compactum and S_X is the family of all closed-and-open subsets of X then in place of $\mathrm{Cen}(S_X)$ we use the notation $\mathrm{Cen}(X)$.

The weight of the compactum $\mathrm{Cen}(S)$ is equal to the cardinality of S. The collection of all finite subfamilies with the finite intersection property is dense in $\mathrm{Cen}(S)$.

Example 11. If a family S has the finite intersection property then the compactum $\mathrm{Cen}(S)$ is homeomorphic to the Cantor cube $D^{|S|}$. If a family S consists of disjoint sets then the compactum $\mathrm{Cen}(S)$ is homeomorphic to the Aleksandrov compactification of a discrete space of cardinality $|S|$.

Theorem 39 (Bell (1985)). *The product of an arbitrary family of compacta of the form* $\mathrm{Cen}(S)$ *is itself a compactum of this type.*

Bell also proved the following subtle result (see Bell (1985)).

Theorem 40. *For any family S of sets the family of all closed-and-open subsets of the compactum* $\mathrm{Cen}(S)$ *is the union of a countable set of antichains.*

Recall that a family of sets is called an *antichain* if none of the elements of this family is a subset of another.

Proposition 10. *A compactum* $\mathrm{Cen}(S)$ *is a Fréchet-Urysohn space if and only if all subfamilies of the family S with the finite intersection property are countable.*

Compacta that are continuous images of compacta of the type $\mathrm{Cen}(S)$ are called *centered.*

Examples of compacta that are not centered may be obtained from the following result.

Theorem 41 (Bell (1985)). *The weight of an arbitrary centered compactum is equal to its π-weight.*

In particular the space $\beta(\mathbb{N})$ is not centered.

Theorem 39 together with Example 11 imply that every polyadic compactum is a centered space. It is not true that every centered space is supercompact. For example, the compactum $\mathrm{Cen}(\beta(\mathbb{N}) \setminus \mathbb{N})$ is not supercompact (see Bell (1985)). One possible generalization of results relating to the theory of dyadic compacta is the following result.

Theorem 42 (Bell (1985)). *If the Stone-Čech compactification $\beta(X)$ of a space X is centered, then X is pseudocompact.*

Theorem 43 (Bell (1985)). *Every centered compactum satisfying the first axiom of countability is metrizable.*

The weight of any polyadic compactum is equal to the product of its Suslin number and tightness (Gerlits, see Bell (1985)). It is worth pointing out that this result does not extend to centered compacta (see Bell (1985)). It is true, however, that the spread of a centered compactum is equal to its weight. In a strengthening of a well known theorem of Schapiro, Bell proved that the space $\mathrm{Exp}(D^{\aleph_2})$ of all closed subsets of the Cantor cube $D^{\aleph_2}$ (in the Vietoris topology) is not centered. However $\mathrm{Exp}(D^{\aleph_2})$ is a supercompactum. Consequently, not every supercompactum is centered.

Theorem 44 (see Bell (1985)). *If the character of a centered compactum X at a point $x \in X$ is not less than τ, then there exists a discrete subspace A of X of cardinality τ such that the subspace $A \cup \{x\}$ is homeomorphic to the Aleksandrov compactification of the space A.*

Corollary 17. *Every nonisolated point in a centered compactum is the limit of a nontrivial convergent sequence.*

The last results generalize the well known theorems relating to the theory of dyadic compacta.

It is not known whether every closed G_δ-set in a centered compactum is a centered compactum (Bell, see Bell (1985)). Bell also raised the question whether the compactum $\mathrm{Cen}(\beta(\mathbb{N}) \setminus \mathbb{N})$ is homogeneous.

§6. Compact Extensions

6.1. General Remarks about Compact Extensions. Compactness, in general, is not a hereditary property. The role of compactness in topology is so essential that it is natural to raise the question of the possibility of representing topological spaces as subspaces of compact spaces. The most general and principally the most important question concerning the possibility of embedding a topological space into a compact space can easily be answered in a positive way. Every topological space X can be "extended" to a compact space by adding to it a single point ξ and defining $X \cup \{\xi\}$ as the only neighborhood of ξ. Under this definition all open subsets of X are open in $X \cup \{\xi\}$. Let us assume that $X \neq \emptyset$. The compact space $X \cup \{\xi\}$ containing X as a dense open subspace has one essential deficiency. It does not even satisfy the T_1 separation axiom since none of the sets $\{x\}$, where $x \in X$, is closed in $X \cup \{\xi\}$.

In connection with the above example the general problem of embedding spaces into compact spaces needs to be divided into separate, very concrete and frequently quite subtle questions whether every topological space with some property $\mathcal{P}$ can be embedded into a compact space with the same property. Of particular importance are the cases when in place of $\mathcal{P}$ one considers separation axioms. This is related to the fact that the properties of compact Hausdorff spaces, or compacta, differ dramatically from the properties of arbitrary compact spaces or even compact T_1-spaces.

In what follows the word "embedding" means "homeomorphic mapping into".

Proposition 1. *Every T_1-space X can be embedded in a compact T_1-space and this can be accomplished by adding to X just one point.*

$\triangleleft$ Let $a \notin X$ and let $\mathcal{F}$ be the family of all closed compact subsets of X. Define a topology on the set $\widetilde{X} = X \cup \{a\}$ in the following way: all open subsets of X are also open if $\widetilde{X}$ and a base at the point a is determined by all sets of the form $\widetilde{X} \setminus B$, where $B \in \mathcal{F}$. The space $\widetilde{X}$ is compact, X is an open subspace of it and if X is not compact, then X is dense in $\widetilde{X}$. All compact closed subsets of X are closed in $\widetilde{X}$. Consequently, if X is a T_1-space then $\widetilde{X}$ is also a T_1-space. $\triangleright$

Not every Hausdorff space can be embedded in a compact Hausdorff space since subspaces of a compactum are Tikhonov spaces and a Hausdorff space does not have to be Tikhonov (see Engelking (1977)). Among the regular T_1-spaces there are also non-Tikhonov ones. They cannot be embedded into any compactum. On the other hand, one of the classical theorems of Tikhonov can be formulated in the following way (see Arhangel'skii and Ponomarev (1974)).

Theorem 1. *A space is Tikhonov if and only if it can be embedded into a compact Hausdorff (hence, normal and Tikhonov) space.*

According to Theorem 1, Proposition 1 and previous remarks, the whole theory of embedding of topological spaces into compact spaces is divided into two basic parts. In the first, one considers the most general questions related to embedding of T_1 spaces into compact T_1-spaces. In the second (and the main one) a theory of embedding of Tikhonov spaces into compacta is developed. This theory is presently well advanced and is rich with important theorems (see Arhangel'skii and Ponomarev (1974), Walker (1974)).

Since compactness is hereditary with respect to closed subspaces, it is natural to limit our attention to embeddings of spaces as dense subspaces of compact spaces. Clearly, in this case the properties of the embedded space are more closely related to the properties of the space containing it. The fundamental definition is therefore formulated in the following way.

A space X is called an *extension of a space* Y if Y is a subspace of X and Y is dense in X. If in addition X is compact (and Hausdorff) then X is called a *compact (Hausdorff) extension (compactification) of the space* Y. Sometimes this terminology is slightly altered. If there exists a homeomorphism φ of the space Y onto a dense subspace of some space X, then X is called an extension of Y keeping in mind that Y is naturally identified with the subspace $\varphi(Y)$ of X. Since the extension theory deals simultaneously with spaces and their subspaces, it was decided to denote the closure in the extension of Y by the bar over a set, while the closure of $A \subset Y$ in the space Y is denoted by $\mathrm{Cl}_Y(A)$. Both in the theory of compact T_1-extensions and in the theory of compact Hausdorff extensions an important role is played by questions of the following type. Let φ be a cardinal topological invariant or a dimensional function. Is it true that every T_1-space Y (every Tikhonov space Y) can be embedded into a compact T_1-space X (into a compactum X) such that $\varphi(X) = \varphi(Y)$?

Another group of natural questions is related to the concept of the remainder. We define the *remainder of the space* Y *in its extension* X as the subspace $X \setminus Y$ of the space X. What are the connections between the properties of a space Y and its remainder? When does a space have a compact extension with a remainder having a prescribed property? It is natural to expect that the properties of the remainder will not duplicate the properties of the space, but will be related to them by some form of duality. One of the major goals of the theory of compact (Hausdorff) extensions is to detect and study systematically such relationships.

6.2. Compact T_1-Extensions. The compact T_1-extension of an arbitrary T_1-space described in the proof of Proposition 1 is interesting only from one point of view. In a certain sense it is minimal since its remainder consists of a single point. Of much more complex nature is the construction of the *Wallman extension* of an arbitrary T_1-space. It can be considered the opposite of the minimal extension. We will provide an outline of this construction.

Let X be an arbitrary T_1-space. Denote by $w(X)$ the collection of all maximal families of closed subsets of X with the finite intersection property (the elements of the collection $w(X)$ do not have to be maximal families of sets in X with the finite intersection property). We will identify every point $x \in X$ with the element $\xi_x = \{F \subset X : F$ is closed in X and $x \in F\}$ of the family $w(X)$, that is with the family of all closed subsets of X containing the point x. By this construction X becomes a subset of the set $w(X)$. We will use the symbol $\mathcal{F}(X)$ to denote the family of all closed subsets of X. For $F \in \mathcal{F}(X)$ put $\widehat{F} = \{\xi \in w(X) : F \in \xi\}$ and define a set $H \subset w(X)$ to be closed if it can be represented as the intersection of some family of elements of $\mathcal{P} = \{\widehat{F} : F \in \mathcal{F}(X)\}$. This determines the topology on $w(X)$. The space $w(X)$ is compact, satisfies the T_1 separation axiom and contains the space X

as a dense subspace. In addition we have $\overline{F} = \widehat{F}$, for all $F \in \mathcal{F}(X)$. The proof of compactness of $w(X)$ is based on the following fact: for every maximal family ξ of closed subsets of X with the finite intersection property it is true that $\{\xi\} = \bigcap\{\overline{F} : F \in \xi\} = \bigcap\{\widehat{F} : F \in \xi\}$. The extension $w(X)$ is called the *Wallman extension* of X in honor of Wallman who was the first to give its construction (see Arhangel'skii and Ponomarev (1974)). Its extraordinary properties are expressed in the following two closely related statements.

Proposition 2. *For any disjoint closed sets A and B in a T_1-space X their closures in the Wallman extension $w(X)$ are disjoint.*

◁ By definition, $\xi \in \overline{A}$ if and only if $A \in \xi$. Analogously, $\xi \in \overline{B}$ is equivalent to $B \in \xi$. However, disjoint sets A and B cannot belong to the same family with the finite intersection property. ▷

Theorem 2. *Every continuous mapping of a T_1-space X into an arbitrary compactum Z can be extended to a continuous mapping of $w(X)$ into Z.*

Of particular importance is the following.

Corollary 1. *Every bounded continuous real-valued function on a T_1-space X can be extended to a continuous, real-valued function on $w(X)$.*

From Proposition 2 it follows that the remainder $w(X) \setminus X$ of the Wallman extension typically contains a very large number of points. Moreover, Proposition 2 proves that, in some sense, the compact extension $w(X)$ is maximal. However, infinite spaces do not have any compact T_1 extensions that are maximal with respect to inclusion. Every such space can be represented as a dense subspace of a compact T_1-space of arbitrarily high cardinality.

Proposition 3. *If Y is a closed subspace of a T_1-space X, then the closure of Y in $w(X)$ is canonically homeomorphic to the Wallman extension $w(Y)$ of Y.*

Example 1. What is the Wallman extension of the interval $J = (0,1)$? Clearly not the interval $[0,1]$, which follows easily from Proposition 2 (or from Theorem 2). It is possible to show that the cardinality of $w(J)$ is equal to $2^{\mathfrak{c}}$, where $\mathfrak{c} = 2^{\aleph_0}$. Moreover, the extension $w(J)$ is not metrizable. Similarly, the Wallman extension $w(\mathbb{N})$ of the set of natural numbers is a compact space of cardinality $2^{\mathfrak{c}}$. Indeed, $\mathbb{N}$ can be mapped continuously on to a dense countable subspace of the Tikhonov cube $I^{\mathfrak{c}}$ (see Arhangel'skii and Ponomarev (1974)). By Theorem 2 this mapping can be extended to a continuous mapping of $w(\mathbb{N})$ onto the cube $D^{\mathfrak{c}}$. But $|D^{\mathfrak{c}}| = 2^{\mathfrak{c}}$ and consequently $|w(\mathbb{N})| \geq 2^{\mathfrak{c}}$.

The examples considered above illustrate the fact that even in the case of the simplest topological spaces the Wallman compact extension has quite complicated structure.

Theorem 3. *The Wallman extension $w(X)$ of a T_1-space X satisfies the Hausdorff separation axiom if and only if the space X is normal.*

In particular, if X is metrizable, then $w(X)$ is a compactum. The compactum $w(\mathbb{N})$ has several remarkable properties (see §6.4). In particular, $w(\mathbb{N})$ does not contain a nontrivial convergent sequence. Consequently, $w(\mathbb{N})$ does not have a countable base. As a result, we arrive at one of the first questions about embedding into compact T_1-spaces while preserving the values of cardinal invariants: does every T_1-space X with a countable base have a compact extension $b(X)$ with a countable base? The answer to this question, as we will see in the next section, is negative.

6.3. Embedding of Topological Spaces into Compact T_1-Spaces of Countable Weight. Arhangel'skii proved the following.

Theorem 4. *A T_1-space X with a countable base has a compact T_1-extension with a countable base if and only if X has a uniform base.*

Corollary 1 (Arhangel'skii (1966)). *Every compact T_1-space with a countable base has a uniform base.*

Corollary 2 (Arhangel'skii (1966)). *Every compact T_1-space with a countable base as well as any subspace of such space can be represented as an image of a separable metrizable space under a continuous open mapping with compact inverse images of points.*

Corollary 3 (see Nedev (1971)). *Every compact T_1-space with a countable base as well as any subspace of such space is symmetrizable with a metric satisfying Cauchy's condition.*

◁ This follows from Corollary 1 (or Corollary 2). ▷

The proof of Theorem 4 is based on the following general result (see Arhangel'skii (1966)).

Theorem 5. *Let X be a T_1-space of weight $\tau \geq \aleph_0$. Then the following conditions are equivalent:*
(a) there exists a compact T_1-extension of weight τ of the space X,
(b) the space X has a network of cardinality τ consisting of closed sets,
(c) every closed subset of X is an intersection of at most τ open subsets.

Example 2 (see Arhangel'skii (1966)). Let $\mathcal{T}^*$ be the minimal topology on the real line $\mathbb{R}$ that contains all standard open sets and the set of irrational numbers. The space $(\mathbb{R}, \mathcal{T}^*)$ is Hausdorff and has a countable base. It does not have a countable network of closed sets since the set of rational numbers is closed in $\mathbb{R}$ and is not an intersection of a countable family of open subsets of $(\mathbb{R}, \mathcal{T}^*)$. Consequently, by Theorem 5, the space $(\mathbb{R}, \mathcal{T}^*)$ cannot be embedded in a compact T_1-space with countable base.

Theorem 4 should be compared to the next result.

Proposition 4. *Every T_1-space with a network of cardinality $\leq \tau$ can be embedded in a compact T_1-space with a network of cardinality $\leq \tau$.*

◁ This follows from the second part of Proposition 1. ▷

The question concerning embedding into compact T_1-spaces with preservation of the other cardinal invariants has not been studied very extensively. Let us mention that the density of a space is always at least equal to density of any of its extensions, while the Suslin number of a space is always equal to the Suslin number of any extension of this space.

6.4. Compact Hausdorff Extensions, Relation of Subordination. The fundamental fact of the theory of compact Hausdorff extensions was already established: Tikhonov spaces and only such spaces have such extensions (Theorem 1). In this section we consider the properties of the family of all compact Hausdorff extensions of an arbitrary Tikhonov space X and we describe the most important compact Hausdorff extension of a Tikhonov space, namely its Stone-Čech compactification. Particular attention is paid to the construction of compact Hausdorff extensions.

Extensions $b_1(X)$ and $b_2(X)$ of a space X are called *equivalent* if there exists a homeomorphism $f : b_1(X) \to b_2(X)$, such that $f(b_1(X)) = b_2(X)$ and $f(x) = x$ for all $x \in X$. Such a homeomorphism maps one remainder onto the other. Since the space X is dense in any of its extensions, every homeomorphism f with the properties described above, if it exists, is determined uniquely. It is natural to consider equivalent extensions identical. Since the cardinality of every compact Hausdorff space does not exceed $2^{2^{|X|}}$, it makes sense to talk about the set $\mathcal{B}(X)$ of all compact Hausdorff extensions of an arbitrary Tikhonov space X and perform all set-theoretic operations on arbitrary subfamilies of $\mathcal{B}(X)$. The following construction is based on these remarks.

Construction 1. Let $\gamma = \{b_\alpha(X) : \alpha \in A\}$ be an arbitrary nonempty family of compact Hausdorff extensions of a Tikhonov space X. Assign to every point $x \in X$ the point $\widehat{x} = \{x_\alpha : \alpha \in A\}$, where $x_\alpha = x$, belonging to the product $B = \prod\{b_\alpha(X) : \alpha \in A\}$. The subspace $\widehat{X} = \{\widehat{x} : x \in X\}$ of the space B is canonically homeomorphic to the space X, and the whole product B is a compactum. Let us identify X with $\widehat{X}$ and denote by $b(X)$ the closure of X in B. In this way we obtain a compact Hausdorff extension $b(X)$ of the space X. Under the natural projection π_α of B onto the α-coordinate $b_\alpha(X)$ the extension $b(X)$ is mapped continuously onto the extension $b_\alpha(X)$ in such way that X is mapped identically onto X and the remainder $b(X) \setminus X$ is mapped onto the remainder $b_\alpha(X) \setminus X$. It is reasonable, in such situation, to say that for every $\alpha \in A$ the extension $b_\alpha(X)$ is subordinate to the extension $b(X)$. This leads to the following definition.

Let $b'(X)$ and $b''(X)$ be any two extensions of a space X. We will say that the extension $b''(X)$ is *subordinate to* the extension $b'(X)$ (notation $b''(X) \leq b'(X)$) if there exists a continuous mapping $f : b'(X) \to b''(X)$ such that $f^{-1}(x) = x$ for all $x \in X$. Note that if $b''(X)$ and $b'(X)$ are compact Hausdorff extensions, then $f(b'(X)) = b''(X)$ since X is dense in $b''(X)$ and a compact set is always closed in a Hausdorff space.

Using the fact that the family $\mathcal{B}(X)$ of all compact Hausdorff extensions of X is a set, we will write it in the form $\mathcal{B}(X) = \{b_\alpha(X) : \alpha \in A\}$ and apply Construction 1 to it. As a result we obtain a compact Hausdorff extension $b(X)$ of X such that every compact Hausdorff extension of X is subordinate to it. The relation of subordination $\leq$ introduced above in the case of Hausdorff extensions is a (partial) order. It is reflexive, transitive and the conditions $b'(X) \leq b''(X)$ and $b''(X) \leq b'(X)$ imply that $b'(X)$ coincides with $b''(X)$ (up to a topological equivalence). In particular, the relation of subordination is a partial order on the set $\mathcal{B}(X)$ of all compact Hausdorff extensions of a Tikhonov space X. The facts established earlier can now be formulated in the following way.

Theorem 6. *For every Tikhonov space X the set $\mathcal{B}(X)$ of all of its compact Hausdorff extensions is partially ordered by the relation of subordination. For each family $\gamma = \{b_\alpha(X) : \alpha \in A\} \subset \mathcal{B}(X)$ in $\mathcal{B}(X)$ there exists a least upper bound in $\mathcal{B}(X)$ (denoted $\sup\{b_\alpha(X) : \alpha \in A\}$). In particular in $\mathcal{B}(X)$ there exists a largest element, a compact Hausdorff extension (unique up to a topological equivalence and denoted by $\beta(X)$) of the Tikhonov space X that can be mapped onto any other extension of X by a mapping that is the identity on the points of X.*

The extension $\beta(X)$ is called the *Stone-Čech compactification* of the space X (see Čech (1937), Stone (1937)).

If a space X is normal then the extension $\beta(X)$ admits a more explicit description.

Theorem 7 (see Engelking (1977)). *If a T_1-space X is normal then the Stone-Čech compactification $\beta(X)$ of the space X coincides (up to a topological equivalence) with its Wallman extension $w(X)$.*

◁ According to Theorem 3, $w(X)$ is a compact Hausdorff extension of a normal space X. By Theorem 2, the identity mapping of X can be extended to a continuous mapping of the extension $w(X)$ onto an arbitrary compact Hausdorff extension $b(X)$ of this space. ▷

In particular $w(\mathbb{N})$ coincides with $\beta(\mathbb{N})$ for the discrete space $\mathbb{N}$ and $w(X) = \beta(X)$ for every metrizable space X.

In the case of an arbitrary Tikhonov space X the description of its Stone-Čech compactification is provided by the following.

Construction 2. Consider the diagonal $\Delta(\mathcal{F})$ of the set $\mathcal{F}$ of all continuous functions of a Tikhonov space X into the interval $[0, 1]$. The mapping $\psi = \Delta(\mathcal{F}) : X \to I^{\mathcal{F}}$ assigns to every $x \in X$ the point $\{x_f : f \in \mathcal{F}\}$ of the cube $I^{\mathcal{F}}$, where $x_f = f(x) \in I$ for all $f \in \mathcal{F}$. The mapping ψ is a homeomorphism of X onto a subspace $\psi(X)$ of $I^{\mathcal{F}}$. The composition of ψ with the projection π_f of the cube $I^{\mathcal{F}}$ onto the f-axis I coincides with the mapping f. Therefore, by identifying the space X with the subspace $\psi(X)$ of the cube $I^{\mathcal{F}}$ we identify an arbitrary mapping $f \in \mathcal{F}$ with the restriction to $\psi(X)$ of the projection π_f of $I^{\mathcal{F}}$ onto I. Consequently, every continuous mapping f of the space $X = \psi(X) \subset I^{\mathcal{F}}$ into the interval I can be extended to a continuous mapping of the whole cube $I^{\mathcal{F}}$, namely to π_f. Therefore every bounded continuous real-valued function f on X can be extended to a continuous real-valued function defined on the closure $\overline{X}$ of the space X in $I^{\mathcal{F}}$. Since $I^{\mathcal{F}}$ is a compactum, $\overline{X}$ is a compact Hausdorff extension of X.

Thus we have proved

Proposition 5. *Every Tikhonov space X has a compact Hausdorff extension $b(X)$ such that every bounded real-valued continuous function on X can be extended to a real-valued continuous function on $b(X)$.*

The next result is an essential generalization of the previous one.

Proposition 6. *Let X be a Tikhonov space and let $b(X)$ be the compact Hausdorff extension of X described in Proposition 5. Then every continuous mapping f of the space X into an arbitrary compactum B can be extended to a continuous mapping of the whole space $b(X)$ into this compactum.*

$\triangleleft$ We may assume that B is a subset of the product $Z = \prod\{I_\alpha : \alpha \in A\}$ of some family of intervals $I_\alpha = [0, 1]$. The composition $g_\alpha = \pi_\alpha \circ f$ of f with the projection $\pi_\alpha : Z \to I_\alpha$ is a continuous mapping of the space X into the interval $I_\alpha = [0, 1]$. By Proposition 5, the mapping g_α can be extended to a continuous mapping $\widetilde{g}_\alpha : b(X) \to I_\alpha$. The diagonal $\widetilde{f} = \Delta\{\widetilde{g}_\alpha : \alpha \in A\}$ of the mappings $\widetilde{g}_\alpha$ is a continuous mapping of the space $b(X)$ into the compactum B that is an extension of the mapping f. $\triangleright$

Proposition 6 implies that if a compact Hausdorff extension $b(X)$ of a space X satisfies the assumptions of Proposition 5, then every compact Hausdorff extension of the space X is subordinate to $b(X)$, that is $b(X)$ is the Stone-Čech compactification of X. Therefore, we have the following.

Theorem 8. *The Stone-Čech compactification $\beta(X)$ of a Tikhonov space X is the unique compact Hausdorff extension of X with the property that every bounded continuous real-valued function on X can be extended to a continuous real-valued function defined on the whole space $\beta(X)$.*

We conclude that Construction 2 described above produces the Stone-Čech compactification of the space X.

The subsets A and B of a Tikhonov space X are called *functionally separated* if there exists a continuous, real-valued function on X assuming the value 0 on the points of A and the value 1 on the points of B. Theorem 8 implies that the closures, in the Stone-Čech compactification $\beta(X)$, of functionally separated subsets of X do not intersect. This property, in fact, characterizes the Stone-Čech compactifications.

Theorem 9. *A compact Hausdorff extension $b(X)$ of a Tikhonov space X is its Stone-Čech compactification if and only if the closures in $b(X)$ of any two functionally separated subsets of X are disjoint.*

Corollary 4. *The Stone-Čech compactification of a normal T_1-space X can be characterized as the unique compact Hausdorff extension of X in which the closures of any two disjoint closed subsets are disjoint.*

Corollary 4 implies the following.

Corollary 5. *If X is a normal T_1-space and Y is a closed subset of X, then the closure of Y in the Stone Čech compactification $\beta(X)$ of X is the Stone-Čech compactification of Y.*

This result cannot be generalized to arbitrary Tikhonov spaces. It is associated with the fact that a continuous, real-valued function defined on a closed subset of a Tikhonov space, in general cannot be extended to a real-valued continuous function defined on the whole space.

We now consider the question of the existence of the smallest extension among all compact Hausdorff extensions of X.

6.5. Compact Extensions of Locally Compact Hausdorff Spaces. Let us begin with the following result.

Theorem 10. *A Tikhonov space X has the smallest (in the sense of the subordination) compact Hausdorff extension if and only if it is locally compact, that is, if every point of X has a neighborhood whose closure is compact.*

◁ Indeed, let $b(X)$ be a compact Hausdorff extension of the space X and let $|b(X) \setminus X| > 1$. Consider two different points $y_1, y_2 \in b(X) \setminus X$ and identify them. The quotient space obtained from $b(X)$ in this way is a compact Hausdorff extension of X that is subordinate to $b(X)$, but not equivalent to it. Consequently, the extension $b(X)$ can be minimal only when the remainder $b(X) \setminus X$ is empty or consists of a single point. But in the last case the space X is locally compact.

Conversely, if X is locally compact Hausdorff space that is not compact, then by adding to X a single new point a and defining the topology on $X \cup \{a\}$ in exactly the same way as in the proof of Proposition 1, we obtain a compact Hausdorff extension $b(X) = X \cup \{a\}$ of the space X whose remainder consists only of the point a. This extension is subordinate to every compact Hausdorff extension of the space X. ▷

In the theory of compact extensions locally compact spaces are characterized also by another condition.

Proposition 7. *A Tikhonov space X is locally compact if and only if it is an open subset of some (and thus every) compact Hausdorff extension of X.*

Proposition 7 leads to an important generalization of the concept of local compactness. A Tikhonov space X is called *Čech-complete* if it is a G_δ-set in some (and thus every) compact Hausdorff extension of X. For metrizable spaces Čech-completeness is equivalent to metrizablity by a complete metric, however Čech completeness is not limited to this case.

6.6. Duality Between Properties of a Space and of Its Remainder. We say (see Henricksen and Isbell (1958)) that a Tikhonov space X *has a topological property P at infinity* if the remainder $b(X) \setminus X$ of some compact extension $b(X)$ of the space X has the property P. Of particular interest are the *perfect properties P*, that is the properties that are invariants and inverse invariants of perfect mappings of Tikhonov spaces.

Recall that a continuous mapping is called *perfect* if it is closed and the inverse image of every point is a compactum. It is precisely such mappings between remainders that arise in connection with the relation of subordination.

Let $b(X)$ be a compact Hausdorff extension of a Tikhonov space X. There exists (see §6.4) a continuous mapping of the Stone-Čech compactification $\beta(X)$ of X onto $b(X)$ such that $f(x) = x$ for all $x \in X$. In addition $f(\beta(X) \setminus X) = b(X) \setminus X$ and the restriction $f|_{\beta(X) \setminus X}$ is a perfect mapping of the remainder $\beta(X) \setminus X$ onto the remainder $b(X) \setminus X$. This is true since f itself is perfect and $f^{-1}|_{(b(X) \setminus X)} = \beta(X) \setminus X$ (see Henricksen and Isbell (1958)). Consequently, we obtain the following result.

Proposition 8. *If P is a perfect property, then for every Tikhonov space X the following conditions are equivalent:*

(a) there exists a compact Hausdorff extension $b(X)$ such that $b(X) \setminus X$ has property P,

(b) the remainder of X in every compact Hausdorff extension of X has property P.

Among perfect properties are paracompactness, Lindelöfness, compactness, σ-compactness, Čech-completeness and many others (but not metrizability!).

A systematic study of the question whether a space has a particular property P at infinity originated with the classical work of Henricksen and Isbell (see Henricksen and Isbell (1958)). We will mention some of the typical results of this type that can be found in their work.

Proposition 9. *A Tikhonov space X is empty at infinity if and only if X is a compactum.*

Proposition 10. *A Tikhonov space is compact at infinity if and only if it is locally compact.*

Proposition 11. *A Tikhonov space is σ-compact at infinity if and only if it is Čech-complete.*

A space X is called a *space of countable order* (see Arhangel'skii (1965b)) if for every compactum $\Phi \subset X$ there exists a compactum $F \subset X$ such that $\Phi \subset F$ and $\chi(F, X) \leq \aleph_0$. The class of spaces of countable order contains all metrizable spaces and even all p-spaces (see Arhangel'skii (1965a)) (consequently, all locally compact and all Čech complete spaces are of countable order).

Theorem 11. *A Tikhonov space X is Lindelöf at infinity if and only if it is of countable order.*

The question which spaces are countable at infinity, that is, which spaces have compact Hausdorff extensions with countable remainder was also considered (in particular by Dimov and Hoshina). Extensions with a pointlike remainder (that is the remainder in which there are no nontrivial connected subsets) were considered by Sklyarenko (see (Sklyarenko (1962)).

6.7. Compact Extensions and Cardinal Invariants. Let us begin with a typical result.

Theorem 12 (see Engelking (1977)). *Every Tikhonov space X has a compact Hausdorff extension $b(X)$ whose weight is equal to the weight of X.*

◁ Assume that the weight of X does not exceed $\tau \geq \aleph_0$. Using the fact that X is Tikhonov and beginning with a base $\mathcal{B}$ of cardinality $\leq \tau$ of the space X we can construct a family F of continuous functions on X with values in the interval $[0, 1]$, such that $|\mathcal{F}| \leq \tau$, and for every point $x \in X$ and every closed set $A \subset X$ not containing x there exists a function $f \in \mathcal{F}$ equal to zero at x and equal to one for all points of A. Then the diagonal $\Delta\mathcal{F}$ of all the mappings in F maps X homeomorphically onto some subspace X' of the Tikhonov cube $I^{\mathcal{F}}$. Since the cube $I^{\mathcal{F}}$ is compact and its weight does not exceed $|\mathcal{F}| \leq \tau$, the closure of the set X' in the space $I^{\mathcal{F}}$ is a desired extension of the space X (as always, we identify X with X'). ▷

Statements similar to Theorem 12 are also true for many other, but not all, cardinal invariants. It is important to identify the properties of cardinal functions φ that determine whether every Tikhonov space X has a compact Hausdorff extension $b(X)$ such that $\varphi(b(X)) \leq \varphi(X)$.

Since for compacta network weight is equal to weight, not every Tikhonov space with a countable network has a compact Hausdorff extension with a countable network. Moreover, there exist countable Tikhonov spaces that cannot be embedded into any compactum of cardinality $\leq 2^{\aleph_0}$. Such is, in particular, every countable dense subspace of the Tikhonov cube $I^{\mathfrak{c}}$ of weight $\mathfrak{c} = 2^{\aleph_0}$ (see Efimov (1975)). Let us add to the discrete space $\mathbb{N}$ any point p of the remainder $\beta\mathbb{N} \setminus \mathbb{N}$. The subspace $\mathbb{N} \cup \{p\}$ of the space $\beta\mathbb{N}$ obtained in this way is countable and has exactly one nonisolated point p from the remainder $\beta\mathbb{N} \setminus \mathbb{N}$. The space $\mathbb{N} \cup \{p\}$ also does not have any compact Hausdorff extension of cardinality $\leq 2^{\aleph_0}$. Przymusiński proved that there exists a countable Tikhonov space that is Fréchet-Urysohn, contains a unique nonisolated point and every compact Hausdorff extension of which contains a topological copy of $\beta\mathbb{N}$ (see Arhangel'skii (1978)). Consequently, this countable space does not have any compact Hausdorff extension of countable tightness. Not every first countable Lindelöf space has a compact Hausdorff extension that is first countable. This is clear from another example given by Przymusiński, namely a first countable Tikhonov space with a countable network, every compact Hausdorff extension of which contains a copy of $\beta\mathbb{N}$ and consequently has uncountable tightness (see Arhangel'skii (1978), Arhangel'skii and Ponomarev (1974)). However, we have the following.

Theorem 13 (see Aleksandrov and Pasynkov (1977)). *Every Tikhonov space X has a compact Hausdorff extension of the same weight and the same dimension* dim *as X.*

◁ This result can be derived from Theorem 12, the theorem about the possibility of extending continuous mappings of a space into compacta to the whole Stone-Čech compactification of this space, and from Mardešić's factorization theorem (see Aleksandrov and Pasynkov (1977)). ▷

If a cardinal function φ is such that not every Tikhonov space has an extension that is a compactum and for which the value of φ is the same as for X, then we face the problem of characterizing Tikhonov spaces admitting such an extension. This problem is not solved in most cases. Ul'yanov found sufficient conditions for a first countable spaces to admit an embedding into a compactum that is also first countable (see Ul'yanov (1975)). It is not known, even in the case of countable spaces, when does a space of countable tightness have a compact Hausdorff extension of countable tightness. One of the reasons why such extensions might not exist is highlighted by the following theorem of Malykhin: the product of a compactum of countable tightness with a Tikhonov space of countable tightness is a space of countable tightness (see Arhangel'skii (1979b)). At the same time, the tightness of the product of a space of countable tightness with a countable space can be uncountable (as an example we can take a countable and an uncountable Fréchet-Urysohn fans, see Arhangel'skii (1979b)).

Theorem 14 (Arhangel'skii (1978)). *Every Tikhonov space of countable tightness has a countably compact Hausdorff extension of countable tightness.*

The following question is closely related to the problems considered earlier: when does a Tikhonov space X have a compact Hausdorff extension belonging to a specific class $\mathcal{P}$? For example, we may ask: when does a space have a compact extension that is a dyadic compactum?

Theorem 15 (Efimov (1965)). *A Tikhonov space satisfying the first axiom of countability has a compact Hausdorff extension that is a dyadic compactum if and only the space has a countable base.*

The last statement cannot be generalized to spaces of countable tightness (take for example the $\sum$-product of intervals in a Tikhonov cube).

Sometimes in the process of taking a compact Hausdorff extension of a space, it is possible not only to preserve the value of a cardinal function, but even to lower it. It is clear that this can happen only if this function is not monotone with respect to dense subspaces. Among the functions of that type the density of a space (that is the minimum of cardinalities of dense subsets of this space) is the most important. A space does not have to be separable to have a separable compact Hausdorff extension.

Example 3 (Comfort, see Engelking (1977)). In the Tikhonov cube I^c of the weight $c = 2^{\aleph_0}$ consider any nonseparable, dense subspace Y (for example, Y could be a $\sum$-product of intervals).

The density of compact extensions of a space has the following unexpected property.

Proposition 12 (see Engelking (1977)). *Any two compact Hausdorff extensions of an arbitrary Tikhonov space have the same density.*

For weight the analogous statement is not true. The question about the behavior of density of a space under the operation of compact Hausdorff extension is reduced, due to Proposition 12, to the comparison of the densities of a space and its Stone-Čech compactification. Van Douwen proved that there exists a nonseparable Čech-complete space X whose Stone-Čech compactification βX is separable (see van Douwen (1977)). The density $d(\beta(X))$ was characterized by van Douwen in terms of the topology of the space X itself in the following way (see van Douwen (1977)).

Theorem 16. *The condition $d(\beta(X)) \leq \tau$ is satisfied if and only if the topology T of the space X is τ-centered, that is, if it is the union of a family of cardinality $\leq \tau$ of families with the finite intersection property.*

6.8. Compact Hausdorff Extensions and Perfect Mappings. Continuous mappings are one of the fundamental means of comparing topological spaces. For that reason the question whether a continuous mapping of a space can be extended to a compact extension of this space is of fundamental importance. The most convenient, from this point of view, is the class of perfect mappings of Tikhonov spaces. We will begin with a general result following from Proposition 6.

Proposition 13. *An arbitrary continuous mapping $f : X \to Y$ of a Tikhonov space X into a Tikhonov space Y can be extended to a continuous mapping $\tilde{f} : \beta(X) \to \beta(Y)$ of the Čech-Stone compactifications of these spaces.*

The mapping $\tilde{f}$ in Proposition 13 is determined uniquely since X is dense in $\beta(X)$. Perfect mappings are characterized by the following property.

Proposition 14. *A continuous mapping f of a Tikhonov space X onto a Tikhonov space Y is perfect if and only if the remainder $\beta(X) \setminus X$ is mapped by $\tilde{f}$ into the remainder $\beta(Y) \setminus Y$.*

The conclusion of Proposition 14 can be expressed in the following way: $\tilde{f}^{-1}(\beta(Y) \setminus Y) = \beta(X) \setminus X$. Therefore, if the mapping $f : X \to Y$ is perfect, then the restriction of $\tilde{f} : \beta(X) \to \beta(Y)$ to the remainder $\beta(X) \setminus X$ of X is a perfect mapping of this remainder into $\beta(Y) \setminus Y$. This makes it possible to apply the theory of compact Hausdorff extensions to the proofs of the invariance of certain topological properties under perfect mappings. For example, a Tikhonov space X is of countable type if and only if the space $\beta(X) \setminus X$ is Lindelöf. Taking into account the fact that a continuous image of a Lindelöf space is Lindelöf we obtain

Corollary 6 (Arhangel'skii (1965b)). *Let X and Y be Tikhonov spaces and let $f : X \to Y$ be a perfect mapping such that $f(X) = Y$. Then, if X is of countable type, so is Y.*

Proposition 14 implies, for example, that perfect mappings of Tikhonov spaces preserve (in both directions) the property of having a compact Hausdorff extension whose remainder is paracompact (or is a paracompact p-space). It is not known whether these last properties can be expressed naturally in terms of the topology of a given space. Clearly the preservation (in both directions) of local compactness by perfect mappings can also be considered as a consequence of Proposition 14.

6.9. Properties of the Stone-Čech Extension. From the earlier results it is clear that we can consider the assignment to a Tikhonov space of its Stone-Čech compactification as a functor from the category of Tikhonov spaces and their continuous mappings to the category of compacta and their continuous mappings. This functor is not exact. Spaces X and Y that are not homeomorphic can have homeomorphic Stone-Čech compactifications. We have, however, the following remarkable result of Čech (see Čech (1937)).

Theorem 17. *If X and Y are first countable Tikhonov spaces whose Stone-Čech compactifications $\beta(X)$ and $\beta(Y)$ are homeomorphic, then the spaces X and Y are homeomorphic.*

Theorem 17 and its generalization can be obtained from the following Proposition 15 that is also of independent interest.

A point x of a space X is called a *strong κ-point* if there exists a sequence $\xi = \{U_n : n \in \mathbb{N}^+\}$ of nonempty, open subsets of X that is convergent to x. The last statement means that every neighborhood of x contains almost all terms of the sequence ξ. In a first countable space all points are strong κ-points.

Proposition 15. *No point of the remainder $\beta(X) \setminus X$ of the Stone-Čech compactification of any Tikhonov space X is a strong κ-point of the space $\beta(X)$.*

◁ Let $x^* \subset \beta(X) \setminus X$ and let $\{U_n : n \in \mathbb{N}^+\}$ be a sequence of nonempty open subsets of $\beta(X)$ convergent to x^*. It is easy to construct a sequence $\{V_n : n \in \mathbb{N}^+\}$ of mutually disjoint, nonempty open subsets of X that is convergent to x^*. For every $n \in \mathbb{N}^+$ select a point $x_n \in V_n$ and let f_n be a continuous function on X with values in the interval $I = [0,1]$, for which $f_n(x_n) = 1$ and $f_n(X \setminus V_n) \subset \{0\}$. Define a function $g : X \to I$ in the following way: g coincides with f_n for even values of n and $g(x) = 0$ for $x \in X \setminus \bigcup\{V_{2n} : n \in \mathbb{N}^+\}$. The function g is continuous, bounded and assumes the value 0 at every point of the set $A = \{x_{2n+1} : n \in \mathbb{N}^+\}$ and the value 1 at every point of the set $B = \{x_{2n} : n \in \mathbb{N}^+\}$. Since $x^* \in \overline{A} \cap \overline{B}$, the function g cannot be extended to a continuous real-valued function defined on $\beta(X)$. But this gives a contradiction with the fundamental property of the Čech-Stone compactification. $\triangleright$

If X is a Tikhonov space every point of which is a strong κ-point in X, then X is called a *Efimov space*. Čech's theorem is a special case of the following general result.

Theorem 18. *If X and Y are Efimov spaces whose Čech-Stone compactifications are homeomorphic, then X and Y are homeomorphic.*

$\triangleleft$ Under a homeomorphism strong κ-points are mapped to strong κ-points. Since X is dense in $\beta(X)$ and $\beta(X)$ is regular, every point of X is a strong κ-point in $\beta(X)$. Similarly, Y is the set of all strong κ-points of compactum $\beta(Y)$. Consequently, under a homeomorphism of $\beta(X)$ onto $\beta(Y)$ the space X is mapped homeomorphically onto Y. $\triangleright$

All bisequential Tikhonov spaces are Efimov spaces (see Arhangel'skij (1979a)). Thus we have the following

Corollary 7 (Arhangel'skii (1979a)). *If the Čech-Stone compactifications of bisequential spaces X and Y are homeomorphic, then the spaces X and Y are homeomorphic.*

Corollary 7 applies also to all (Tikhonov) spaces that are perfect images of first countable spaces.

Corollary 8. *If the Stone-Čech compactifications of metrizable spaces X and Y are homeomorphic, then also X and Y are homeomorphic.*

Example 4. $\triangleleft$ Let $X = \mathbb{Q}$ be the space of rational numbers. Take any subspace Y of the space $\beta(X)$, such that $X \subset Y \subset \beta(X)$ and $X \neq Y \neq \beta(X)$. It is obvious that $\beta(X) = \beta(Y)$. No point $y \in Y \setminus X$ is a strong κ-point of Y (see Proposition 9). Consequently, Y is not a Efimov space and moreover X and Y are not homeomorphic. This argument proves that Čech's theorem does not generalize to the class of all Tikhonov spaces. $\triangleright$

Čech's theorem cannot be generalized to Tikhonov spaces that are Fréchet-Urysohn. Reznichenko gave a construction of nonhomeomorphic, monolithic, Tikhonov spaces that are Fréchet-Urysohn and have homeomorphic Stone-Čech compactifications.

The question when $\beta(X \times Y) = \beta(X) \times \beta(Y)$ turns out to be quite subtle. Glicksberg (see Engelking 1977)) proved the following.

Theorem 19. *For infinite Tikhonov spaces X and Y the formula $\beta(X \times Y) = \beta(X) \times \beta(Y)$ is true if and only if the space $X \times Y$ is pseudocompact.*

Recall that the product of two pseudocompact spaces is not always pseudocompact (see Engelking (1977)). Theorem 19 can be generalized with certain additional assumptions to the case of infinite products (see Engelking (1977)).

A special variety of locally compact spaces consists of those, whose Stone-Čech compactification has a one-point remainder.

Example 5. ◁ Let I^τ be the Tikhonov cube, where $\tau > \aleph_0$, $x \in I^\tau$ and $X = I^\tau \setminus \{x\}$. Then $\beta(X) = I^\tau$ and $\beta(X) \setminus X = \{x\}$. This follows from the fact that every continuous real-valued function on the space X depends only on a countable number of coordinates. ▷

Example 6. ◁ Let $X = T(\omega_1)$ be the space of ordinal numbers that are less than the first uncountable ordinal ω_1 in the order topology. Then $\beta(X) = T(\omega_1 + 1) = T(\omega_1) \cup \{\omega_1\}$ and the remainder $\beta(X) \setminus X = \{\omega_1\}$ consists of a single point. This follows from the fact that every continuous real-valued function on the space $X = T(\omega_1)$ is eventually constant, that is beginning with some ordinal it assumes the same value. ▷

The Stone-Čech compactification $\beta\mathbb{N}$ of the set of $\mathbb{N}$ of all natural numbers has been the subject of particularly careful study. Even though this space is separable, its remainder $\beta\mathbb{N} \setminus \mathbb{N}$ is not. It follows that there does not exist an operator of extension of continuous, real- valued functions from the compactum $\beta\mathbb{N}\setminus\mathbb{N}$ to the whole compactum $\beta\mathbb{N}$ that is continuous in the topology of pointwise convergence or in the topology of uniform convergence. The compactum $\beta\mathbb{N} \setminus \mathbb{N}$ does not contain any nontrivial convergent sequences and it is not homogeneous (see Arhangel'skii (1978), Frolik (1977)).

The following result shows that, in a certain sense, the compactum $\beta\mathbb{N}$ is the largest among all separable compacta.

Theorem 20. *Every separable compactum B is a continuous image of the compactum $\beta\mathbb{N}$.*

◁ Let S be a countable, dense subset of B. Consider an arbitrary mapping f of $\mathbb{N}$ onto S and extend f to a continuous mapping $\tilde{f}$ of $\beta\mathbb{N}$ into B. Then $\tilde{f}(\beta\mathbb{N}) = B$. ▷

In particular $\beta\mathbb{N}$ can be mapped continuously onto $D^{\mathfrak{c}}$ and $I^{\mathfrak{c}}$.

Another direction in the study of compact Hausdorff extensions is illustrated by the following result

Proposition 16 (see Arhangel'skii and Ponomarcv (1974), Parovichenko (1963)). *Every compactum Y of weight $\aleph_1$ can be represented as the remainder of some compact Hausdorff extension of the discrete space $\mathbb{N}$.*

We will give one application of Proposition 16.

Theorem 21 (Parovichenko (1963)). *Every compactum B of weight $\leq \aleph_1$ is a continuous image of $\beta\mathbb{N} \setminus \mathbb{N}$.*

◁ Take a compact Hausdorff extension $b(\mathbb{N})$ of the space $\mathbb{N}$ whose remainder $b(\mathbb{N}) \setminus \mathbb{N}$ is homeomorphic to B. The identity mapping of the space $\mathbb{N}$ onto itself can be extended to a continuous mapping of the compactum βN onto the compactum $b(N)$. Then $g(\beta N \setminus N) = b(N) \setminus N = B$. $\triangleright$

Corollary 9 (Parovichenko (1963)). *If the Continuum Hypothesis $2^{\aleph_0} = \aleph_1$ is true, then compacta of weight $\leq 2^{\aleph_0}$ and only such compacta are continuous images of $\beta N \setminus N$.*

$\triangleleft$ In addition to Theorem 21 it is necessary to use the fact that continuous mappings of compacta onto Hausdorff spaces do not increase weight (see Arhangel'skii and Ponomarev (1974)). $\triangleright$

Corollary 9 shows that, under the Continuum Hypothesis, the compactum $\beta N \setminus N$ plays in the class of all compacta of weight $\leq \mathfrak{c} = 2^{\aleph_0}$ the same role as the Cantor set in the class of all metrizable compacta. However, it is consistent with ZFC to assume that $\aleph_2 < 2^{\aleph_0}$ and the space $T(\omega_2 + 1)$ of all ordinal numbers not exceeding ω_2 is not a continuous image of $\beta N \setminus N$ (Kunen, see Juhász (1980)).

Under the indicated assumptions a perfectly normal compactum does not have to be separable. However, the following statement can be proved "naively".

Theorem 21 (Przymusiński, see Przymusiński (1982)). *Every perfectly normal compactum can be represented as a continuous image of the compactum $\beta N \setminus N$.*

The question whether every first countable compactum is a continuous image of the compactum $\beta N \setminus N$ remains open. According to Corollary 9 the answer is positive if $2^{\aleph_0} = \aleph_1$.

There exists a characterization of compacta that can be embedded into $\beta N \setminus N$. Zero-dimensionality and the absence of nontrivial convergent sequences are necessary conditions for the existence of such embedding (see van Mill (1984)).

There are many delicate, set-theoretic and topological questions related to the study of different natural orders on the set $\beta N \setminus N$ of all free ultrafilters on N (see van Mill (1984), Rudin (1971)).

6.10. Closing Remarks Concerning Compact Hausdorff Extensions. Many papers (see Arhangel'skii and Ponomarev (1974), Engelking (1977)) were dedicated to different descriptions and constructions of compact Hausdorff extensions of a Tikhonov space X in terms reflecting the nature of the space X itself.

Many attempts to represent each compact Hausdorff extension of a space X as an extension of Wallman type whose elements are maximal families of closed subsets of X with the finite intersection property belonging to some fixed family satisfying natural conditions, resulted in a negative result obtained V.M. Ul'yanov (see Ul'yanov (1977)).

A description of compact Hausdorff extensions using the concept of a proximity on a space was obtained by Smirnov. It is based on a general concept of a proximity space proposed by Efremovich (see Engelking (1977)). There exists a canonical one-to-one relationship between compact Hausdorff extensions and proximities. In one direction this relationship can be described in the following way: if $b(X)$ is a compact Hausdorff extension of X, then the subsets A and B of X are near if and only if their closures in $b(X)$ intersect (see Engelking (1977)). A similar "connection" exists between the theory of uniform spaces and the theory of compact Hausdorff extensions, however in this case there is no one-to one correspondence. Different uniformities on a Tikhonov space X can correspond to the same proximity and consequently to the same compact Hausdorff extension (see Arhangel'skii and Ponomarev (1974), Engelking (1977)).

Sklyarenko (see Sklyarenko (1962)) created the theory of *perfect compact Hausdorff extensions* of Tikhonov spaces. They are characterized by the following property: $O(U \cup V) = O(U) \cup O(V)$ for any disjoint open subsets U and V of X, where $O(U)$ is the largest open subset of the extension such that $O(U) \cap X = U$. A compact Hausdorff extension $b(X)$ is perfect if the Stone-Čech compactification $\beta(X)$ can be mapped continuously onto $b(X)$ in such way that inverse images of all points are connected (Sklyarenko, see Sklyarenko (1962)).

Inasaridze (see Inasaridze (1966)) considered the remainders of higher order of Tikhonov spaces. If X is a Tikhonov space and $b(X)$ is its compact Hausdorff extension, then $Y = b(X) \setminus X$ is referred to as the remainder of the first order, $Z = \overline{Y} \setminus Y$ the remainder of the second order, $\overline{Z} \setminus Z$ of the third order, etc. (closures are considered in $b(X)$). He clarified, in particular, when a sequence of remainders defined above, terminates. This is associated with the degree to which a space resembles a locally compact space.

§7. Compactness and Spaces of Functions

While the foundations of general topology, and thus also of the theory of compact extensions, belong to the area of set theory and mathematical logic, the credit for the creation and growth of general topology must be given to the theory of functions. Precisely this theory provided (and is still providing) the enlightenment and energy responsible for the development of modern general topology.

One of the sources of general topology, the theory of metric spaces, is a vital part of theory of functions. On the other hand, precisely the ideas of functional analysis are responsible for the growth of general topology beyond the class of metric spaces. These ideas include concepts of pointwise convergence, weak topology, compact-open topology, uniform structure and many others.

It is not surprising, therefore, that many important applications of topological concepts and constructions belong to the area of functional analysis and that several fundamental principles, associated with general topology, have a twofold, topologically-functional or linear-topological, character. Such results include the theorems of Stone-Weierstrass, Krein-Milman, Arzela-Ascoli, Alaoglu, Markov-Kakutani and many others.

In this section we provide a review of such principles and discuss their role in general topology and its applications.

7.1. Natural Topologies on Spaces of Functions. We will denote by $\mathbb{R}^X$ and $C(X)$, respectively, the sets of all real-valued and the set of all continuous real-valued functions on a topological space X. There are several different natural ways of introducing a topology on $\mathbb{R}^X$ and $C(X)$ and they all follow the general approach indicated below.

Let S be a family of subsets of X. We assign to each finite sequence $P_1, \ldots, P_k$ of elements of S and to each sequence (of the same length) $U_1, \ldots, U_k$ of open subsets of $\mathbb{R}$ the set $O'\langle P_1, \ldots, P_k; U_1, \ldots, U_k \rangle$ consisting of all $f \in \mathbb{R}^X$ for which $f(P_i) \subset U_i$, for all $i = 1, \ldots, k$. Put $O\langle P_1, \ldots, P_k; U_1, \ldots, U_k \rangle = C(X) \cap O'\langle P_1, \ldots, P_k; U_1, \ldots, U_k \rangle$. Sets of this form are a base of some topology on $\mathbb{R}^X$, respectively on $C(X)$, called the topology of convergence with respect to elements of S. Under this construction the space $C(X)$ is a subspace of the space $\mathbb{R}^X$. The family S is typically assumed to be quite rich, for example, one can require that it be a network of X. This condition assures sufficiently good separation properties of $\mathbb{R}^X$ and $C(X)$. The most important topologies obtained in this way are the following: the *topology of pointwise convergence* corresponding to the case when S consists of all one-point (or all finite) subsets of X, and the *compact-open topology* arising when S consists of all compact subsets of the space X. If S consists of all bounded subsets of X (recall, that those are the sets $A \subset X$ on which every continuous, real-valued function on X is bounded) then the topology of convergence with respect to S is called the *bounded-open topology*. The following result is of fundamental importance.

Proposition 1 (see Bourbaki (1969a)). *For every Tikhonov space X the compact-open topology, the topology of pointwise convergence and the bounded-open topology are compatible with the natural linear structure on the space $C(X)$ and they induce on it the structure of a locally convex linear topological space over $\mathbb{R}$.*

We will denote by $C_k(X), C_0(X)$ and $C_p(X)$, respectively, the set $C(X)$ equipped with the compact-open topology $\mathcal{T}_k$, bounded-open topology $\mathcal{T}_0$ and the topology of pointwise convergence $\mathcal{T}_p$.

The topologies of *uniform convergence* on $\mathbb{R}^X$ and $C(X)$ are described in a slightly different way. Their bases consist of the sets $O'_\epsilon(f) = \{g \in \mathbb{R}^X : \sup_{x \in X} |f(x) - g(x)| < \epsilon\}$ and $O_\epsilon(f) = O'_\epsilon(f) \cap C(X)$, respectively. However, if a Tikhonov space X is not pseudocompact, then the topology of uniform convergence is not compatible with linear operations on $C(X)$. Indeed, if f is an unbounded, continuous function on X, then $0 \cdot f = \theta$ is a function that is identically equal to zero on X, but for no nonzero $\lambda \in \mathbb{R}$ the function $\lambda \cdot f$ belongs to the neighborhood $O_1(\theta) = \{g \in C(X) : \sup_{x \in X}|g(x)| < 1\}$.

For this reason the topology of uniform convergence is defined traditionally not on all of $C(X)$ but on the set $C^0(X)$ of all bounded continuous real-valued functions on X. In this case it is compatible with the linear structure on $C^0(X)$.

In the study of spaces of functions the question of compactness of some sets of functions in one of the indicated topologies is of great importance. The topology of pointwise convergence is included in the compact-open topology and the last one, in turn, is a subset of the bounded-open topology. Thus, since the topology of pointwise convergence is Hausdorff, the largest family of compact sets is created if the space $C(X)$ is equipped with the topology of pointwise convergence. More precisely, we have the following.

Proposition 2. *If a set $F \subset C(X)$ of continuous, real-valued functions on X is compact in the compact-open topology, the topology of uniform convergence or the bounded-open topology, then this topology coincides with the one induced on F by the topology of pointwise convergence.*

Below we will explain first when a family of real-valued functions is compact in the topology of pointwise convergence and later we will describe a more narrow class of families of functions that are compact in the compact-open topology.

We will use the following notation: if $F \subset \mathbb{R}^X$ and $x \in X$, then $F[x] = \{f(x) : f \in F\} \subset \mathbb{R}$.

Proposition 3. *A set $F \subset \mathbb{R}^X$ is compact in the topology of pointwise convergence if and only if F is closed in $\mathbb{R}^X$ relative to that topology and $F[x]$ is bounded (in $\mathbb{R}$) for every $x \in X$.*

A family F of real-valued functions on a Tikhonov space X is called *equicontinuous* if for every point $x \in X$ and every $\epsilon > 0$ there exists a neighborhood O_x of the point x such that $|f(y) - f(x)| < \epsilon$ for every $y \in O_x$ and every $f \in F$. The closure of every family of equicontinuous functions in $\mathbb{R}^X$ in the topology of pointwise convergence is again an equicontinuous family of functions (and, moreover, is contained in $C(X)$).

Proposition 4. *The topology of pointwise convergence coincides with the compact-open topology on every equicontinuous family $F \subset C(X)$ of real-valued functions on X.*

Recall that a space X is called a $k_{\mathbb{R}}$-*space* if for every real-valued function on X the function is continuous if and only if its restriction to every compactum in X is continuous. The class of $k_{\mathbb{R}}$-spaces contains all locally compact Hausdorff spaces, all Čech-complete spaces and all Hausdorff k-spaces. The following result is known as *Ascoli's Theorem*.

Theorem 1. *Let X be a Hausdorff $k_{\mathbb{R}}$-space and let F be a subspace of $C_k(X)$ of all continuous real-valued functions on X in the compact-open topol-*

ogy. Then F is compact if and only if F is closed in $C_k(X)$, the set $F[x]$ is bounded in $\mathbb{R}$ for each $x \in X$, and for each compactum $\Phi \subset X$ the family $\{f|_\Phi : f \in F\}$ of functions on Φ is equicontinuous.

Corollary 1. *Let X be a compactum, $F \subset C(X)$ and let F be equipped with the compact-open topology. Then F is compact if and only if F is closed in $C_k(X)$, $F[x]$ is bounded in $\mathbb{R}$ for each $x \in X$, and F is equicontinuous.*

Other versions of Ascoli's theorem deal with families of functions with values not necessarily in the space of real numbers, but also with values in an arbitrary uniform or Tikhonov space. Ascoli's theorem is often used in proofs of existence theorems in mathematical analysis. As a classical example one can take the proof of existence of a solution of a differential equation $y' = f(x, y)$, where f is a continuous function.

7.2. Joint Continuity and the Compact-Open Topology. In this subsection X will always denote a Tikhonov space and F will denote a subset of $C(X)$. Let us assign to each point (f, x) of the product $F \times X$ the point $f(x) \in \mathbb{R}$. The indicated mapping $F \times X \to \mathbb{R}$ will be denoted by ψ. A topology on F will be called *jointly continuous* if ψ is continuous in the product topology of $F \times X$. The discrete topology on F is always jointly continuous while the topology of pointwise convergence on F is almost never jointly continuous. If a topology is jointly continuous, then every larger topology is also jointly continuous. However, the smallest jointly continuous topology on F usually does not exist (see Kelley (1955)). In connection with this fact the following modification of the concept of joint continuity was proposed.

A topology on a set $F \subset C(X)$ is called *k-jointly continuous* if for every compactum $\Phi \subset X$ the restriction of the mapping ψ to the subspace $F \times \Phi$ of the space $F \times X$ is a continuous, real-valued function on $F \times \Phi$. One of the greatest advantages of the compact-open topology is expressed by the following result.

Theorem 2 (see Kelley (1955)). *The compact-open topology on F is the smallest k-jointly continuous topology on F.*

7.3. Stone-Weierstrass Theorem. The theorem of Stone-Weierstrass deals with the possibility of a uniform approximation of an arbitrary continuous real-valued functions on a compactum by functions of special type. The following is one of its formulations

Theorem 3 (see Engelking (1977)). *Let X be a compactum and let P be a subring of $C(X)$ of continuous real-valued functions on X that contains all constant functions and separates the points of X (the last statement means that for any two different points $x_1, x_2 \in X$ there exists a function $f \in P$ such that $f(x_1) \neq f(x_2)$). Then for every $\epsilon > 0$ there exists a function $f \in P$, such that $|g(x) - f(x)| < \epsilon$ for all $x \in X$.*

Theorem 3 implies the classical *theorem of Weierstrass* which states that every continuous real-valued function on an interval $[a, b] \subset \mathbb{R}$ is the uniform limit of a sequence of polynomials. The *Weierstrass theorem on approximation of periodic continuous functions by trigonometric polynomials* (see Engelking (1977)) is also based on Theorem 3.

Of great importance in functional analysis is the complex analog of the Stone-Weierstrass theorem in which one additionally assumes that together with each $f \in P$, the complex conjugate of f also belongs to P.

Theorem 3 can be obtained from the following "lattice" version proved by Kakutani and M.G. Krein (see Engelking (1977)). Let X be a compactum and let $P \subset C(X)$. Assume that the constant function $\mathbb{1}$ belongs to P, and that for every $f, g \in P$ and every $\alpha, \beta \in \mathbb{R}$ the functions $\alpha f + \beta g, f \vee g = \max\{f, g\}, f \wedge g = \min\{f, g\}$ also belong to P. If in addition P separates the points of X, then every function in $C(X)$ can be uniformly approximated by functions from P.

The Stone-Weierstrass theorem is frequently used inside general topology. In particular, with its help one can obtain a theorem about factoring continuous real-valued functions on products of compacta (see below). The proof of countability of the Lindelöf's number of the space $C_p(X)$ of continuous real-valued functions on an arbitrary Eberlein compact space provides another brilliant application of the Stone-Weierstrass theorem.

Let us point out that all bounded, continuous real-valued functions on X form a subring $P = C^0(X)$ in $C(X)$ satisfying the assumptions of Theorem 3. Since only bounded functions can be approximated uniformly by the functions of this subring, we conclude that if a Tikhonov space satisfies the Stone-Weirstrass theorem, then it is pseudocompact. Indeed, it is then a compactum, as was demonstrated by Hewitt (see Hewitt (1947)). The set of all continuous real valued functions on $\mathbb{R}$ that are constant outside of some interval satisfies the assumptions of the Stone-Weierstrass theorem but the bounded function $\sin x$ cannot be approximated uniformly by functions from this set. Thus, also in the case of bounded real-valued functions the Stone-Weierstrass theorem is not true even on the real line which is a locally compact, σ-compact, metrizable space. Nevertheless, by using the compact-open topology we can deduce from Theorem 3 an important corollary applicable in the case of Tikhonov spaces.

Corollary 2. *Let X be a Tikhonov space and let us assume that a subring $P \subset C(X)$ separates points of X and contains all constant functions. Then P is dense in $C(X)$ in the compact-open topology (that is in $C_k(X)$).*

◁ For a proof it is enough to consider the restrictions of functions in P to an arbitrary compactum and apply Theorem 3. ▷

Theorem 4 (Mibu, see Engelking (1977)). *An arbitrary continuous real-valued function f on the product $X = \prod\{X_\alpha : \alpha \in A\}$ of compacta depends only on a countable number of coordinates. More precisely, there exists* *a countable set $A_0 \subset A$ and a continuous function $f_0 : \prod\{X_\alpha : \alpha \in A_0\} \to \mathbb{R}$ such that f is a composition of the projection π_{A_0} of the product $\prod\{X_\alpha : \alpha \in A\}$ onto the face $\prod\{X_\alpha : \alpha \in A_0\}$ and of f_0, that is $f = f_0 \circ \pi_{A_0}$.*

◁ Denote by P the set of all continuous real-valued functions on X, for which Theorem 4 is true. They form a subring in $C(X)$ satisfying the assumptions of the Stone-Weierstrass theorem. In addition the limit of a uniformly convergent sequence of elements of P belongs to P. This follows from the fact that for every countable family of countable faces of X there exists a countable face containing all of them. Now the equality $P = C(X)$ follows from Theorem 3. ▷

Theorem 4 has several generalizations that are not associated directly with the concept of compactness. In particular, it can be generalized to dense subspaces of products.

Theorem 5 (Arhangel'skii (1984b), Arhangel'skii and Ponomarev (1974)). *Let Y be a dense subspace of the product $X = \prod\{X_\alpha : \alpha \in A\}$ of Tikhonov spaces and assume that for every countable set $B \subset A$ the product $\prod\{X_\alpha : \alpha \in B\}$ is hereditarily separable. Then for every continuous mapping f of Y onto a first countable Tikhonov space Z there exist a countable set $A^* \subset A$ and a continuous mapping g of the projection $\pi_{A^*}(Y)$ of the space Y (considered as a subspace of the face $\prod\{X_\alpha : \alpha \in A^*\}$ of X) onto Z such that $f = g \circ \pi_{A^*}$.*

Corollary 3 (Arhangel'skii (1984b)). *The conclusion of Theorem 5 is true for continuous real-valued functions on dense subspaces of arbitrary products of Tikhonov spaces with a countable network.*

The last statement is applied in studying continuous real-valued functions on spaces $C_p(X)$ and on Banach spaces with the weak topology (see Arhangel'skii (1984b)).

Other useful formulations of this factorization theorem can be found in the papers of Engelking (see Engelking (1977)) and Tkachenko (see Tkachenko (1981)).

7.4. Convex Compact Sets and the Krein-Milman Theorem. From the point of view of applications, some important principles dealing with the notions of linearity and topology are related to convex compacta. Among them the most important are the Krein-Milman theorem and the fixed point theorems.

We will use the term *convex compactum* to describe a compact, convex subset of a locally convex Hausdorff linear topological space L over the field $\mathbb{R}$ of real numbers.

A point x of a subset A of a linear topological space L over $\mathbb{R}$ is called an *extreme point* of A if A does not contain points y, z different from x such that the interval joining y and z is completely contained in A and contains x. Let us point out that a closed ball in an infinite dimensional Banach space can have the empty set of extreme points (see Bourbaki (1969a)). We have, however, the following.

Theorem 6 (Krein-Milman, see Bourbaki (1969a)). *Every convex compactum in a locally convex Hausdorff linear topological space is the closed convex hull of the set of its extreme points.*

As always, by the closed convex hull of a set A in L we understand the smallest closed, convex set in L containing A.

The Krein-Milman Theorem has many applications (see Haydon (1976)). One of them is related to the theory of irreducible representations of locally compact groups. The characters on locally compact Abelian groups are interpreted as extreme points of some convex compacta and in this way one can show the existence of sufficiently many characters (for more details see §8).

From the point of view of general topology the following question is of particular interest: what can be the topological properties of a topological space consisting of the extreme points of a convex compactum? This space does not have any linear structure and it can be studied only by methods of general topology. Let us notice that, even in the case when K is a convex compactum in a finite dimensional space L, the set of all extreme points of K does not have to be closed in K and thus does not have to be compact. We will use the notation $E(K)$ to denote the space of all extreme points of a convex compactum K and we will call it the *extreme set* of K. What is the relationship between the topological properties of $E(K)$ and K? What properties of $E(K)$ imply metrizability of K? One of the results related to this question was obtained by Haydon (see Haydon (1974)).

Theorem 7. *If the extreme set $E(K)$ of a convex compactum K is a space with a countable network, then K is metrizable.*

A fundamental step in the proof of Theorem 7 consists in constructing a countable network in K. Let us point out that the finite linear combinations of vectors from K do not represent all of K, they form only a dense subset of K.

It is not known whether the network weight of a convex compactum K is always equal to the network weight of its extreme set.

Theorem 7 was improved by Debs, who obtained the following result (see Debs (1980)).

Theorem 8. *If the extreme set $E(K)$ of a convex compactum K is a Lindelöf space with a G_δ-diagonal, then K is metrizable.*

The following question is natural: which Tikhonov spaces can be represented as sets of extreme points of a convex compactum? We will call such spaces *extreme*. The class of extreme spaces contains all spaces that are metrizable by a complete metric and all locally compact Hausdorff spaces. This class is closed with respect to the operation of the topological product. Not all Tikhonov spaces are extreme. We have the following (see Debs (1980))

Proposition 5. *Every extreme space X has the Baire property: the intersection of every countable family of dense, open subsets of X is dense in X.*

One of the most interesting unsolved questions concerning the topology of convex compacta is the following (Jayne): is every perfectly normal, convex compactum metrizable? It is not even clear whether every perfectly normal convex compactum is separable.

7.5. Theorem of Alaoglu and Convex Hulls of Compacta. There exists a standard way of constructing the convex hull of an arbitrary compactum X. Let Y be an arbitrary normed linear space over $\mathbb{R}$ and let Y^* be the space of all continuous real functionals on Y with the standard norm $\|f\| = \sup\{|f(y)| : \|y\| \le 1\}$. The topology of pointwise convergence T_p on Y^* is traditionally called the *w^*-topology*. A set $F \subset Y^*$ is called *pointwise bounded* (or *w^*-bounded*) if the set $\{f(y) : f \in F\}$ is bounded for each $y \in Y$. Note that the space Y^* in the w^*-topology is a subspace of the space $\mathbb{R}^Y$ of all functions on Y in the product topology. In general the space Y^* is not closed in $\mathbb{R}^Y$. However, the unit ball $U^0 = \{f \in Y^* : \|f\| \le 1\}$ of the space Y^* is a closed subset of $\mathbb{R}^Y$ in the product topology on $\mathbb{R}^Y$. Clearly, the ball U^0 is pointwise bounded since for every $y \in Y$ and every $f \in U^0$, $|f(y)| \le \|y\|$. Denoting by I_y the interval $[-\|y\|, \|y\|]$ on the real axis we conclude that the unit ball U^0 in the w^*- topology is a closed subset of the product $\prod\{I_y : y \in Y\}$. Since the last set, according to Tikhonov's theorem, is a compactum, we conclude that the unit ball U^0 is compact in the w^*-topology. Thus we have the following

Theorem 9 (Alaoglu, see Bourbaki (1969a)). *The unit ball of the space Y^* dual to a normed linear space Y is compact in the topology of pointwise convergence (that is in the w^*-topology).*

However, the unit sphere $S^0 = \{f \in Y^* : \|f\| = 1\}$ does not have to be compact. Typically it is dense in the ball U^0 in the w^*-topology.

Theorem 9 implies the following.

Corollary 4. *The dual of a normed linear space is σ-compact in the w^*-topology (and thus is Lindelöf and paracompact).*

Note that the normed space Y^* does not always have a countable base and thus is not always a Lidelöf space.

The unit ball in (Y^*, T_p) is not only compact but also convex. The following applications of the construction described above are associated with this fact.

Let X be a Tikhonov space, and let $Y = C^0(X)$ be the linear space of all bounded continuous real-valued functions on X with the usual norm: $\|f\| = \sup\{|f(x)| : x \in X\}$. Introduce the topology T_p of pointwise convergence on the dual Y^* of the normed space Y and consider a canonical mapping ψ of X into Y^* that assigns to each $x \in X$ the function $\psi(x) = g_x$, where $g_x(f) = f(x)$, for all $f \in Y$. The image $\psi(X)$ of X is contained in the unit sphere $S^0 = \{g \in Y^* : \|g\| = 1\}$ and thus in the unit ball U^0 of the space Y^*.

The mapping ψ is a homeomorphism between X and the subspace $\psi(X)$ of the ball U^0. Every $g_x = \psi(x)$ is an extreme point of the convex compactum U^0 in the linear space Y^* and U^0 is the smallest convex compactum in Y^* containing $\psi(X)$ and its antipodal copy $-\psi(X) = \{-g : g \in \psi(X)\}$. The set $\psi(X)$ is closed in the ball U^0 if and only if X is a compactum. However, if X is not a compactum then by taking the closure of $\psi(X)$ in U^0 we obtain a compact Hausdorff extension of $\psi(X)$, which turns out to be the Stone-Čech compactification of X (we identify X with $\psi(X)$).

The closure $\overline{\psi(X)}$ consists precisely of these points g of the unit ball U^0 that assume the value 1 on the function $\mathbb{1}$ that is identically equal to one.

The constructions described above lead to the following result.

Theorem 10 (see Kelley (1955)). *Every compactum X can be represented as the extreme set of some convex compactum K in a linear topological space L. In addition, this can be done in such a way that every continuous real-valued function on X can be extended to a continuous, linear, real-valued functional on K (by means of an extension operator that is norm continuous).*

$\triangleleft$ Consider the embedding of X into the unit ball U^0 of the space $(C(X))^*$ by the homeomorphism ψ. The extreme set of U^0 is the direct sum of two spaces homeomorphic to X: $E(U^0) = \psi(X) \cup (-\psi(X))$. The closed convex hull of $\psi(X)$ in $(C(X))^*$ is the desired compactum K. $\triangleright$

In this way we easily obtain the following fundamental *Banach-Stone theorem* (see Stone (1937)).

Theorem 11. *Let X and Y be compacta such that the normed spaces $C(X)$ and $C(Y)$ are linearly isometric. Then X and Y are homeomorphic.*

There is also a slightly different way to view the compactum K described in Theorem 10. More specifically, K consists of all probability (that is countably additive, nonnegative, Borel) regular measures on X assuming the value 1 on X, in the weak topology (see Fedorchuk (1981)). With this interpretation K is traditionally denoted by $P(X)$.

Of considerable interest is the question of the relationship between the properties of X and $P(X)$.

Since X is the extreme set of compactum $P(X)$, metrizability of X implies metrizability of $P(X)$ (for example, by Theorem 7). In works related to functional analysis the following facts found extensive applications. If X is an Eberlein compact space, so is $P(X)$. If X is Gul'ko compact, so is $P(X)$ (for more details see R. Pol (1982)). Haydon constructed, assuming the Continuum Hypothesis, an example of a compactum X of countable tightness such that $P(X)$ has uncountable tightness (see R. Pol (1982)).

Example 1. Let $2^{\aleph_0} = 2^{\aleph_1}$. Then there exists a nonseparable compactum X satisfying the first countability axiom, and admitting a regular probability measure μ (Haydon, see R. Pol (1982)). The point μ of $P(X)$ then belongs to the closure of the set $A = \{\lambda \in P(X) : \text{the support of } \lambda \text{ is finite}\}$. At the same time for every countable set $B \subset A$, $\mu \notin \overline{B}$, since the support of μ is not separable. Consequently $t(P(X)) > \aleph_0$.

It is not clear whether it is possible to construct, within ZFC, a compactum X of countable tightness for which $t(P(X)) > \aleph_0$. If a compactum X is scattered and if the tightness of X is countable then the tightness of $P(X)$ is also countable (see R. Pol (1982)). R. Pol raised the following questions. Let X be a compactum of countable tightness, $x \in X \subset P(X), A \subset P(X)$ and $a \in \overline{A}$. Does there then exist a countable set $B \subset A$ such that $x \in \overline{B}$ (that is, is the tightness of $P(X)$ at the point x countable)? Let X be a compactum and $t(P(X)) \leq \aleph_0$. Is it then true that $t(P(X \times X)) \leq \aleph_0$?

7.6. Fixed-point Theorems for Continuous Mappings of Convex Compacta. The following *Schauder-Tikhonov theorem* (see Bourbaki (1969a)) is one of the principles related to the theory of linear topological spaces that found important applications, primarily in the theory of differential equations.

Theorem 12. *Let K be a nonempty convex compactum in a locally convex Hausdorff linear topological space L (over $\mathbb{R}$). Then every continuous mapping φ of K into itself has a fixed point, that is a point $x \in K$ such that $\varphi(x) = x$.*

In particular, Theorem 12 applies to the Hilbert cube Q and to finite-dimensional Euclidean cubes. It is a generalization of the classical *Brouwer's fixed-point theorem* (see Aleksandrov and Pasynkov (1977)).

Another fundamental principle is related to families of continuous mappings of a convex compactum into itself. A mapping u of a convex set $K \subset L$ is called *affine* if for every $\alpha, \beta \in \mathbb{R}$ such that $\alpha + \beta = 1$, and every $x, y \in K$ the following condition is satisfied: $u(\alpha x + \beta y) = \alpha u(x) + \beta u(y)$.

Theorem 13 (Markov-Kakutani, see Bourbaki (1969a)). *Let K be a nonempty convex compactum in a linear topological Hausdorff space L over the reals $\mathbb{R}$, and let $\mathcal{A}$ be a family of continuous affine mappings of K into itself that commute with each other. Then there exists a common fixed-point in K for all mappings of $\mathcal{A}$, that is a point $x \in K$ such that $f(x) = x$, for all $f \in \mathcal{A}$.*

Using Theorem 13 one can show, for example, the existence of an invariant mean on an Abelian topological group. Let G be such a group and let $L = C^0(G)$ be the linear topological space of all bounded continuous real-valued functions on G. With the norm $\|f\| = \sup\{|f(x)| : x \in G\}$ the space L is a Banach space. In the space L^* of all linear real functionals on L consider the set K consisting of all $g \in L^*$ that assume the value 1 on the function identically equal to one. The set K is convex and, if equipped with the topology of pointwise convergence, compact.

The group G acts by means of "shifts" on K (with respect to "shifts" compactum K is invariant), the "shifts" commute with each other and in addition are continuous and linear. Thus we can use Theorem 13 to obtain the following.

Corollary 5. *Every Abelian topological group admits an invariant mean, that is a linear functional on $C^0(G)$ invariant under shifts and equal to 1 on the function identically equal to one.*

In particular, an invariant mean exists on the discrete group $\mathbb{R}$ of real numbers. By identifying the subsets of $\mathbb{R}$ with their characteristic functions we obtain from Corollary 5 the proof of the existence of the so called *Banach measure* on $\mathbb{R}$ (see Bourbaki (1969a)).

Corollary 6. *There exists a nonnegative, finitely additive real-valued function m defined on the set of all subsets of $\mathbb{R}$ that is shift invariant and such that $m(\mathbb{R}) = 1$.*

As is well known, it is impossible to construct a countably additive function on the set of all subsets of $\mathbb{R}$ with these properties (see Arhangel'skii and Ponomarev (1974)).

7.7. Milyutin Compact Spaces. In the previous section we mentioned the following theorem of Banach and Stone: if the Banach spaces $C(X)$ and $C(Y)$ of all continuous real-valued functions on compacta X and Y are isomorphic (that is linearly isometric), then X and Y are homeomorphic. It is natural to ask whether the assumptions of this theorem can be weakened by assuming only that the Banach spaces $C(X)$ and $C(Y)$ are linearly homeomorphic. The answer is quite unexpected. It follows from it that Banach spaces of continuous real-valued functions on an interval and on a square are linearly homeomorphic (Milyutin, see Pełczyński (1968)). The following is the general result of Milyutin.

Theorem 14. *If X and Y are uncountable metrizable compacta, then the Banach spaces $C(X)$ and $C(Y)$ are linearly homeomorphic.*

This theorem should be compared with the following, no less remarkable, result of Kadec (see Pełczyński (1968)).

Theorem 15. *All separable Banach spaces are homeomorphic.*

It is also true that Banach spaces of the same density are homeomorphic (see Negrepontis (1984)).

The following concepts of the extension and the mean operators and, corresponding to them, classes of Dugundji and Milyutin compact spaces, are of importance in linear classification of spaces of continuous functions on compacta

Let X and Y be compacta and let $\varphi : X \to Y$ be a continuous mapping such that $\varphi(X) = Y$. A linear mapping $u : C(X) \to C(Y)$, continuous with respect to the norm, is called a *linear mean operator* for φ if for every $g \in C(Y)$ the following condition is satisfied: $u(g \circ \varphi) = g$. If, in addition, $\|u(f)\| \le \|f\|$, for all $f \in C(X)$ and $u(1_X) = 1_Y$, then u is called a *regular mean operator*. We will use 1_X to denote the constant function identically equal to 1 on X.

Let us introduce the following dual concept. Assume that X and Y are compacta and that $Y \subset X$. A linear mapping $u : C(Y) \to C(X)$, continuous with respect to the norm, is called a *linear extension operator* if for every function $g \in C(Y)$ the restriction to Y of the function $u(g) \in C(X)$ coincides with g, that is $u(g)|_Y = g$ for all $g \in C(Y)$. If, in addition, the mapping u does not increase norm and $u(1_Y) = 1_X$, then u is called a *regular extension operator*.

The following theorem was obtained by Ditor (see Shchepin (1981)): every compactum can be represented as a continuous image of a zero-dimensional compactum under a mapping admitting a regular mean operator.

Let us recall that D^τ is the product of τ copies of D, where D is a two-point, discrete space.

A compactum X is called a *Milyutin compact space* if for some $\tau \geq \aleph_0$ there exists a continuous mapping of D^τ onto X admitting a regular mean operator. By weakening the assumption of regularity in the previous definition to the assumption of linearity and continuity we obtain the notion of an *almost Milyutin compact space* (see Pełczyński (1968)).

The class of Milyutin compact spaces is closed with respect to arbitrary products. Milyutin first proved (see Pełczyński (1968)) that this class contains the unit interval $I = [0, 1]$. It also contains all metrizable compacta and all compact absolute neighborhood retracts (see Pełczyński (1968)). Generalized Cantor discontinua D^τ and Tikhonov cubes I^τ are Milyutin compact. However, not every continuous image, nor a closed subspace of a Milyutin compact space is Milyutin compact. This is implied by the following

Proposition 6. *Let U and V be disjoint open subsets of a Milyutin compact space X. Then, there exist disjoint σ-compact sets $\widetilde{U}$ and $\widetilde{V}$ in X such that $U \subset \widetilde{U}$ and $V \subset \widetilde{V}$.*

Example 2. ◁ Consider the space $X = D^\tau \times \{0, 1\}$ homeomorphic to D^τ, where $\tau > \aleph_0$. This space is Milyutin compact. Fix an arbitrary point $a \in D^\tau$ and put $U = (D^\tau \times \{0\}) \setminus \{(a, 0)\}$, $V = (D^\tau \times \{1\}) \setminus \{(a, 1)\}$. The compactum Y obtained by identifying the points $(a, 0)$ and $(a, 1)$ in X is not Milyutin compact. Indeed, the subsets U and V are open and disjoint in Y but cannot be enlarged to disjoint F_σ-sets in Y. If this were possible then the compactum D^τ would satisfy the first countability axiom at the points $(a, 0)$ and $(a, 1)$. Notice that Y is a dyadic compactum. Accordingly, not every dyadic compactum is Milyutin compact. ▷

Several applications of Milyutin compact spaces are related to the fact that every compact topological group and every quotient group of a compact group with respect to a closed subgroup is Milyutin compact (for more details see Pełczyński (1968)).

The conclusion of Theorem 14 is no longer true if the topologies induced by the norms on $C(X)$ and $C(Y)$ are replaced by topologies of pointwise convergence. In particular, if X and Y are compacta for which the spaces

$C_p(X)$ and $C_p(Y)$ are linearly homeomorphic, then the dimensions dim of compacta X and Y are equal (Arhangel'skii and Zambahidze, see Efimov (1975), Arhangel'skii (1987a)). Pestov (see Pestov (1982)) generalized this result to the case of arbitrary Tikhonov spaces and Gul'ko proved that Pestov's theorem is true also for uniform homeomorphisms. On the other hand Pavlovskij showed that if X and Y are compact polyhedra, and if $0 < \dim X = \dim Y < \infty$, then $C_p(X)$ is linearly homeomorphic to $C_p(Y)$ (see Pavlovskij (1980)). The following problem remains open. Let X and Y be compacta and let $C_p(X)$ be homeomorphic to $C_p(Y)$. Is it true then, that $\dim X = \dim Y$? Is it true for arbitrary Tikhonov spaces X and Y?

Notice that from the fact that $C_p(X)$ and $C_p(Y)$ are linearly homeomorphic, where X and Y are compacta, it follows that also the Banach spaces $C(X)$ and $C(Y)$ are linearly homeomorphic (see Pavlovskij (1982)), but the converse, as demonstrated by Theorem 14 and by the examples presented above, is not true.

Questions related to preservation of properties related to compactness by homeomorphisms and linear homeomorphisms of function spaces with the topology of pointwise convergence were considered by Arhangel'skii, Gul'ko and Khmyleva, Okun'ev, Pytkeev, Pavlovskij, Uspenskij and many others. In particular, Arhangel'skii and Pavlovskij proved that if X is a compactum and Y is a Tikhonov space such that the spaces $C_p(X)$ and $C_p(Y)$ are linearly homeomorphic, then Y is also a compactum (see Arhangel'skii (1982, 1987a)). Gul'ko and Khmyleva constructed a homeomorphism between the spaces $C_p(I)$ and $C_p(\mathbb{R})$, where $I = [0, 1]$ is an interval and $\mathbb{R}$ is the real line (see Gul'ko and Khmyleva (1986)). Okun'ev proved the following theorem (see Arhangel'skii (1987a)): if X and Y are Tikhonov spaces for which the spaces $C_p(X)$ and $C_p(Y)$ are homeomorphic and if X is σ-compact, then Y is also σ-compact.

7.8. Dugundji Compact Spaces. A compactum Y is called a *Dugundji compact space* (*almost Dugundji compact space*) if for every compactum X containing Y there exists a regular (linear) extension operator $\varphi : C(Y) \to C(X)$. If Y is embedded into a Tikhonov cube I^τ and if there exists a regular extension operator $\varphi : C(Y) \to C(I^\tau)$, then Y is Dugundji compact. Every compact group and every quotient group of a compact group with respect to a closed subgroup are Dugundji compact (see Pełczyński (1968)). All Dugundji compact spaces are also Milyutin compact (Haydon, see Haydon (1974)), but the converse is not true (Shchepin, see Shchepin (1981)). From results of Ditor and Haydon it follows that a compactum of weight $\leq \aleph_1$ is Milyutin compact if and only if it is Dugundji compact (see Shchepin (1981)). The class of Dugundji compact spaces is closed with respect to products. Consequently, all metrizable compacta and their products (in particular, compacta D^τ and I^τ) are Dugundji compact. Every compact absolute neighborhood retract is Dugundji compact.

Haydon (see Haydon (1976)) gave a description of Dugundji compact spaces as limits of continuous well ordered inverse systems of compacta beginning with a point, in which all projections $p_\alpha^{\alpha+1}$ are continuous open mappings with metrizable kernels. In this formulation, a mapping $f : X \to Y$ between compacta X and Y is said to have a *metrizable kernel* if there exists a continuous mapping $g : X \to M$ onto a metrizable compactum M whose restriction to the inverse image $f^{-1}(y)$ of every point $y \in Y$ is one-to-one. The last condition is equivalent to the assumption that the diagonal mapping $f \triangle g : X \to Y \times M$ is injective.

Dugundji compact spaces Y can also be characterized by the following property: for every embedding of a compactum Y into a compactum X, every continuous mapping of Y into a convex subspace Z of Tikhonov cube I^τ can be extended to a continuous mapping of X into Z (see Shchepin (1981)).

A compactum Z is called an *absolute extensor in dimension 0* (abbreviated $Z \in AE(0)$) if for every paracompact space X and every closed subspace Y of X such that $\dim X = 0$, and every continuous mapping $f : Y \to Z$ there exists a continuous mapping $\widetilde{f} : X \to Z$ extending f, that is $\widetilde{f}|_Y = f$. Hewitt proved the following result (see Haydon (1974)).

Theorem 16. *Dugundji compact spaces and only such spaces are the absolute extensors in dimension 0.*

Hoffman (see Hoffman (1979)) found a property of continuous surjections that allows another point of view on Dugundji compact spaces. We will call a continuous mapping $f : X \to Y$ invertible in dimension 0, if for any zero-dimensional compactum B and for every continuous mapping $f : B \to Y$, it is possible to "lift" g to a continuous mapping, that is, it is possible find a continuous mapping $g_1 : B \to X$ such that $f \circ g_1 = g$.

Theorem 17 (Hoffman, see Hoffman (1979)). *Dugundji compact spaces and only such spaces are the images of generalized Cantor discontinua D^τ under continuous mappings that are invertible in dimension 0.*

For spaces of weight $\aleph_1$ we have the following result of Ditor and Haydon (see Shchepin (1981)).

Theorem 18. *Let X be a nonmetrizable compactum. Then the space $P(X)$ is an absolute retract if and only if X is a Dugundji compact space of weight $\aleph_1$.*

Shchopin characterized the Cantor discontinua D^τ in the following way (see Shchepin (1981)).

Theorem 19. *A nonmetrizable compactum X is homeomorphic to D^τ for some $\tau > \aleph_0$ if and only if X is a zero-dimensional Dugundji compact space in which any two points have the same character.*

In a similar way Shchepin characterized also the Tikhonov cubes.

Results of Haydon, Ditor, Pełczyński were further developed in the papers of Shchepin, Fedorchuk, Ivanov, Dranishnikov, Chigogidze and others. In particular, Shchepin introduced the class of κ-metrizable compacta and characterized them in terms of inverse limits with open projections.

Recall that *canonically closed sets* are defined as closures of open sets. A nonnegative real-valued function $\rho(x, C)$, where $x \in X$, and C is a canonically closed subset of X, is called a κ-*metric* on X if the following conditions are satisfied:

1) $\rho(x, C) = 0 \iff x \in C$,

2) if $C_1 \subset C_2$, then $\rho(x, C_1) \geq \rho(x, C_2)$, for all $x \in X$,

3) for a fixed C, $\rho(x, C)$ is a continuous function of x,

4) $\rho(x, \overline{\bigcup\{C_\alpha : \alpha \in A\}}) = \inf\{\rho(x, C_\alpha) : \alpha \in A\}$ for any chain $\{C_\alpha : \alpha \in A\}$ of canonically closed subsets of X and any $x \in X$.

The scope of the class of all κ-metrizable spaces is described by the following results (see Shchepin (1973)). Every metrizable space is κ-metrizable since it is enough to take in place of $\rho(x, C)$ the distance from x to C. The product of any family of κ-metrizable (in particular, metrizable) spaces is κ-metrizable. All Dugundji compact spaces are κ-metrizable. It became clear that in several situations κ-metrics are quite a useful tool in the study of topological properties (see Shchepin (1973)). The class of κ-metrizable compacta in many situations is similar to the class of dyadic compacta, however, none of these classes contains the other. For example, the compactum $\mathrm{Exp}(D^{\aleph_2})$ is κ-metrizable but not dyadic. It is worth pointing out that compacta of all three classes: Milyutin, Dugundji and κ-metrizable, are very similar, from the point of view of topological properties, to dyadic compacta (see Shchepin (1973, 1976)).

A unified approach to Dugundji compact spaces and κ-metrizable spaces was developed by Shirokov (see Shirokov (1982)). We will say that a compactum $Y \subset I^\tau$ is K_1-*positioned in the Tikhonov cube* I^τ, if there exists a mapping φ of the family $\mathcal{T}$ of all open subsets of Y into the family of all open subsets of the compactum I^τ such that the following conditions are satisfied: $U = \varphi(U) \cap U$ for all $U \in \mathcal{T}$, and if $U, V \in \mathcal{T}$ are such that $U \cap V = \emptyset$, then $\varphi(U) \cap \varphi(V) = \emptyset$ (see Shirokov (1982)). If in addition $\varphi(U \cap V) = \varphi(U) \cap \varphi(V)$ for all open subsets U, V of Y, then we call the compactum Y K_0-*positioned in the cube* I^τ.

Theorem 20 (Shirokov (1982)). *A compactum $Y \subset I^\tau$ is Dugundji compact if and only if it is K_0-positioned in I^τ.*

Theorem 21 (Shirokov (1982)). *A compactum $Y \subset I^\tau$ is κ-metrizable if and only if it is K_1-positioned in I^τ.*

A remarkable connection between κ-metrizable compacta and Dugundji compact spaces is revealed through the concept of *superextension* introduced by de Groot (see van Mill (1977)). We will describe here the construction of the superextension. A family of sets is called a *linked* if every two sets in this family have a nonempty intersection. Let X be a T_1-spaces. A linked system of closed subsets of X is called maximal if it is not contained in any other linked system of closed subsets of X. Denote by $\lambda(X)$ the space whose points are all maximal linked systems of closed subsets of X and whose closed subsets are defined as the intersections of the families of the type $\lambda(F) = \{\xi \in \lambda(X) : F \in \xi\}$, where F is a closed subset of X. The space X can be naturally embedded in $\lambda(X)$ by assigning to a point x the system $\lambda(x)$. If X is a compactum, so is $\lambda(X)$ (see van Mill (1977)).

Theorem 22 (see Ivanov (1981), Shchepin (1981)). *A compactum X is κ-metrizable if and only if $\lambda(X)$ is Dugundji compact.*

We note that every compactum X is K_1-positioned in $\lambda(X)$. Since the class of all Dugundji compact spaces contains all Tikhonov cubes, but does not coincide with the class of all compacta, the property of being Dugundji compact is not hereditary with respect to closed subspaces. Moreover, we have the following.

Theorem 23. *If X is a compactum and every compactum in X is Dugundji compact, then X is metrizable.*

The converse is also true.

◁ Since every Dugundji compact space is dyadic, its Suslin number is countable. Consequently, every discrete subspace of X is countable. Therefore X has countable tightness (see Arhangel'skii (1978)), but every dyadic compactum of countable tightness is metrizable (see Arhangel'skii and Ponomarev (1974)). ▷

In connection with Dugundji compactness, Arhangel'skii and Choban introduced the following concept. A subspace $Y \subset X$ is called *t-embedded* (*l-embedded*) in X if there exists a continuous (linear) operator $\varphi : C_p(Y) \to C_p(X)$ such that $\varphi(f)|_Y = f$ for all $f \in C_p(Y)$. A Tikhonov space Y that is l-embedded (t-embedded) in every Tikhonov space X is called *extral* (*t-extral*).

Theorem 24 (Arhangel'skii, Choban). *Every t-extral space is a compactum.*

Extral compacta have properties that are similar to those of dyadic compacta. For example, the tightness and the weight of such spaces coincide. However, not every dyadic compactum (even a Dugundji compact space) is extral. Moreover, the subspace D^τ of the cube I^τ for $\tau > \aleph_0$ is not l-embedded in I^τ. Consequently, the compactum D^τ is not extral. The question of "internal" characterization of extral compacta remains open.

Theorem 25 (Arhangel'skii, Choban.) *If X is a compactum and if every compactum contained in X is extral, then the compactum X is metrizable.*

The concept of Dugundji compact space is of crucial importance in topological algebra, in particular, in connection with questions related to continuous actions of topological groups on compacta (see Uspenskij (1987)). Below we will present some results of Uspenskij.

A topological group G is called $\aleph_0$-*bounded* (see Arhangel'skii (1981)) if it satisfies one of the following equivalent conditions: (a) for every neighborhood U of the unit element there exists a countable set $A \subset G$, such that $A \cdot U = G$, (b) G is topologically isomorphic to a subgroup of the product of some family of groups with countable base. A topological group G is called $\aleph_0$-*balanced* if it is topologically isomorphic to a subgroup of a product of metrizable groups. All Lindelöf groups, as well as all groups with countable Suslin number are $\aleph_0$-bounded. In particular, all σ compact and all pseudocompact groups are $\aleph_0$-bounded. Every $\aleph_0$-bounded group is $\aleph_0$-balanced.

Theorem 26 (Uspenskij (1987)). *If X is a compactum admitting a transitive and continuous action of some $\aleph_0$-bounded topological group, then X is Dugundji compact.*

Theorem 27 (Uspenskij (1987)). *Let G be an $\aleph_0$-balanced topological group and let H be a closed subgroup of it. Then every compactum of type G_δ in the quotient space G/H is Dugundji compact.*

Proofs of Theorems 26 and 27 are based on Haydon's theorem on the representation of Dugundji compact spaces as limits of special types of inverse systems.

Theorems 26 and 19 imply that every zero-dimensional compactum of weight τ admitting a continuous and transitive action of an $\aleph_0$-bounded topological group is homeomorphic to the Cantor discontinuum D^τ (see Uspenskij (1987)). Uspenskij proved also that if a compactum X is a retract of some topological group, then X is Dugundji compact.

Applications of the mean and the extension operators in linear topological classification of function spaces are based on the following results (see Pełczyński (1968)).

Proposition 7. *If a continuous mapping f of a compactum X onto compactum Y admits a linear mean operator, then the Banach space $C(Y)$ is a divisor of the Banach space $C(X)$, that is, there exists a Banach space B such that $C(X)$ is homeomorphic to $C(Y) \times B$.*

Proposition 8. *If a compactum Y is embedded in a compactum X in such a way that there exists a linear extension operator $U : C(Y) \to C(X)$, then the Banach space $C(Y)$ is a divisor of the Banach space $C(X)$ (more precisely, the subspace $u(C(Y))$ is complemented in $C(X)$.)*

Proposition 9. *If a Banach space B is a divisor of a Banach space $C(D^\tau)$ and if $C(D^\tau)$ is a divisor of B then B is linearly homeomorphic to $C(D^\tau)$ (see Pełczyński (1968)).*

By combining Propositions 8 and 9 it is possible to prove the following.

Theorem 28 (Pełczyński (1968)). *Let X be an infinite compactum that is either almost Dugundji compact or almost Milyutin compact and that contains* *a compactum homeomorphic to D^τ, where τ is equal to the weight of X. Then the Banach spaces $C(X)$ and $C(D^\tau)$ are linearly homeomorphic.*

Every uncountable metrizable compactum contains a topological copy of the Cantor perfect set (Aleksandrov, see Arhangel'skii and Ponomarev (1974)). Thus Theorem 28, under the assumption that $\tau = \aleph_0$, implies Theorem 14 of Milyutin.

§8. Algebraic Structures and Compactness. A Review of the Most Important Results

8.1. Compacta and Ideals in Rings of Functions. A fundamental connection between algebraic structures and compacta is realized through the concept of an ideal. This connection is most clearly visible in the case of mutual correspondence between an arbitrary Tikhonov space X and the ring $C^0(X)$ of all bounded, continuous real-valued functions on X. Denote by $\mathcal{H}^0(X)$ the set of all nontrivial homomorphisms of the ring $C^0(X)$ into the ring $\mathbb{R}$ of real numbers and introduce the topology of pointwise convergence in $\mathcal{H}^0(X)$. To each $x \in X$ there corresponds the evaluation homomorphism $w_x : C^0(X) \to \mathbb{R}$ described by $w_x(f) = f(x)$, for all $f \in C^0(X)$. The mapping of X into the space $\mathcal{H}^0(X)$ obtained in this way is denoted by μ. Every homomorphism $w \in \mathcal{H}^0(X)$ is onto, that is $w(C^0(X)) = \mathbb{R}$, and its kernel $\ker(w)$ is a maximal ideal of $C^0(X)$. Conversely, to every maximal ideal of the ring $C^0(X)$ there corresponds a homomorphism of $C^0(X)$ onto $\mathbb{R}$, that is an element of the set $\mathcal{H}^0(X)$. Thus the set $\mathcal{H}^0(X)$ can be canonically identified with the set of all maximal ideals of the ring $C^0(X)$. The mapping μ of the space X into the space $\mathcal{H}^0(X)$ described by $x \to w_x$ is a homeomorphism of X onto a subspace of the space $\mathcal{H}^0(X)$. The following result is known as the *Gel'fand-Kolmogorov theorem* (see Naimark (1956)).

Theorem 1. *If X is a compactum, then μ is a homeomorphism of X onto $\mathcal{H}^0(X)$.*

Since the topological space $\mathcal{H}^0(X)$ is completely determined by the ring $C^0(X)$ which so far has been considered as purely algebraic object, Theorem 1 implies the following

Corollary 1 (see Naimark (1956)). *If X and Y are compacta and if the rings $C(X)$ and $C(Y)$ of all continuous real-valued functions on X and Y respectively are (algebraically) isomorphic, then the spaces X and Y are homeomorphic.*

Corollary 1 raises the possibility of characterizing topological properties of compacta by purely algebraic properties of rings of functions and applying well developed algebraic tools in the study of compacta.

If X is a noncompact Tikhonov space, then by identifying X with $\mu(X) \subset \mathcal{H}^0(X)$, we represent $\mathcal{H}^0(X)$ as an extension of X. It turns out that this is the Stone-Čech compactification of X.

If in place of $C^0(X)$ we consider the ring $C(X)$ of all continuous real-valued functions on a Tikhonov space X, then the space $\mathcal{H}(X)$ of all nontrivial ring homomorphisms $C(X) \to \mathbb{R}$ in the topology of pointwise convergence can be interpreted, through evaluation, as the Hewitt extension of X. Every continuous real-valued function on X can be extended to a continuous real-valued function defined on this extension. If a space X is *complete in the sense of Hewitt* (that is, if X is homeomorphic to a closed subspace of the space $\mathbb{R}^\tau$ for some τ), then X is canonically homeomorphic to $\mathcal{H}(X)$. Therefore, we obtain the following theorem of Hewitt (see Hewitt (1948)).

Theorem 2. *Two spaces X and Y complete in the sense of Hewitt are homeomorphic if and only if the rings of continuous real-valued functions $C(X)$ and $C(Y)$ are (algebraically) isomorphic.*

8.2. Spectrum of a Ring. Zariski Topology. The connection between rings of functions and compacta described in the previous section has a far-reaching generalization. It turns out that it is possible to assign to each commutative ring with a unit a certain compact space, called its spectrum. It is true that the last space does not even have to be a T_1-space. We will describe briefly its construction.

Let A be a commutative ring with a unit. The *spectrum* $\operatorname{spec}(A)$ *of the ring* A consists of all the prime ideals of A. For each ideal α of ring A let us denote by $P(\alpha)$ the set $\{\beta \in \operatorname{spec}(A) : \alpha \subset \beta\}$. Open subsets of $\operatorname{spec}(A)$ are defined as complements of the sets $P(\alpha)$, that is the sets of the form $\operatorname{spec}(A) \setminus P(\alpha)$. The sets $V(f) = \{\beta \in \operatorname{spec}(A) : f \notin \beta\}$, where f is any element of A, form a base of the space $\operatorname{spec}(A)$. The topology on $\operatorname{spec}(A)$ described above is called the *Zariski topology*.

Example 1. The spectrum of the ring $\mathbb{Z}$ of integers is homeomorphic to an infinite countable space X in which, apart from X, only finite subsets are closed. This space is compact, satisfies the T_1-separation axiom, but is not Hausdorff. Any two nonempty open subsets of X intersect.

Proposition 1. *The spectrum of any commutative ring with a unit is a compact T_0-space. In addition, the sets $V(f)$ that form its base, where $f \in A$ (see above), are compact (even though they are not necessarily closed) sets.*

The topological properties of the compact space $\operatorname{spec}(A)$ are closely related to the algebraic properties of the ring A. Therefore it is possible to use ideas associated with compactness in a study of purely algebraic questions related to A. This approach is used in algebraic geometry (see Bourbaki (1969b)).

8.3. The Space of Maximal Ideals of a Commutative Banach Algebra. Let X be a compactum. The space $C(X, \mathbb{R})$ of all continuous real-valued functions on X with the natural operations of addition and multiplication and with the natural norm $\|f\| = \text{Max}\{|f(x)| : x \in X\}$ is a Banach algebra over $\mathbb{R}$. In a similar way we define the Banach algebra $C(X, \mathbb{C})$, over the field $\mathbb{C}$ of complex numbers, of all continuous complex-valued functions on compactum X. The Banach algebra $C(X, \mathbb{C})$ is commutative and has a unit.

It is also possible to realize a construction leading in the opposite direction, namely to assign a compactum to every commutative Banach algebra with a unit.

Let K be an arbitrary Banach algebra with a unit over the field $\mathbb{C}$. Let us introduce the topology on K generated by the norm. All maximal ideals in K are closed (for arbitrary topological rings this statement is not true). Every maximal ideal M of K induces a canonical homomorphism of K onto the field of complex numbers. This is related to the famous *Gel'fand-Mazur theorem*: if every element of a Banach algebra A over $\mathbb{C}$ has an inverse in A, then the algebra A is isometrically isomorphic with the field of complex numbers $\mathbb{C}$ (see Naimark (1956)). For every $x \in K$ denote by $x(M)$ the image of x under the homeomorphism $\psi_M : K \to \mathbb{C}$ induced by a maximal ideal M. Thus the elements of K can be interpreted as complex-valued functions on the set $\text{Max}(K)$ of all maximal ideals of the Banach algebra K. The set $\text{Max}(K)$ equipped with the weakest topology with respect to which all these functions are continuous, is a compact Hausdorff space. The compactum $\text{Max}(K)$ is called the *space of the maximal ideals* of the Banach algebra K or its *spectrum*.

An algebra K is called *semisimple* if its radical is trivial, that is if the intersection of all maximal ideals of K consists of the zero element only. The construction described above leads to the following result.

Theorem 3 (see Naimark (1956)). *Every semisimple commutative Banach algebra K with a unit over the field $\mathbb{C}$ is canonically isomorphic (in general, without preservation of the norm) with some subalgebra of the Banach algebra $C(X, \mathbb{C})$ of all continuous, complex-valued functions on some compactum X, specifically on the space $\text{Max}(K)$ of all maximal ideals of K.*

The compactum $\text{Max}(K)$ can also be described as the space of all multiplicative linear functionals on K (with values in $\mathbb{C}$) in the topology of pointwise convergence.

Example 2. Let K be the algebra of all continuous complex-valued functions defined on the unit disc $\{z \in \mathbb{C} : |z| \leq 1\}$ that are analytic in the interior of this disc and such that $f'(0) = 0$. Then the unit disc is canonically identical with the space $\text{Max}(K)$ of the maximal ideals of this algebra.

Example 3. Let K_n be the Banach algebra of functions on the interval $[0, 1]$ that are n times continuously differentiable, with the norm $\|f\| = \sum_{k=0}^{n} \text{Max} |f^{(k)}(t)|$. The spectrum $\text{Max}(K_n)$ of this algebra coincides with the interval $[0, 1]$, and the canonical isomorphism described in Theorem 3 is the identity embedding of the set K_n into the Banach algebra $K = C([0, 1], \mathbb{C})$. This embedding is not norm preserving since the norm of an arbitrary function f in K is the maximum of its absolute value. In this norm the algebra K_n is not complete and thus is not a Banach space. This example shows that the isomorphism mentioned in Theorem 3 does not have to be an isometry.

A Banach algebra K is called an *algebra with a uniform convergence* if the isomorphism mentioned in Theorem 3 is a homeomorphism of the space K onto a subspace of the space $C(X, \mathbb{C})$.

The following natural question can be asked in connection with Theorem 3: when is a Banach algebra K isomorphic with the Banach algebra $C(X, \mathbb{C})$ for some compactum X?

A Banach algebra K over the field $\mathbb{C}$ is called an *algebra with an involution* if there exists an operation $a \mapsto a^*$ on K such that $(\lambda a + \mu b)^* = \overline{\lambda} a^* + \overline{\mu} b^*, (a^*)^* = a, (ab)^* = b^* a^*$ for all $a, b \in K$ and all $\lambda, \mu \in \mathbb{C}$. A Banach algebra K with an involution is called a *C^*-algebra* if $\|x^* x\| = \|x\|^2$ for every $x \in K$. The algebras $C(X, \mathbb{C})$ serve as examples of C^*-algebras where the involution is defined as taking a complex conjugate of a function.

Theorem 4 (see Naimark (1956)). *An arbitrary commutative C^*-algebra K with a unit is isometrically isomorphic (with preservation of the involution) with the Banach algebra $C(X, \mathbb{C})$, where X is a certain compactum (specifically, $X = \mathrm{Max}(K)$ is the space of maximal ideals of the algebra K).*

For Banach algebras over the reals we have the following theorem (see Hewitt (1948)).

Theorem 5. *Let K be a Banach algebra with a unit e over the field $\mathbb{R}$. In order for K to be isometrically isomorphic with a Banach algebra $C(X, \mathbb{R})$ for some compactum X it is necessary and sufficient that the following conditions be satisfied: (a) for every $x \in L$ there exists the inverse $(x^2 + e)^{-1}$ of the element $x^2 + e \in K$, (b) no element of the form $\|x^2\| e - x^2$, where $x \in K$ has an inverse in K.*

Example 4. Let Y be a Tikhonov space and let $K = C^0(Y, \mathbb{R})$ be the Banach algebra of all bounded, continuous real-valued functions on Y. It is easy to see that K satisfies the assumptions of Theorem 5. In the light of Theorem 5 the compactum X corresponding to the algebra K is, up to a canonical interpretation of Y as a subspace of X, the Čech-Stone compactification of Y.

The algebraic properties of a Banach algebra K are closely related to the topological properties of its space of maximal ideals. Specifically, *Shilov's theorem on idempotents*, in the case of a commutative Banach algebra K with a unit, states that if the spectrum $\mathrm{Max}(K)$ is represented as two disjoint closed subsets A and B, then there exists an idempotent $z \in K$ such that $z(x) = 1$ for all $x \in A$, and $z(x) = 0$ for all $x \in B$.

An important topological object in the study of commutative Banach algebra with a unit is its Shilov boundary. Consider all closed subsets A of the spectrum $X = \mathrm{Max}(K)$ of K, such that $\max\{|a(x)| : x \in A\} = \max\{|a(x)| : x \in X\}$ for all $a \in K$. It turns out that among such sets $A \subset X$ there exists the smallest one in the sense of inclusion. It is called the *Shilov boundary* or the *ring boundary* of the spectrum $\mathrm{Max}(K)$ of K (for more details see Naimark (1956), Khelemskij (1986)). In the works of Khelemskij and other authors homological properties of Banach algebras were related to topological properties of their spectra. In particular, Khelemskij showed that if X is a compactum, F is a closed subset of X and I is the ideal in $C(X, \mathbb{C})$ consisting of all the functions $f \in C(X, \mathbb{C})$ assuming value 0 at all points of F, then this ideal is projective as a module over the Banach algebra $K = C(X, \mathbb{C})$ if and only if the space $X \setminus F$ is paracompact (see Khelemskij (1986)).

8.4. The Stone Space of a Boolean Algebra. The concept of an ideal serves also as a link between Boolean algebras and certain compacta, namely their Stone spaces.

We will use the symbol $\mathbb{Z}_2$ to denote the discrete field of integers mod 2.

A *Boolean ring* is a ring $(B, +, \cdot)$ in which $a \cdot a = a$ and $a + a = 0$ for all $a \in B$. Every Boolean ring is commutative. There exists a canonical one-to-one correspondence between the maximal ideals of a Boolean ring B and the nontrivial homomorphisms of B into $\mathbb{Z}_2$ (the kernel of such homomorphism is a maximal ideal).

As an example of a Boolean ring with a unit we can consider the algebra of all subsets of a set X with the operations of symmetric difference Δ (where $A \Delta B = (A \setminus B) \cup (B \setminus A)$) and intersection. However, not every Boolean ring with a unit is isomorphic to the algebra of all subsets of some set. As an example we can take any countable, infinite Boolean ring. In connection with the above, the following construction based on the ideas described in previous paragraphs is of particular interest.

Let $(B, +, \cdot)$ be a Boolean ring and let S be the set of all nontrivial homomorphisms of the ring B into the discrete field $\mathbb{Z}_2$. Assume that S is equipped with the topology of pointwise convergence. The topological space S_B obtained in this way is called the *Stone space* of the Boolean ring B. From this definition it follows that S_B is a subspace of the Tikhonov product $\mathbb{Z}_2^B$. It is clear that S_B is a Hausdorff space with a base consisting of sets that are closed and open. The space S_B is always locally compact and if B has a unit, S_B is a compactum. We have the following representation theorem of M. Stone (see Stone (1937)).

Theorem 6. *Every Boolean ring with a unit is isomorphic to the algebra of all closed and open subsets of some zero-dimensional compactum (more specifically, this compactum is the Stone space of this Boolean ring).*

Conversely, to every zero-dimensional compactum there corresponds a Boolean ring with a unit formed by all of its subsets that are both closed and open. This indicates that the algebraic theory of Boolean rings is particularly suitable for the study of zero-dimensional compacta. It is convenient to study many topological properties of such compacta in this way.

There exists a natural partial order on every Boolean ring B. For $x, y \in B$ we put $x \leq y$ if and only if $x \wedge y = x$. A Boolean ring B is called *complete* if every set $A \subset B$ has a least upper bound $\sup A$ and a greatest lower bound $\inf A$ in $(B, \leq)$. A Boolean ring B with a unit is complete if and only if its Stone space is an extremally disconnected compactum. A very interesting Boolean ring corresponds to the remainder $\beta \mathbb{N} \setminus \mathbb{N}$ of the Stone-Čech compactification of the discrete space $\mathbb{N}$.

The Boolean ring whose Stone space is the Cantor cube D^τ (that is, the algebra of all closed-and-open subsets of D^τ) plays an important role in mathematical logic. It is used in the construction of boolean-valued models of an axiomatic set theory (see Malykhin (1983)).

8.5. Pontryagin's Duality Theory. Let $\mathbb{T}$ be the group of complex numbers of absolute value 1 in the usual topology. Thus topologically $\mathbb{T}$ is a circle. The role of $\mathbb{T}$ in topological algebra is associated with the following results. It follows from the work of Pontryagin (see Pontryagin (1973)) that for an arbitrary (Hausdorff) compact Abelian group G there exist sufficiently many continuous homomorphisms of G into $\mathbb{T}$ and that they completely determine the algebraic and topological structures of the group G. Accordingly, the group $\mathbb{T}$ plays the same role in the class of all compact Abelian groups as the interval $[0, 1]$ in the class of compacta. However, it is not this analogy (even though it is sufficiently important) that determines the role of $\mathbb{T}$ and continuous homomorphisms into it. Truly deep and fundamentally important facts related to it were established by Pontryagin. He showed that it is possible to assign to every compact Abelian group G a purely algebraic object, namely the discrete Abelian group $\widehat{G}$ (the group of continuous characters of G) that determines it completely (up to a homeomorphism) according to a rule that establishes a duality between these two groups. This duality states that the group G is "generated" by the group $\widehat{G}$ as the group of its characters in the topology of pointwise convergence. This fact is a key result in the *duality theory* created by Pontryagin and applicable to all locally compact, Hausdorff, Abelian groups (see Pontryagin (1973)).

Here are some basic definitions. Let G be a topological group. A *character of the group G* is defined as an arbitrary homomorphism of G into $\mathbb{T}$. Of particular importance are continuous homomorphisms of G into $\mathbb{T}$ called *continuous characters*.

The *group of characters* of a topological group G is the group G^0 of all continuous characters of G equipped with the compact-open topology. The group G^0 is an Abelian Hausdorff topological group. Pontryagin proved the following *duality theorem* (see Pontryagin (1973)).

Theorem 7. *If G is a locally compact Hausdorff Abelian group, then its group of characters G^0 is also a locally compact Hausdorff Abelian group and there exists a natural topological isomorphism of the groups G and G^{00}, the group of characters of G^0, that is given by the evaluation mapping.*

It is impossible to consider only compact Abelian groups in Theorem 7: a compact Abelian group has a discrete group of characters and a discrete Abelian group has a compact group of characters.

The group of characters of the compact group $\mathbb{T}$ is the discrete group of all integers. The group of characters of any discrete finite group is isomorphic to the original group. Finally, the character group of the group $\mathbb{R}$ of real numbers with the standard topology is topologically isomorphic to $\mathbb{R}$.

Let us mention that to every continuous homomorphism of topological groups there corresponds a continuous homomorphism of their character groups, going in the opposite direction.

According to Pontryagin's duality theorem topological properties of a compact Abelian group G should correspond to purely algebraic properties of its group of characters. A compact Abelian group G is metrizable if and only if its group of characters G^0 is countable. A group G is connected if and only if G^0 is torsion free (that is when G^0 is divisible). A compact Abelian group G is finite-dimensional if and only if its group of characters G^0 has finite rank. The above results were obtained by Pontryagin and they are a part of the so-called duality theory (see Pontryagin (1973)). Connected, locally connected, compact, Hausdorff, Abelian groups G have the property that their group of characters G^0 are discrete, torsion free and all subgroups of G^0 of finite rank are free. The following theorem was also proved by Pontryagin.

Theorem 8. *A finite dimensional compact Abelian group G is locally connected if and only if its group of characters G^0 is finitely generated.*

This result allows us to use well known theorems about the algebraic structure of finitely generated Abelian groups in studying the structure of locally compact groups. In particular, exactly in this way we can obtain the following result.

Theorem 9 (Pontryagin (1973)). *Let G be a locally Euclidean Abelian topological group. Then G is topologically isomorphic with the group $\mathbb{R}^n \times \mathbb{T}^m \times A$, where A is a discrete group and n, m are nonnegative integers.*

Duality theory found extensive application in the theory of locally compact groups and in algebraic topology, in particular, in proofs of theorems about the isomorphism of corresponding homology groups of compacta contained in manifolds and cohomology groups of their complements.

For every Abelian group G the characters form a family of homomorphisms into $\mathbb{T}$ separating the points of G. The diagonal of all characters is an isomorphism of G onto some subgroup H of T^τ, for a suitable τ. Since T^τ is compact, the group H is totally bounded (see Comfort (1984)). If the group G is infinite, so is H, and consequently H is not discrete. Thus we obtain the following result

Theorem 10 (Comfort (1984)). *Every infinite Abelian group G admits a non discrete topology relative to which it is a totally bounded topological group.*

Not every infinite Abelian group G can be equipped with a compact topology consistent with its algebraic structure. If a topological group G is countable and not discrete, then it is not compact because of Baire's property for compact spaces.

Duality theory makes it possible to prove explicitly that every compact Abelian group is *dyadic*, that is, a continuous image of D^τ (see Ivanovskij (1958), Kuz'minov (1959)). Let us mention that continuous homomorphic images of the topological group D^τ do not include all compact Abelian groups since all such images are zero-dimensional.

The duality theorem has been generalized in many directions. For example, there exists a version of this theorem for non-abelian, locally compact groups. Topological groups for which the duality theorem is true are called *reflexive*. Among such groups are, for example, all Banach spaces considered as topological groups. An important step in the proof of Theorem 7 is based on the following fact: for every non-zero element x of a locally compact Abelian (Hausdorff) group G there exists a continuous character $\varphi : G \to T$ such that $\varphi(x) \neq 1$. Not every Abelian topological group has this property, and this fact already is an obstruction to further generalizations of duality theory.

An unexpected result concerning the existence of continuous characters was obtained by Cotlar and Ricabarra (see Cotlar and Ricabarra (1954)). Let G be an Abelian topological group and let x be a non-zero element in G. Assume that there exists a neighborhood U of the zero element of G such that $G = F + U$ for some finite set F and $x \notin 6 \cdot U = U + U + U + U + U + U$. Then there exists a continuous character $\varphi : G \to \mathbb{T}$ for which $\varphi(x) \neq 1$. An elementary example of an infinite Abelian topological group G with a single continuous character, identically equal to 1, was described by Prodanov (see Comfort (1984)).

Totally bounded Abelian topological groups as well as free Abelian topological groups of Tikhonov spaces admit sufficiently many continuous characters. Pestov found conditions under which a free Abelian topological group $A(X)$ of a Tikhonov space X is reflexive, that is, satisfies the duality principle of Pontryagin (see Pestov (1986)). In particular, it follows from his results that if X is zero-dimensional in the sense of dim (that is in the sense of coverings) and metrizable, then the topological group $A(X)$ is reflexive.

Let us mention that it was Pontryagin who pointed out the unique role of the group $\mathbb{T}$ in duality theory. For no other locally compact Abelian (Hausdorff) group is the duality theorem true (see Pontryagin (1973)).

8.6. Compact Extensions of Topological Groups. Almost Periodic Functions.
Every topological subgroup H of a compact topological group is totally bounded. Indeed, we can cover H with a finite number of shifts of any open, nonempty subset of H. Total boundedness of a topological group H happens to be a sufficient condition for H to be topologically isomorphic to a subgroup of a compact group. Indeed, since H is totally bounded, H has a Weil completion with respect to the right uniform structure of H. This completion $\tilde{H}$ is

itself a topological group. The group $\widetilde{H}$ is totally bounded and Weil complete, and consequently it is compact. By strengthening the above reasoning slightly we can obtain the following result (see Comfort (1984)).

Theorem 11. *A topological group G can be represented as a dense subgroup of a compact topological group if and only if G is totally bounded. The compact group mentioned earlier is determined uniquely up to natural isomorphism.*

Let us mention that not every Hausdorff topological group has a Weil completion (in the class of groups).

The class of totally bounded topological groups contains all pseudocompact groups, that is those groups on which every continuous real-valued function is bounded. We have the following result (see Comfort (1984)).

Theorem 12. *Let K be a compact topological group and let H be a dense subgroup of it. Then H is pseudocompact if and only if every nonempty G_δ-set in K intersects H.*

From Theorem 11 it follows directly that the product of any family of totally bounded topological groups is a totally bounded group. The following result of Comfort and Ross (see Comfort and Ross (1966)) is much more surprising.

Theorem 13. *The product of any family of pseudocompact topological groups is a pseudocompact topological group.*

In the class of topological spaces pseudocompactness is not multiplicative. A square of a pseudocompact Tikhonov space does not have to be pseudocompact (see Engelking (1977)). Van Douwen and Malykhin showed that Theorem 13 can not be extended to countably compact groups.

Theorems 11 and 12 are nicely complemented by the following remarkable result.

Theorem 14. (see Comfort (1984)). *Let K be a compact topological group and let H be a dense subgroup of it. Then H is pseudocompact if and only if the compactum K is the Stone-Čech compactification of the space H, that is, if $K = \beta(H)$.*

Van Douwen proved (see van Douwen (1980)) that every Tikhonov space X whose Stone-Čech compactification is topologically homogeneous is pseudocompact. By combining this result with Theorems 14 and 11 he deduced that if for a topological group G the space $\beta(G)$ is topologically homogeneous, then the space $\beta(G)$ is a topological group with respect to algebraic operations that extend the algebraic operations in G. Van Douwen also gave an example of a pseudocompact, topologically homogeneous space that does not have topologically homogeneous compact extensions.

With the help of the following construction it is possible to assign to every topological group a compact group and its dense subgroup. Consider a set $\mathcal{H}$ of continuous homomorphisms of the group G into compact topological groups such that every continuous homomorphism of G into a compact group is equivalent, in the obvious sense, to a homomorphism $h \in \mathcal{H}$. Thus $\mathcal{H}$ can be considered as the set of "all" continuous homomorphisms of the group G into compact groups. Let $\mathcal{H} = \{h_\alpha : \alpha \in A\}$, where $h_\alpha : G \to H_\alpha$ and the group H_α is compact. The diagonal $\psi = \Delta\mathcal{H} = \Delta\{h_\alpha : \alpha \in A\}$ of the homomorphisms h_α is a continuous homomorphism of G into the compact group $K = \prod\{H_\alpha : \alpha \in A\}$. The closure $\overline{\psi(G)}$ of the image $\psi(G)$ in K is the desired compact group, and $\psi(G)$ is a dense subgroup of it. The compact group $\overline{\psi(G)}$ is denoted by $\alpha[G]$. From this construction it is easy to see that for any continuous homomorphism φ of G into a compact group H there exists a continuous homomorphism μ of the group $\alpha[G]$ into H such that $\varphi = \mu \circ \psi$. For every fixed group G this condition, together with the requirement of compactness of $\alpha[G]$ and the fact that $\psi(G)$ is dense in $\alpha[G]$, determines the pair $(\psi, \alpha[G])$ uniquely up to a natural equivalence.

We note that an arbitrary continuous homomorphism of a topological group G into a topological group H induces a natural continuous homomorphism of the group $\alpha[G]$ into the group $\alpha[H]$.

The compact group $\alpha[G]$ is associated in a natural way with the family of almost periodic functions on G. We will describe briefly this association.

For every bounded real-valued function f of the group G and for every $x \in G$ we will denote by $L_x(f)$ the left shift by x of the function f, that is the function on G defined by the formula: $L_x(f)(y) = f(x^{-1}y)$. As always we assume that for bounded functions f and g the distance $d(f, g)$ is defined as $d(f, g) = \sup\{|f(x) - g(x)| : x \in G\}$. The closure of the set $\{L_x(f) : x \in G\}$ in the topology generated by this metric is called the left orbit of the function f. A bounded real-valued function f on a group G is called *left almost periodic* if its left orbit is a compactum.

We will denote by A_G the set of all continuous left almost periodic functions on G. Every left shift L_x maps the set A_G into itself. Consider the topology of pointwise convergence on the set of all mappings of the set A_G into itself and denote by $\alpha'[G]$ the closure, in this topology, of the set of all left shifts L_x. It can be shown that $\alpha'[G]$ is a compact group and that the mapping L of G into $\alpha'[G]$ that sends an arbitrary $x \in G$ to L_x is a continuous homomorphism onto a dense subgroup of the group $\alpha'[G]$. It turns out that there exists a topological isomorphism of the compact groups $\alpha[G]$ and $\alpha'[G]$ that maps the homomorphism ψ to the homomorphism L. From the construction of $\alpha'[G]$ described above we obtain the following theorem (see Kelly (1955)).

Theorem 15. *The Banach algebra A_G of all continuous left almost periodic functions on a topological group G is isometrically isomorphic to a Banach algebra of all continuous real-valued functions on the compactum $\alpha[G]$.*

Accordingly, $\alpha[G]$ can be viewed as the space of all maximal ideals of the Banach algebra of all almost periodic continuous functions on a topological group G.

The homomorphism $\psi : G \to \alpha[G]$ is injective if and only if arbitrary two elements of the group G can be separated be a continuous homomorphism of the group G into some compact group. This is the case when, for example, G has a separating family of characters. However, even in this case ψ does not have to be a topological isomorphism between the groups G and $\psi(G)$ since the group $\psi(G)$ is always totally bounded while G does not have to be (one can take, for example, $\mathbb{R}$ in place of G).

8.7. Compacta and Namioka's Theorem on Joint Continuity of Separately Continuous Functions. Let us begin with the following result.

Theorem 16. *Let X and Y be compacta and let $f : X \times Y \to \mathbb{R}$ be a real-valued function that is continuous separately with respect to each variable. Then X contains a dense G_δ-subset A such that f is jointly continuous (that is, continuous with respect to the product topology on $X \times Y$) at every point of $A \times Y$.*

This remarkable theorem was proved by Namioka (see Namioka (1974)). One of its important corollaries is the theorem which states that if Y is a compactum and Z is a compactum contained in $C_p(Y)$, then Z contains a dense G_δ-subset M with the property that the topology induced on Z by the topology of uniform convergence coincides with the topology of pointwise convergence at all points of M (that is, with the topology induced from $C_p(Y)$). It follows easily from the last result that, for example, every Eberlein compact space with countable Suslin number is metrizable. Namioka's theorem was substantially generalized in the papers of Talagrand, Troallic, Christensen, Lie, Piotrowski and the others (see Christensen (1981), Talagrand (1985)). In particular, it remains true if one considers in place of X an arbitrary Čech complete space.

Theorem 16 implies, in particular, the following result of Ellis (see Ellis (1953)).

Theorem 17. *Let a group G be equipped with a locally compact (Hausdorff) topology $\mathcal{T}$, with respect to which all left and right shifts are continuous. Then $(G, \mathcal{T})$ is a topological group.*

Using Theorem 16 it is also easy to obtain the following deep result of Glicksberg (see Christensen (1981)).

Theorem 18. *Let G be a locally compact Abelian group, Γ its group of characters and $\mathcal{T}_\rho$ the weakest topology on G with respect to which all characters $g \in \Gamma$ are continuous. If $\Phi \subset G$ and if Φ is compact relative to $\mathcal{T}_\rho$ then Φ is a compact subspace of G itself.*

Several applications of Namioka's theorem in functional analysis were presented in the papers of Kenderov (see Choban and Kenderov (1985)).

8.8. Fragmentable and Strongly Fragmentable Compacta and Radon-Nikodým Compact Spaces. The results of Namioka presented in the previous section lead naturally to the concepts of fragmentable compactum and Radon-Nikodým compact space. Quite recently Reznichenko introduced an intermediate concept of a strongly fragmentable compactum. The concept of fragmentability is associated with a particular type of relationship, different from subordination, between a topology on a set and a metric.

Let $(X, \mathcal{T})$ be a topological space and let ρ be a metric on X. We will say, following Jayne and Rogers, that the metric ρ *fragments* X if for every nonempty subspace $Y \subset X$ and every $\epsilon > 0$ there exists a nonempty, open subset V of Y such that the diameter of V (with respect to the metric ρ) does not exceed ϵ. An equivalent condition is obtained when one requires Y to be closed in X.

Example 5. (see Namioka (1974)). If X is a compactum contained in a Banach space E equipped with the weak topology, then the metric induced by the norm fragments X.

A metric ρ on a topological space X is called *lower semicontinuous* if from the fact that $x, y \in X$ and $\rho(x, y) > \epsilon$ it follows that there exist neighborhoods O_x and O_y of the points x and y respectively, such that $\rho(x', y') > \epsilon$ for all $x' \in O_x$ and all $y' \in O_y$.

Example 6. If X is a scattered compactum, then the metric ρ such that $\rho(x, y) = 1$, for every two different points x, y of the set X, fragments X and is lower semicontinuous.

One of the equivalent definitions of the *Radon-Nikodým compact spaces* is the following: these are the compacta that are fragmented by some lower semicontinuous metric. Accordingly, all scattered compact spaces are among Radon-Nikodým compacta. Also Eberlein compact spaces belong to this class, however, as demonstrated by Reznichenko, not every Talagrand compact space (and a fortiori, not every Corson compact space) is Radon-Nikodým compact. In addition we have the following theorem of Reznichenko.

Theorem 20. *A compactum X is Eberlein compact if and only if it is both Corson and Radon-Nikodým compact.*

On the other hand, every Gul'ko compact space is fragmentable.

We will say that a metric ρ *strongly fragments* a space X if it fragments X and for any two different points $x, y \in X$ there exist neighborhoods O_x and O_y of x and y respectively such that $\rho(O_x, O_y) > 0$. A space admitting a strongly fragmenting metric is called *strongly fragmentable* (Reznichenko). From those definitions it is clear that every Radon-Nikodým compact space is strongly fragmentable. It is not known whether the converse is true. However, every strongly fragmentable Corson compact space is Eberlein compact.

At present the answer to the following question of Namioka is not known: is a continuous image of a Radon-Nikodým compact space also Radon-Nikodým compact? However, as was established by Reznichenko, a continuous image of a strongly fragmentable compactum is a strongly fragmentable compactum. The following result was obtained by Baturov and Reznichenko.

Theorem 21. *Let X be a strongly fragmentable compactum. Then the following conditions are equivalent: (a) X is Preiss-Simon compact (that is, all pseudocompact subspaces in X are closed), (b) X is Fréchet-Urysohn compact.*

This statement can not be generalized to fragmentable compacta.

The concept of fragmentability and its different versions proved useful, as indicated by the above results, in the study of topological properties of compacta related to Eberlein compact spaces. Let us note that Reznichenko characterized *Eberlein compact* spaces as hereditarily metaLindelöf strongly fragmentable compacta and proved that every hereditarily metaLindelöf fragmentable compactum is Corson compact.

8.9. Hilbert Modules over C^*-Algebras of Continuous Functions on Compacta. Let A be a fixed C^*-algebra with a unit. A right A-module H is called a *pre-Hilbert A- module* if H is equipped with a scalar product with values in A, that is a function $\langle \cdot, \cdot \rangle$ transforming $H \times H$ into A and satisfying for all $h_1, h_2, h_3 \in H$ and all $a \in A$, the following conditions:

1) $\langle h_1 + h_2, h \rangle = \langle h_1, h \rangle + \langle h_2, h \rangle$;
2) $\langle h_1, h_2 a \rangle = \langle h_1, h_2 \rangle a$;
3) $\langle h_1, h_2 \rangle^* = \langle h_2, h_1 \rangle$;
4) $\langle h, h \rangle \geq 0$, and if $\langle h, h \rangle = 0$ then $h = 0$.

The norm on H is defined by: $\|h\| = \sqrt{\|\langle h, h \rangle\|}$, for all $h \in H$. If a pre-Hilbert A-module H is complete with respect to this norm, it is called a *Hilbert A-module*.

Every Hilbert A-module H is a Banach A-module and for every A-module H there exists the dual Banach A-module H^* whose elements are A-linear continuous morphisms from H to A. The action of A on H^* is defined in the following way: $(h'a)(h) = a^* h'(h)$. Consequently, there are also defined the Banach modules H^{**}, H^{***}, etc. The modules H, H^* and H^{**} are related through a canonical norm preserving A-morphisms (which are, moreover, injective). These A-morphisms allow the assumption that $H \subset H^{**} \subset H^*$. It turns out that the following dual Banach modules do not provide additional information since the Banach A-modules H^* and H^{***} are canonically isomorphic and so are the Banach A-modules H^{**} and H^{****} (see Paschke (1973)). This fact adds interest to the following question: when are some of the inclusions in the chain $H \subset H^{**} \subset H^*$ equalities? In connection with this question we introduce the following definitions.

A Hilbert A-module H is called *reflexive* if $H = H^{**}$, *antireflexive* if $H^{**} = H^*$ and *self-dual* if $H = H^*$ (see Mishchenko (1984)).

Mishchenko gave a particularly detailed account of the case when $H = H^{**}$. A compactum X is called *Mishchenko compact*, or *M-compact* if every countably generated Hilbert module H over the C^*-algebra $C(X)$ of all continuous complex-valued functions on X is reflexive. Trofimov proved that a compactum X is Mishchenko compact if for no point $x \in X$ the compactum X is the Stone-Čech compactification of the space $X - \{x\}$. In particular, all first countable and all bisequential compacta, as well as all Eberlein compact spaces are M-compact.

On the other hand a compactum X is extremally disconnected if and only if every countably generated Hilbert module H over the C^*-algebra $C(X)$ is antireflexive, that is it satisfies $H^* = H^{**}$.

If an M-compact space X is extremally disconnected then every countably generated Hilbert module H over $C(X)$ satisfies the condition $H = H^*$, that is, it is self-dual. But then one can show that the compactum X is finite (see Trofimov (1987)). Consequently, no infinite, extremally disconnected compactum is Mishchenko compact (see Mishchenko (1984), Trofimov (1987)).

It seems that the general problem of explaining connections between the topological properties of a compactum X and the structure of Hilbert modules over the C^*-algebra $C(X)$ posed by Mishchenko is quite interesting but far from being completely solved. Let us point out that the Hilbert A-modules have many applications, in particular in K-theory, in the theory of extensions of C^*-algebras and in the operator theory.

8.10. Compact Subsets of Topological Fields. Recall that a *division ring* is defined as an associative ring in which non-zero elements form a group with respect to multiplication. A division ring in which multiplication is commutative is called a field. A *topological division ring (field)* is a division ring (field) T with a topology $\mathcal{T}$ such that with respect to addition T is a topological group, and the set of all non-zero elements of T is a topological group with respect to multiplication. The role of compactness in the theory of topological division rings is determined primarily by the following remarkable result of Pontryagin (see Pontryagin (1973)).

Theorem 22. *There exist exactly three connected locally connected topological division rings: the field of real numbers, the field of complex numbers and the division ring of quaternions.*

A topological space is called *totally disconnected* if it does not contain any nontrivial connected subspaces. There are many totally disconnected locally compact division rings but they are quite well known. Every such division ring is a finite extension of either the field of p-adic numbers or the field of infinite series over the field of integers mod p (where p is a prime number) (Pontryagin, see Pontryagin (1973)).

The main variety of topological fields is outside of the realm of locally compact spaces. Kiltinen proved (see Shakhmatov (1983)), that every infinite field P admits $2^{2^{|P|}}$ mutually different topologies with respect to which P is a Hausdorff topological field. This leads naturally to the general problem of studying the topological structure of topological fields. Here are two more specific questions related to this subject. Which Tikhonov spaces can be obtained as subspaces (closed subspaces) of topological fields? Which compacta can be embedded into topological fields?

Important results in this direction were obtained by Shakhmatov (see Shakhmatov (1983)). Here we will present some of them.

Theorem 23. *Every pseudocompact subspace of an arbitrary topological field is metrizable.*

Theorem 24. *Assume that a Tikhonov space X admits a one-to-one continuous mapping onto some metrizable space. Then X can be embedded as a subspace in a connected topological field P of an arbitrary characteristic.*

Theorems 23 and 24 together imply the following result.

Corollary 2. *A compactum can be embedded into some topological field if and only if this compactum is metrizable.*

Let us point out that every compactum X can be embedded into a topological ring, for example, into $C_p C_p(X)$. It is even possible to embed X into a topological integral domain (see Shakhmatov (1983)).

Theorem 23 and Corollary 2 cannot be generalized to locally compact subspaces of topological fields. A locally compact closed subspace of a topological field not only does not have to be metrizable, but even could fail to be paracompact (Shakhmatov, see Shakhmatov (1983)). The following addition to Theorem 24 is quite important.

Theorem 25. *If a space X admits a continuous one-to-one mapping onto a metrizable compactum, then X can be embedded into a Tikhonov topological field P as a closed subspace.*

Since there exist non normal, Tikhonov spaces that admit a continuous one-to-one mapping onto a metrizable compactum, the last theorem implies that a topological field does not have to be a normal topological space (Shakhmatov, see Shakhmatov (1983)).

8.11. Locally Compact Topological Groups and Paracompactness. In the section devoted to Pontryagin's duality theory we mentioned the prominent role of local compactness in the theory of topological groups. The theory of locally compact groups is the subject of separate presentations. In this section we will mention only one important topological feature of the structure of locally compact topological groups.

Theorem 26 (see Kelley (1955)). *The space of each (Hausdorff) locally compact topological group G is paracompact. More precisely, the space of G can be represented as a union of mutually disjoint σ-compact open subspaces, and thus, G is strongly paracompact.*

This theorem is fundamental in the proof of the equality of dimensions ind and dim for locally compact groups (see Aleksandrov and Pasynkov (1977)). A more complicated approach, using spectral methods, proves that for locally compact groups all three dimensions ind, dim and Ind coincide (Pasynkov, see Aleksandrov and Pasynkov (1977)).

8.12. Final Remarks. The theme "compactness in algebra and functional analysis" will be continued in the articles, Topology and Algebraic Structures, and Topological Spaces of Functions presented in other issues. Compactness plays a great role in probability theory and in general measure theory on topological spaces, but to discuss these issues one needs separate surveys reaching far beyond the limits of general topology. The same can be said about the role of compactness in the theory of differential equations (in particular, in questions related to correctness in the sense of Tikhonov). It is impossible to present all the important applications of compactness in mathematics in one survey. In this presentation we limited ourselves to a description of fundamental results of the theory of compact spaces and to a detailed description of the most important principles associated with some applications of this theory.

References*

Aleksandrov, P.S. (1939): On bicompact extensions of topological spaces. Mat. Sb., Nov. Ser. **5**(4), 403–423 (Russian). Zbl.22,412

Aleksandrov, P.S., Urysohn, P.S. (1971): Memoir on Compact Topological Spaces. Nauka, Moscow, 3rd ed. (Russian). Zbl.228.54017

Aleksandrov, P.S., Pasynkov, B.A. (1973): Introduction to Dimension Theory. Nauka, Moscow (Russian). Zbl.272.54028

Alster, K. (1979): Some remarks on Eberlein compacts. Fundam. Math. **104**(1), 43–46. Zbl.426.54015

Arhangel'skii, A.V. (1966): On embeddings of T_1-spaces in bicompact T_1-spaces of the same weight. Bull. Acad. Pol. Sci., Ser. Math., Astron., Phys **14**(7), 361–166. Zbl.144,217

Arhangel'skii, A.V. (1965a): On a class of spaces containing all metric and all locally bicompact spaces. Mat. Sb., Nov. Ser. **67**(1), 55–88 (Russian). Zbl.127,131

Arhangel'skii, A.V. (1965b): Bicompact sets and topology of spaces. Tr. Mosk. Mat. O-va. 13, 3–55 (Russian). Zbl.149,404

Arhangel'skii, A.V. (1976): On some topological spaces arising in functional analysis. Usp. Mat. Nauk **31**(5), 17–32. Zbl.344.46058; English translation: Russ. Math. Surv. 31(5), 14–30 (1976)

Arhangel'skii, A.V. (1978): Structure and classification of topological spaces and cardinal invariants. Usp. Mat. Nauk **33**(6), 29–84. Zbl.414.54002; English translation: Russ. Math. Surv. 33(6), 33–96 (1978)

Arhangel'skii, A.V. (1979a): On invariants related to character and weight. Tr. Mosk. Mat. O-va **38**, 3–23. Zbl.439.54005; English translation: Trans. Mosc. Mat. Soc. 2, 1–23 (1980)

---

* For the convenience of the reader, references to reviews in Zentralblatt für Mathematik (Zbl.), compiled using the MATH database, and Jahrbuch über die Fortschritte der Mathematik (Jrb.) have, as far as possible, been included in this bibliography.

Arhangel'skii, A.V. (1979b): Spectrum of frequencies of topological spaces and products. Tr. Mosk. Mat. O-va **40**, 171–206. Zbl.447.54004; English translation: Trans. Mosc. Mat. Soc. 2, 163–200 (1981)

Arhangel'skii, A.V. (1980a): On some properties of radial spaces. Mat. Zametki **27**(1), 95–104. Zbl.434.54004; English translation: Math. Notes 27, 50–54 (1980)

Arhangel'skii, A.V. (1980b): The method of stars, new classes of spaces and countable compactness. Dokl. Akad. Nauk SSSR **251**(5), 1033–1037. Zbl.485.54020; English translation: Sov. Math., Dokl. 21, 550–554 (1980)

Arhangel'skii, A.V.(1981): Classes of topological groups. Usp. Mat. Nauk **36**(3), 127–146. Zbl.468.22001; English translation: Russ. Math. Surv. 36(3), 151–174 (1981)

Arhangel'skii, A.V. (1982): On linear homeomorphisms of spaces of functions. Dokl. Akad. Nauk SSSR **264**(6), 1289–1292. Zbl.522.54015; English translation: Sov. Math., Dokl. 25, 852–855 (1982)

Arhangel'skii, A.V. (1984a): Spaces of mappings in the topology of pointwise convergence and compacta. Usp. Mat. Nauk **39**(5), 11–50. Zbl.568.54016; English translation: Russ. Math. Surv. 39(5), 9–56 (1984)

Arhangel'skii, A.V. (1984b): Continuous mappings, factorization theorems and spaces of mappings. Tr. Mosk. Mat. O-va **47**, 3–21. Zbl.588.54018; English translation: Trans. Mosc. Math. Soc. 1985, 1–22 (1985)

Arhangel'skii, A.V. (1986): On R-quotient mappings of spaces with a countable base. Dokl. Akad. Nauk SSSR **287**(1), 14–17. Zbl.616.54008; English translation: Sov. Math. Dokl. 33, 302–305 (1986)

Arhangel'skii, A.V. (1987a): A survey of C_p-theory. Quest. Answers Gen. Topology **5**(1), 1–109. Zbl.634.54012

Arhangel'skii, A.V. (1987b): Topological homogeneity. Topological groups and their continuous images, Usp. Mat. Nauk **42**(2), 69–105. Zbl.642.54017; English translation: Russ. Math. Surv. 42(2), 83–131 (1987)

Arhangel'skii, A.V., Ponomarev, V.I. (1974): Principles of General Topology in Problems and Questions. Izd. Nauka, Moscow. Zbl.287.54001; English translation: Reidel 1984

Arhangel'skii, A.V., Tkachuk, B.V. (1985): Spaces of Functions and Topological Invariants. MGU

Asanov, M.O. (1977): Products of normal countably compact spaces. In: Set Theory and Topology, Izhevsk, 1, 1–6. Zbl.491.54016

Balogh, Z. (1987): On compact Hausdorff spaces of countable tightness, Baku Int. Top. Conf. (Abstracts), Part 1, Baku

Balogh, Z. (1989): On compact Hausdorff spaces of countable spaces. Proc. Amer. Math. Soc. **105**, 755–764

Bandlov, Yu., Ponomarev, V.I. (1980): Cardinal invariants related to tightness, absolutes of spaces and continuous mappings. Usp. Mat. Nauk **35**(3), 23–30. Zbl.461.54002. English tansl.: Russ. Math. Surv. 35(3), 25–33 (1980)

Bell, M. (1980): Compact ccc non-separable spaces of small weight. Topology Proceedings **5**, 11–25. Zbl.464.54017

Bell, M. (1985): Generalized dyadic spaces. Fundam. Math. **125**(1), 47–58. Zbl.589.54019

Biryukov, P.A. (1981): Orders of families of sets and properties of topological spaces, Dokl. Akad. Nauk SSSR **257**(4), 777–779. Zbl.499.54004. English translation: Sov. Math. Dokl. 23, 315–318 (1981)

Borges, C.J.R. (1966): On stratifiable spaces. Pac. J. Math. **17**(1), 1–16. Zbl.175,198

Bourbaki, N. (1969a): Espaces Vectorielles Topologiques. Paris, Hermann. Zbl.50,107

Bourbaki, N. (1969b): Algebra Commutative, Ch. 1 at 2 (3-ème ed.). Paris, Hermann. Zbl.205,60 (Zbl.108,40)

Bourgin, J. (1978): Some remarks on compact sets of the first Baire class. Bull. Soc. Belg., Ser. B **30**(1), 3–10. Zbl.414.54011

Bregman, Yu., Shapirovskij, B., Shostak, A. (1984): On partitions of topological spaces. Caš. Pestovani Mat. **108**(1), 27–53. Zbl.539.54005

Broverman, S., Weiss, W. (1981): Spaces co-absolute with $\beta N \setminus N$. Topology and Appl. **12**(2), 127–133. Zbl.455.54020

Bula, W., Turzański, M.A. (1986): A characterization of continuous images of compact ordered spaces and the Hahn-Mazurkiewicz problem. Topology Appl. **22**, 7–17. Zbl.587.54053

Burke, D.K. (1984): Covering properties. in Handbook of Set-Theoretic Topology, Amsterdam, North Holland, 347–422. Zbl. 569.54022

Burke, D.K., Davis, S.W. (1981): Compactifications of symmetrizable spaces. Proc. Am. Math. Soc. **81**, 647–651. Zbl. 451.54021

Čech, E. (1937): On bicompact spaces. Ann. Math. **38**,, II. Ser. 823–844. Zbl.17.428

Chaber, J. (1976): Conditions which imply compactness in countably compact spaces. Bull. Acad. Pol. Sci. Ser. Sci. Math. **24**, 993–998. Zbl. 347.54013

Charin, V.S. (1966): Topological Groups. in Itogi Nauki VINITI, Algebra, 123–160 (Russian). Zbl.209,59

Chigogidze, A. Ch. (1982): On κ-metrizable spaces. Usp. Mat. Nauk **37**(2), 241–242. Zbl.494.54008; English translation: Russ. Math. Surv 37(2) 209–210 (1982)

Choban, M.M. (1976): Topological structure of subsets of topological groups and their quotient spaces. Dokl. Akad. Nauk SSSR **228**(1), 52–55. Zbl.346.22007; English translation: Sov. Math. Dokl. 17, 669-673 (1976)

Choban M.M., Dodon, N.K. (1979): Theory of $\mathcal{P}$–Scattered Spaces. Shtiinitsa, Kishiniev, 100 pp.(Russian). Zbl.506.54029

Choban, M.M., Kenderov, P.S. (1985): Dense Gâteaux differentiability of the sup-norm in $C(T)$ and the topological properties of T. Dokl. Bulg. Acad. Sci. **38**(12), 1603–1604. Zbl.608.46012

Christensen, J.P. (1981): Joint continuity of separately continuous functions. Proc. Am. Math. Soc. **82**(3), 455–461. Zbl.472.54007

Comfort, W.W. (1984): Topological groups. In: Handbook of Set-Theoretic Topology, Amsterdam, North Holland, 1143–1263. Zbl.604.22002

Comfort, W.W., Ross, R.A. (1966): Pseduocompactness and uniform continuity in topological groups. Pac. J. Math. **16**, 483–496. Zbl.214,285

Corson, H.H. (1959): Normality in subsets of product spaces. Am. J. Math. **81**(3), 785–796. Zbl.95,373

Corson, H.H., Lindenstrauss, J. (1966): On weakly compact subsets of Banach spaces. Proc. Am. Math. Soc. **17**(2), 407–412. Zbl.186,447

Cotlar, M., Ricabarra, R. (1954): On existence of characters in topological groups. Am. J. Math. **76**, 375–388. Zbl.55,257

Debs, G. (1980): Quelques propriètés des espaces α-favorables et applications aux convexes compacts. Ann. Inst. Fourier **30**(2), 29–43. Zbl.432.54027

Debs, G. (1986): Points de continuité d'une fonction séparémant continue. Proc. Am. Math. Soc. **97**(1), 167–176. Zbl.592.54012

van Douwen, E.K. (1977): Density of compactifications. In: Set-Theoretic Topology, New York, Acad. Press, 97–110. Zbl.379.54006

van Douwen, E.K. (1980): Homogeneity of $\beta(G)$ when G is a topological group. Colloq. Math. **41**(2), 193–199. Zbl.454.22001

van Douwen, E.K., van Mill, J. (1978): Parovichenko's characterization of $\beta(\omega) \setminus \omega$ implies CH. Proc. Am. Math. Soc. **72**(2), 539–541. Zbl.393.54004

Dranishnikov, A.N. (1984): Absolute extensors in dimension n and n-soft mappings. Usp. Mat. Nauk **39**(5), 55–95. Zbl.572.54012; English translation: Russ. Math. Surv. 39(5), 63–111 (1984)

Efimov, B.A. (1965): Dyadic bicompacta. Tr. Mosk. Mat. O-va. 14, 211–241. Zbl.163,172; English translation Trans. Mosc. Math. Soc. 14, 229–267 (1967)

Efimov, B.A. (1975): On the cardinality of extensions of dyadic spaces. Mat. Sb., Nov. Ser. **96**(4), 614–632. Zbl.312.54005; English translation: Math. USSR, Sb. 25, 579–583 (1976)

Ellis, R. (1953): Continuity and homeomorphism groups. Proc. Am. Math. Soc. **4**(6), 969–973. Zbl.52,396

Engelking, R. (1977): General Topology, Warszawa, PWN. Zbl.373.54001

Fedorchuk, V.V. (1976): Completely closed mappings and consistency of some theorems of general topology with axioms of set theory. Mat. Sb., Nov. Ser. **99**(1), 3–33. Zbl.357.54018; English translation: Math. USSR, Sb. 28, 1–26 (1977)

Fedorchuk, V.V. (1980): The method of splitting spectra and completely closed mappings in general topology. Usp. Mat. Nauk **35**(3), 112–121. Zbl.529.54008; English translation: Russ. Math. Surv. 35(3), 131–143 (1980)

Fedorchuk, V.V. (1981): Covariant functors in the category of compacta, absolute retracts and Q manifolds. Usp. Mat. Nauk **36**(3), 177–198. Zbl.495.54008; English translation: Russ. Math. Surv. 36(3), 211–233 (1981)

Fillipov, V.V. (1969): Quotient spaces and rank of a base. Mat. Sb., Nov. Ser.**80**(4), 521–532. Zbl.199,573; English translation: Math. USSR, Sb. 9, 487–496 (1969)

Fillipov, V.V. (1969): On perfectly normal compacta. Dokl. Akad. Nauk SSSR **189**, 736–739. Zbl.198,555; English translation: Sov. Math. Dokl. 10, 1503–1507 (1970)

Frolik, Z. (1977): Homogeneity problems for extremally disconnected spaces. Comment. Math. Univ. Carol. **8**(4), 757–763. Zbl.163,445

Gillman, L., Jerison, M. (1960): Rings of Continuous Functions. Princeton, Van Nostrand. Zbl.93,300

Gleason, A.M. (1958): Projective topological spaces, Ill. J. Math. **2**, 482–489. Zbl.83,174

Godefroy, G. (1958): Compacts de Rosenthal. Pac. J. Math. **91**(2), 293–306. Zbl.475.46003

Grothendieck, A. (1953): Sur les applications linéaire faiblement compactes d'espaces du type $C(K)$. Can. J. Math. **5**(2), 129–173. Zbl.50,109

Gruenhage, G. (1984): Generalized metric spaces. In: Handbook of Set-Theoretic Topology, Amsterdam, North Holland, 423–502. Zbl.555.54015

Gruenhage, G. (1986): Games, covering properties and Eberlein compacts. Topology and Appl. **23**(3), 291–297. Zbl.604.54022

Gruenhage, G., Nyikos, P. (1978): Spaces with the base of countable rank. Gener. Topology Appl. **8**(3), 233–257. Zbl.412.54034

Gryzlov, A.A. (1980): Two theorems on cardinality of topological spaces. Dokl. Akad. Nauk SSSR **251**(3), 780–783. Zbl.449.54005; English translation: Sov. Math. Dokl. 21, 506–509 (1980)

Gul'ko, S.P. (1979): On the structure of spaces of continuous functions and their hereditary paracompactness. Usp. Mat. Nauk **34**(6), 33–40. Zbl.421.46020; English translation: Russ. Math. Surv. 34(6), 36–44 (1979)

Gul'ko, S.P., Khmyleva, T.E. (1986): Compactness is not preserved by the relation of t-equivalence. Mat. Zametki **39**(6), 895–903. Zbl.629.54007; English translation: Math. Notes 39, 484–488 (1986)

Hagler, J. (1975): On the structure of S and $C(S)$ for S dyadic. Trans. Am. Math. Soc. **214**, 415–428. Zbl.321,46022

Haydon, R. (1974): On a problem of Pełczyński; Milutin spaces, Dugundji spaces and AE (dim 0). Studia Math. **52**(1), 23–31. Zbl.294.46016

Haydon, R. (1976): An extreme point criterion for separability of a dual Banach space and a new proof of a theorem of Corson. Q. J. Math., Oxf. II. Ser. **27**, 379–380. Zbl.335.46012

Heath, R.W., Lutzer, P.L., Zenor, P.L. (1973): Monotonically normal spaces. Trans. Am. Math. Soc. **178**, 481–493. Zbl.269.54009

Henricksen, M., Isbell, J.B. (1958): Some properties of compactifications. Duke Math J. **25**, 85–105. Zbl.81,386

Hewitt, E. (1947): Certain generalizations of the Weierstrass approximation theorem. Duke Math. J. **14**, 419–427. Zbl.29,303

Hewitt E. (1948): Rings of real-valued continuous functions I. Trans. Am. Math. Soc. **64**(1), 45–99. Zbl.32,286

Hewitt, E. (1960): The role of compactness in harmonic analysis. Am. Math. Month. **67**, 499–516. Zbl.101,153

Hewitt, E. (1962): Some applications of compactness in harmonic analysis. General Topology and its Relations to Modern Analysis and Algebra, Proc. Sympos., Prague 1961, 204–210. Zbl.116,329

Hochster, M. (1960): Prime ideal structure in commutative rings. Trans. Am. Math. Soc. **142**, 43–60. Zbl.184,294

Hödel, R. (1984): Cardinal functions. In: Handbook of Set-Theoretic Topology, Amsterdam, North Holland, 1–62. Zbl.559.54003

Hoffman, B. (1979): A surjective characterization of Dugundji spaces. Proc. Am. Math. Soc.. **76**(1), 151–156. Zbl.432.54016

Hušek, M. (1977): Topological spaces without κ-accessible diagonal. Comment. Math. Univ. Carol. **18**(4), 777–788. Zbl.374.54035

Inasaridze, H.N. (1966): On extensions and remainders of finite order of Tikhonov spaces. Dokl. Akad. Nauk SSSR **166**(6), 1043–1045. Zbl.166,184; English translation: Sov. Math. Dokl. 7, 229–231 (1966)

Isbell, J.R. (1964): Uniform Spaces. Providence, Rhode Island, Am. Math. Soc.. Zbl.124,156

Ivanov, A.V. (1978): On bicompacta all of whose finite powers are hereditarily separable. Dokl. Akad. Nauk SSSR **243**(5), 1109–1112. Zbl.409.54028; English translation: Sov. Math. Dokl. 19, 1470–1473 (1978)

Ivanov, A.V. (1981): Superextensions of openly generated bicompacta. Dokl. Akad. Nauk SSSR **259**(2), 275–278. Zbl.476.54019; English translation: Sov. Math. Dokl. 19, 1470–1473 (1978)

Ivanovskij, L.N. (1958): On a conjecture of P.S. Aleksandrov. Dokl. Akad. Nauk SSSR **123**(5), 785–786 (Russian). Zbl.113,258

Jayne, J.E. (1978): Metrization of compact convex sets. Math. Ann. **234**, 109–115. Zbl.409.46014

Juhász, I. (1980): Cardinal functions in topology - ten years later. Math. Center Tracts 123, Amsterdam. Zbl.479.54001

Juhász, I. (1984): Cardinal functions II. In: Handbook of Set-Theoretic Topology, Amsterdam, North Holland, 63–110. Zbl.559.54004

Kalmutskij, L.I. (1986): On extensions of Wallman type of T_0-spaces. In: Studies in General Algebra, Geometry and their Applications, Shtiinitsa, Kishinev, 77–85 (Russian). Zbl.679.54021

Katetov, M. (1940): Über H-abgeschlossene und bikompakte Räume. Čas. Mat. Fus. **69**, 36–49. Zbl.22,412

Keldysh, L.V. (1959): Zero-dimensional open mappings. Izv. Akad. Nauk SSSR, Ser. Mat.SM **23**(2), 165–184 (Russian). Zbl.86,158

Kelley, J.L. (1955): General Topology. Princeton-New York, D. Van Nostrand Company. Zbl.66,166

Khelemskij, A. Ya. (1986): Homology in Banach Topological Algebras. MGU, Moscow, 288 pp. (Russian). Zbl.608.46046

Kimura, T. (1984): Every compact tree-like space is regular supercompact. Quest. Answers Gen. Topology **2**, 69–72. Zbl.547.54019

Koumoullis, G. (1985): Topological spaces containing compact perfect sets and Prohorov spaces. Topology Appl., **21**(1), 59–72. Zbl.574.54041

Kuz'minov, V.I. (1959): On a conjecture of P.S. Aleksandrov in the theory of topological groups. Dokl. Akad. Nauk SSSR **125**(4), 727–729 (Russian). Zbl.133,287

Leiderman, A.G. (1985): On dense metrizable subspaces of Corson compact spaces. Mat. Zametki **38**(3), 440–449. Zbl.602.54024; English translation: Math. Notes 38, 751–755 (1985)

Leiderman, A.G., Sokolov, G.A. (1984): Adequate families of sets and Corson compacts. Comment. Math. Univ. Carol. **25**(2), 233–246. Zbl.586.54022

Lindenstrauss, J. (1972): Weakly compact sets – their topological properties and the Banach spaces they generate.Ann. Math. Stud. **69**, 235–273. Zbl.232.46019

Malykhin, V.I. (1982): On cardinality of bicompact spaces satisfying the weak form of the first axiom of countability. Mat. Zametki **32**(2), 261–268. Zbl.505.54006; English translation: Math. Notes 32, 610–614 (1983)

Malykhin, V.I. (1983): Topology and forcing. Usp. Mat. Nauk **38**(1), 69–118. Zbl.515.54003; English translation: Russ. Math. Surv. 38(1), 77-136 (1983)

Malykhin, V.I. (1987a): Nonpreservation of properties of topological groups under the operation of squaring. Sib. Mat. Zh. **28**(4), 639–645. Zbl.636.54007; English translation: Sib. Math. J. 28(4), 154–161 (1987)

Malykhin, V.I. (1987b): A Fréchet-Urysohn bicompactum without points of countable character. Mat. Zametki **41**(3), 365–376. Zbl.624.54003; English translation: Math. Notes 41, 210–216 (1987)

Malykhin, V.I., Ponomarev, V.I. (1975): General Topology (Set-Theoretic Approach), Itogi Nauki Tekh., VINITI, Ser. Algebra, Topology, Geometry 13, 149–229. Zbl.434.54001; English translation: J. Sov. Math. 7, 587–629 (1977)

Mardešič, S. (1970): Mapping products of ordered compacta onto products of more factors. Gla. Mat. **5**, 163–170

Matveev, M.V. (1984): On properties similar to pseudocompactness and countable compactness. Vestn. Mosk. Univ., Ser. I Mat. Mekh., 1984(2), 24–27. Zbl.556.54016; English translation: Mosc. Univ. Math. Bull. 39, No. 2, 32–36 (1984)

Michael, E. (1966): $\aleph_0$-spaces. J. Math. Mech. **15**(6), 923–1002. Zbl.148,167

Michael, E. (1972): A quintuple quotient quest. General Topol. Appl. **2**(2), 91–138. Zbl.238.54009

Michael, E. (1973): On k-spaces. k_R-spaces and $k(X)$, Pac. J. Math. **47**(2), 487–498. Zbl.262.54017

van Mill, J. (1977): Supercompactness and Wallman spaces. Math. Center Tracts **85**. Zbl.407.54001

van Mill, J. (1982): A homogeneous Eberlein compact space which is not metrizable. Pac. J. Math. **101**(1), 141–146. Zbl.495.54020

van Mill, J. (1984): An introduction to $\beta(\omega)$. In: Handbook of Set-Theoretic Topology, Amsterdam, North Holland, 503–568. Zbl.555.54004

van Mill, J., Mills, C. (1978): On the character of supercompact spaces. Topol. Proc. **3**, 227–236. Zbl.413.54031

Mishchenko, A.S. (1962): On spaces with point countable base. Dokl. Akad. Nauk SSSR **144**(6), 985–986. Zbl.122,173; English translation: Sov. Math. Dokl. 3, 855–858 (1962)

Mishchenko, A.S. (1984): Representations of compact groups in Hilbert modules over C^*-algebras. Tr. Mat. Inst. Steklova **166**, 161–176. Zbl.562.46038; English translation: Proc. Steklov Inst. Math 166, 179–185 (1986)

Mukhin, Yu. N. (1982): Topological groups. Itogi Nauki Tekhniki, VINITI, Ser. Algebra,Topology. Geometry. 20, 3–64 (Russian). Zbl.518.22001

Nagami, K. (1969): $\sum$-spaces. Fundam. Math. **61**, 169–192. Zbl.181,507

Naimark, M.A. (1956): Normed Rings. Gostekhizdat', Moscow, 487 pp. (Russian). Zbl.73,89

Nakhmanson, L.B., Yakovlev, N.N. (1981): On bicompacta in σ-products. Commentat. Math. Univ. Carol. **22**(4), 705–719. Zbl.499.54006

Namioka, I. (1974): Separate continuity and joint continuity. Pac. J. Math. **51**(2), 515–531. Zbl.294.54010

Nedev, C.J. (1971): o-metrizable spaces. Tr. Mosk. Mat. O-va. 24, 201–236. Zbl.244. 54016; English translation: Trans. Mosc. Math. Soc. 24, 213–247 (1974)

Negrepontis, S. (1984): Banach spaces and topology. In: Handbook of Set-Theoretic Topology, Amsterdam, North Holland, 1045–1142. Zbl.584.46007

Nikiel, J. (1986): Some problems on continuous images of compact ordered spaces. Quest. Answers Gen. Topology **4**(2), 117–128. Zbl.625.54039

Noble, N. (1969): Countably compact and pseudocompact products. Czech. Math, J. **19**, 390–397. Zbl.184,477

Nyikos, P.J. (1977): Some surprising properties in topology, II, Set-Theoretic Topology, New York, Acad. Press, 277–305. Zbl.397.54004

Okromeshko, N.G. (1984): Retracts of bicompacta similar to homogeneous ones. Vest. Mosk. Univ. Ser. I Mat. Mekh., 1984(6), 7–10. Zbl.587.54038; English transl: Mosc. Univ. Math. Bull. 39(6), 7–11 (1984)

Parovichenko, I.I. (1963): On a universal bicompactum of weight $\aleph_1$. Dokl. Akad. Nauk SSSR **150**(1), 36–39. Zbl.171,213; English translation: Sov. Math. Dokl. 4, 592–595 (1963)

Paschke, W.L. (1973): Linear product modules over B^*-algebras. Trans. Am. Math. Soc. **182**, 443–468. Zbl.265.46056

Pavlovskij, D.S. (1980): On spaces of continuous functions in the topology of pointwise convergence. Dokl. Akad. Nauk SSSR **253**(1), 38–41. Zbl.489.54011; English translation: Sov. Math. Dokl. 22, 34–37 (1980)

Pavlovskij, D.S. (1982): On spaces with linearly homeomorphic spaces of continuous functions in the topology of pointwise convergence. Usp. Mat. Nauk **37**(2), 185–186. Zbl.501.54009; English translation: Russ. Math. Surv. 37(2), 241–241 (1982)

Pełczyński, A. (1968): Linear Extensions, Linear Averagings and their Applications to Linear Topological Classification of Spaces of Continuous Functions. Rozpr. Mat. **58**. Zbl.165,146

Peregudov, S.A., Shapirovskij, B. Eh. (1976): On a class of bicompacta. Dokl. Akad. Nauk SSSR **230**(2), 279–282. Zbl.364.54012; English translation: Sov. Math. Dokl. 17, 1296–1300 (1977)

Pestov, V.G. (1982): Coincidence of dimension for dim l-equivalent topological spaces. Dokl. Akad. Nauk SSSR **266**(3), 553–556. Zbl.518.54030; English translation: Sov. Math. DOkl. 26, 380–383 (1982)

Pestov, V.G. (1986): Free topological abelian groups and Pontryagin's duality. Vestn. Mat. Univ. Ser. I Mat. Mekh. 1986(1), 3–5. Zbl.599.22003; English translation: Mosc. Univ. Math. Bull. 41(1), 1–4 (1986)

Pol, R. (1982): Note on the spaces $P(S)$ of regular probability measures whose topology is determined by countable subsets. Pac. J. Math. **100**(1), 185–201. Zbl.522.46019

Pol, R. (1980): A theorem on the weak topology of $C(X)$ for compact scattered X. Fundam. Math. **106**(2), 135–140. Zbl.444.54010

Ponomarev, V.I., Shapiro, L.B. (1976): Absolutes of topological spaces and their continuous mappings. Usp. Mat. Nauk **31**(6), 120–136. Zbl.341.54048; English translation: Russ. Math. Surv. 31(5), 138–154 (1976)

Pontryagin, L.S. (1973): Continuous Groups. 3rd edition, Nauka, Moscow (Russian). Zbl.265.22001; see also Selected Scientific Works Val. III (1988 Nauka). Zbl.659.22001

Preiss, D., Simon, P. (1974): A weakly pseudocompact subspace of Banach space is weakly compact. Commentat. Math. Univ. Carol. **15**(4), 603–609. Zbl. 306.54033

Prodanov, I., Stoyanov, L.N. (1984): Every minimal Abelian group is precompact. C.R. Acad. Bulg. **37**(1), 23–26. Zbl.546.22001

Protasov, I.V. (1985): 0-dimensional groups with compact space of subgroups. Mat. Zametki **37**(4), 483–490 (Russian). Zbl.578.22003

Przymusiński, T. (1982): Perfectly normal compact spaces are continuous images of $\beta N \setminus N$. Proc. Am. Math. Soc. **86**(2), 541–544. Zbl.496.54020

Pytkeev, E.G., Yakovlev, N.N. (1980): On bicompacta that are unions of spaces defined by means of coverings. Commentat. Math. Univ. Carol. **21**, 247–261. Zbl.436.54018

Rosenthal, H.P. (1974): The heredity problem for weakly compactly generated Banach spaces. Compos. Math. **28**(1), 88–111. Zbl.298.46013

Rosenthal, H.P. (1977): Pointwise compact subsets of the first Baire class. Am. J. Math. **99**(2), 362–378. Zbl.392.54009

Rudin, M.E. (1971): Partial orders on the types βN. Trans. Am. Math. Soc. **155**, 353-362. Zbl.212,549

Rudin, M.E. (1976): Lectures on Set Theoretic Topology. Providence, Am. Math. Soc., No. **23**. Zbl.318.54001

Samuel, P. (1948): Ultrafilters and compactifications of uniform spaces. Trans. Am. Math. Soc. **64**(1), 100–132. Zbl.32,314

Scott, B.M. (1979): Pseudocompact metacompact spaces are compact. Topology Proc. **4**, 577–587. Zbl.449.54020

Semadeni, Z. (1954): Sur les ensembles clairsemés. Rozpr. Mat. **19**

Semadeni, Z. (1971): Banach Spaces of Continuous Functions, Vol. I, Monogr. Mat. **55**. Zbl.225.46030

Shakhmatov, D.B. (1983): On imbeddings into topological fields and the structure of a field. Commentat. Math. Univ. Carol. **24**(3), 525-540. Zbl.568.54029

Shakhmatov, D.B. (1984): On pseudocompact spaces with a point-countable base. Dokl. Akad. Nauk SSSR **279**(5), 825–829. Zbl.598.54010; English translation: Sov. Math. Dokl. 30, 747–751 (1984)

Shakhmatov, D.B. (1986): A pseudocompact Tychonoff space all countable subspaces of which are closed and C-imbedded. Topology Appl. **22**(2), 139–144. Zbl.586.54020

Shapirovskij, B.E. (1980a): Special types of imbeddings in Tikhonov cubes. Subspaces of $\sum$-products and cardinal invariants. in Colloq. Math. Soc. J. Bolyai, 23. Topology. North-Holland, 1055–1086. Zbl.437.54008

Shapirovskij, B.E. (1980b): On mappings onto Tikhonov cubes. Usp. Mat. Nauk **35**(3), 122–130. Zbl.454.54013; English translation: Russ. Math. Surv. 35(3), 145–156 (1980)

Shchepin, E.V. (1979): On κ-metrizable spaces. Izv. Akad. Nauk SSSR, Ser. Mat. **43**(2), 442–478. Zbl.409,54040; English translation: Math. USSR, Izv. 14, 407–440 (1980)

Shchepin, E.V. (1976): Topology of limit spaces of uncountable inverse spectra. Usp. Mat. Nauk **31**(5), 191–226. Zbl.345.54022; English translation: Russ. Mat. Surv. 31(5), 155–191 (1976)

Shchepin, E.V. (1981): Functors and uncountable powers of compacta. Usp. Mat. Nauk **36**(3), 3–62. Zbl.463.54009; English translation: Russ. Mat. Surv. 36 (3), 1–71 (1981)

Shirokov, L.V. (1982): On internal characterization of Dugundji compact spaces and κ-metrizable bicompacta. Dokl. Akad. Nauk SSSR **263**(3), 1073–1077. Zbl.515.54019; English translation: Sov. Math. Dokl. 25, 507–510 (1982)

Sikorski, R. (1960): Boolean Algebras. Berlin, Springer. Zbl.87,25

Simon, P. (1980): A compact Fréchet space whose square is not Fréchet. Commentat. Math. Univ. Carol. **21**, 749–753. Zbl.466.54022

Sklyarenko, E.G. (1962): Some questions of the theory of bicompact extensions. Izv. Akad. Nauk SSSR, Ser. Mat. 26, 427–452. Zbl.103.390; English translation: II Ser. Am. Math. Soc. 58, 216–244 (1966)

Stone, M.N. (1937): Applications of the theory of Boolean rings to general topology. Trans. Am. Math. Soc. **41**, 375–481. Zbl. 17,135

Stone, M.N. (1948): The generalized Weierstrass approximation theorem. Math. Mag. **21**, 167–184

Talagrand, M. (1979): Espaces de Banach faiblement K-analytiques. Ann. Math., II. Ser. **110**(3), 407–438. Zbl.414.46015

Talagrand, M. (1985): Propriété de Baire et propriété de Namioka. Math. Ann. **270**, 159–164. Zbl.582.54008

Tall, F.D. (1974): The countable chain condition versus separability - Applications of Martin's Axiom. Gen. Topol. Appl. **4**(4), 315–340. Zbl.293.54003

Terpe, F., Flachsmeyer, J. (1977): On some applications of theory of extensions of topological spaces to measure theory. Usp. Mat. Nauk **32**(5), 125–162. Zbl.374. 54017; English translation: Russ. Math. Surv. 32(5), 133–171 (1977)

Tkachenko, M.G. (1981): Some results on inverse spectra II. Commentat. Math. Univ. Carol. **22**(4), 819–841. Zbl.494.54007

Tkachenko, M.G. (1983): On Suslin property in free topological groups on bicompacta. Mat. Zametki **34**(4), 601–607. Zbl.535.22002; English translation: Math. Notes 34, 790–793 (1983)

Todorčevič, S. (1983): Forcing positive partition relations. Trans. Am. Math. Soc. **280**(2), 703–720. Zbl.532.03023

Todorčevič, S. (1984): Trees and linearly ordered sets. In: Handbook of Set-Theoretic Topology, Amsterdam, North Holland, 235–293. Zbl.557.54021

Trofimov, V.A. (1987): Reflexive and selfdual Hilbert modules over some C^*-algebras. Usp. Mat. Nauk **42**(2), 247–248. Zbl.637.46047; English translation: Russ. Math. Surv 42(2), 303–304 (1987)

Tsuda, K. (1986): Metrizability of general ANR. Proc. Am. Math. Soc. **96**(2), 375–378. Zbl.578.54012

Ul'yanov, V.M. (1975): On compactifications satisfying the first axiom of countability. Mat. Sb., Nov. Ser. **98**(2), 223–254. Zbl.316.54024; English translation: Math. USSR, Sb. 27, 199–226 (1977)

Ul'yanov, V.M. (1977): A solution of a problem concerning compactifications of Wallman type. Dokl. Akad. Nauk SSSR **233**(6), 1056–1059. Zbl.371.54048; English translation: Sov. Math. Dokl. 18, 567–571 (1977)

Uspenskij, V.V. (1983): On the theorem of Balcar and Franek about maps of extremally disconnected bicompacta onto the Cantor discontinuum. Commentat. Math. Univ. Carol. **42**(1), 155–165. Zbl.536.54029

Uspenskij, V.V. (1984): Pseudocompact spaces with a σ-point finite base are metrizable. Commentat. Math. Univ. Carol. **25**(2), 261–264. Zbl.574.54021

Uspenskij, V.V. (1987): Compact quotients of topological groups and Haydon spectra. Mat. Zametki **42**(4), 594–602. Zbl.631.22002; English translation: Math. Notes 42(3/4), 827–831 (1987)

Vaughan, J.E. (1984): Countably compact and sequentially compact spaces. In: Handbook of Set-Theoretic Topology, Amsterdam, North Holland, 569–602. Zbl. 562.54031

Wage, M.L. (1980): Weakly compact subsets of Banach spaces. in Surveys on General Topology, New York, Acad. Press, 479–494. Zbl.445.54009

Walker, R.C. (1974): The Stone-Čech Compactification. Berlin Springer. Zbl.292. 54001

Warner, S. (1958): The topology of compact convergence on continuous function spaces. Duke Math. J. **25**, 265–282. Zbl.81,328

Watson, W.S. (1981): Pseudocompact metacompact spaces are compact. Proc. Am. Math. Soc. **81**, 151–152. Zbl.468.54014

Watson, W.S. (1985): A pseudocompact meta-Lindelöf space which is not compact. Topology Appl. **20**, 237–243. Zbl.589.54030

Weil, A. (1940): L'intégration dans les Groups Topologiques et Applications, Actual Sci. Ind

Yakovlev, N.N. (1976): On a theory of g-metrizable spaces. Dokl. Akad. Nauk SSSR **229**(6), 1330–1331. Zbl.345.54037; English translation: Sov. Math. Dokl. 17, 1217–1219 (1977)

Yakovlev, N.N. (1980): On bicompacta in σ-products and related spaces. Commentat. Math. Univ. Carol. **21**(2), 263–283. Zbl.436.54019

Zambahidze, L.G. (1980): On relationships between dimensions of free bases of free topological groups. Soobshch. Akad. Nauk Gruzin. SSR **97**(3), 569–572 (Russian). Zbl.452.54031

Zhou, Hao-Xuan (1982): On the small diagonals. Topology Appl. **13**(3), 283–293. Zbl.495.54028

II. Homology and Cohomology Theories of General Spaces

E.G. Sklyarenko

Translated from the Russian
by Janusz M. Lysko

Contents

Introduction

Homology and cohomology theories are a result of the study of specific problems related to polyhedra and triangulated manifolds. Although described in a combinatorial way they nevertheless reflect the majority of the most typical topological invariants of those objects.

In topology, however, it is often necessary to deal with objects more general than manifolds or polyhedra such as subspaces and quotient spaces, spaces of maps, etc. Similar situations exist in the theory of transformation groups (the study of sets of fixed and periodic points, orbits, spaces of orbits), in the theory of fixed points and points of incidence, in homotopic topology in classification of continuous maps and the study of the structure of some of them, in homological dimension theory and other areas. This requires the development of theories which, by comparison with classical homology and cohomology, are broader and more effectively applicable in such general situations.

However, the indicated reasons justifying the development of more general theories are not the only ones and in general they may even not be the most important. It is fairly typical in topology itself as well as in some other areas of mathematics (differential equations and geometry, functional and complex analysis, algebraic geometry) that the objects of study (manifolds, analytic spaces, algebraic manifolds, etc.), even when they can be triangulated, be described not by a triangulation (which is sometimes difficult to establish), but by how they are made up of their less complicated sub-objects forming in most cases an open (or closed) cover. In this process the coefficient domain for homology or cohomology is introduced separately for each element of the cover and is frequently different for different elements of the cover.

Those circumstances, together with the internal development of homology and cohomology theories contributed to the creation of the apparatus of sheaf theory. In the light of this apparatus and the development of homological algebra, this lead to the revision of the very foundation of these theories, thus completely changing their nature and immensely enriching their content. The suitability of the new methods for application to broad categories of topological spaces is by no means their main advantage (as was sometimes considered). In fact a new theory was created, different in content – the theory, for which classical cohomology happens to be just one of its concrete manifestation and which allows us to look at all such theories from a single point of view. Thanks to this theory, new connections, not visible from previous approaches, were discovered; in particular with such algebraic constructions as functors Tor and Ext, the theory of derived functors, etc. As a typical example we can consider the connection between such a geometric concept as the dimension of a space and the global dimension of a ring in algebra. Poincaré duality is a spectacular example demonstrating the effectiveness of the new methods. Established as a result of purely combinatorial considerations, it acquires its complete form only after clarification of its other essentially different, abstract nature. It turns out to be a result of some general construction in homological algebra describing duality within the theory of derived functors.

Even though geometrically homology is by far more intuitive than cohomology and its original definition is simpler (as a result of which it appeared in topology much earlier than cohomology), cohomology theory of general spaces surpassed it by far in its development and has proved to be essentially fuller, logically more complete and more effective. This explains, in particular, why cohomology has received more attention in this survey.

Chapter 1
Classical Theories

In addition to the fundamental facts reflected in the Steenrod-Eilenberg axioms and their immediate consequences, in homology and cohomology theory of noncompact polyhedra some statements are made, which, even though they do not follow from the axioms, are nevertheless considered either obvious or self explanatory. As such they seldom attract attention, and thus are sometimes forgotten to be taken into account in the development of a theory outside of the realm of the category of polyhedra. The goal of this chapter is to identify some of these essential facts.

§1. Simplicial Homology and Cohomology

1.1. Homology Groups. The definition of homology groups is based on the concept of the *boundary operator* ∂ on a simplex Δ^n with vertices $a_0, a_1, \ldots, a_n$ described by the formula

$$\partial(a_0, \ldots, a_n) = \sum_{i=0}^{n} (-1)^i (a_0, \ldots, \hat{a}_i, \ldots, a_n). \tag{1}$$

The symbol $\hat{\ }$ over a_i indicates that the vertex a_i is omitted. It is essential to point out that each simplex is considered with its *orientation*, defined by a specific order (modulo an arbitrary even permutation) of its vertices. The minus sign in front of a simplex denotes the change of its orientation, and therefore all faces of Δ^n which appear on the right hand side of formula (1) have the same orientation determining the orientation of the boundary of Δ^n. Formula (1) defines the *boundary homomorphism* $\partial : C_n(K; G) \to C_{n-1}(K; G)$ *of the groups of chains* $C_n(K; G)$ *of a complex* K *with coefficients in an Abelian group* G (that is the formal linear combinations with coefficients in G of oriented simplices $\Delta^n \in K$). Chains in the kernel $\operatorname{Ker} \partial$ are called *cycles* while those in the image Im ∂ *boundaries*. Intuitively it is clear that every boundary must be a cycle (a boundary has no boundary). This follows from the identity $\partial\partial = \partial^2 = 0$ (when we apply ∂ to both sides of formula (1) the final result contains only simplices of the form $(a_0, \ldots, \hat{a}_i, \ldots, \hat{a}_j, \ldots, a_n)$, each appearing exactly twice with opposite signs $(-1)^{i+j}$ and $(-1)^{i+j-1}$). The *homology group* $H_n(K; G)$ is defined as the quotient group of the group of cycles with respect to its subgroup of boundaries.

Poincaré was the founder of homology theory. For a long time homology was described only in terms of classes of homological cycles and numerical invariants of finitely generated groups, namely rank (called the *Betti number*) and torsion. Only during the period between 1925 and 1935, following the work of Emmy Noether on Abelian groups, did the homology groups gradually became a widely accepted concept.

The concept of a chain as well as the boundary formula were known for a long time and they appear naturally in many concrete situations. The intensity of the flow of electrical current on the intervals between junction points in an electrical circuit, the intensity of the flow of a fluid in a branched canal are some of the examples of one-dimensional chains. The "boundary" ∂ of such chains represents the change of electrical potential between junction points in the first example and the rate of accumulation of the fluid at branch points in the second. The rate of change of electromagnetic field on the contours of an electrical circuit is an example of a two-dimensional chain, while its "boundary" describes the resulting electrodynamical forces on separate contours. Chains and their boundaries appear in mathematical analysis in questions related to integration. In all of the above examples it is obviously necessary to take into account the orientation of all simplices. It is also obvious that cycles (that is chains without a boundary) are distinguished from arbitrary "chains" by their special physical or geometric meaning.

1.2. Cohomology. In simillar fashion the concept of a coboundary arose from the consideration of concrete problems. If ξ is an arbitrary function defined on the n-dimensional oriented simplices and taking values in a group G (it is assumed that ξ changes its sign if its argument changes orientation), then there exists a function $d\xi$ (the *coboundary* of ξ) defined on the $(n + 1)$-dimensional simplices in the following way:

$$d\xi(a_0, \ldots, a_{n+1}) = \sum_{i=0}^{n+1} (-1)^i \xi(a_0, \ldots, \hat{a}_i, \ldots, a_{n+1}) \qquad (2)$$

(the value of $d\xi$ on a polyhedron is equal to the sum of the values of ξ on properly oriented faces).

For example, if ξ is the value of a potential at a point (zero-dimensional cochain), then $d\xi$ on every (oriented) interval is the difference of the potential between the end points of the interval. If ξ represents the forces acting along the edges of a polyhedron (one-dimensional cochain) then $d\xi$ describes their turning moments. If ξ is the intensity of the flow of a gas through the boundary (two-dimensional cochain), then $d\xi$ is the rate of accumulation of the gas inside the corresponding polyhedra. Kirchhoff's laws for electrical circuits, in Maxwell's form, are represented in terms of the operators ∂ and d. We should bare in mind the fact that in these cases too, it is essential to take into account the orientation of all simplices (polygons, polyhedra) and their boundaries. In any case the *cocycles* (that is, the cochains ξ for which $d\xi = 0$) are clearly distinguished from the remaining cochains by their special physical meaning.

As in the case of chains, it is true that $dd = d^2 = 0$, that is, cochains of the form $d\xi$ (they are called the *coboundaries*) are cocycles. To verify this it is enough to consider $d(d\xi)$ for a cochain ξ that is non-zero on a single simplex $\Delta^n \in K$. For any simplex Δ^{n+2} containing Δ^n as its face the value of $d(d\xi)$ on Δ^{n+2} is zero since it is equal the sum of values of ξ on Δ^n counted twice with opposite signs. In accordance with the above, the *cohomology groups* $H^n(K; G)$ are defined as the quotient groups of n-dimensional cocycles of K with respect to the subgroup of coboundaries.[1]

1.3. Betti Numbers. The fact that $d^2 = 0$ is a consequence of the relation $\partial^2 = 0$ and a simple observation that the homomorphism d is dual to ∂ under the obvious identification $C_n(K; G) = \mathrm{Hom}(C^n(K; \mathbb{Z}), G)$ of cochains with the groups of homomorphisms of free chain groups into G ($\mathbb{Z}$ will be used to denote the group of integers). In the case of a finite K the cochain groups $C^n(K; \mathbb{Z})$ are also free, $C^n(K; G) = \mathrm{Hom}(C_n(K; \mathbb{Z}), G)$ and the homomorphisms ∂ and d are dual to each other. This follows from the symmetry in notation $(d\xi, c) = (\xi, \partial c)$ if we agree to denote the value of a cochain ξ on a chain c in the form of the scalar product (ξ, c). From the universal coefficients formulas (see the end of this paragraph) it follows that for every finite K the groups $H^n(K; \mathbb{Z})$ and $H_n(K; \mathbb{Z})$ have the same rank. It is called the *n-dimensional Betti number* of the compact polyhedron K.

1.4. Homology and Cohomology of the Second Kind. In general it makes no sense to consider *infinite chains*. If a certain $(n-1)$-dimensional simplex is a face of an infinite number of n-dimensional simplices contained in a chain then the boundary of such chain is not defined. For the same reason it does not make sense, in the general case, to consider *finite cochains*, the coboundary of such cochain will be infinite. However, finite cochains (or *cochains with compact supports*, different from zero on a finite number of simplices) as well as arbitrary infinite chains are fully admissible when the simplicial complex K under consideration is locally finite. In this situation the geometric content of the concept of a cycle remains the same as earlier, it is the chain without boundary. For example, in three-dimensional space a closed polygonal line as well as an infinite polygonal line with ends at infinity are one-dimensional

---

[1] Cohomology was introduced for the first time in Kolmogorov (1936a) and Alexander (1935).

cycles. The surface of a polyhedron, a cylindrical surface or a surface in the shape of a paraboloid serve as examples of two dimensional cycles. Homology and cohomology defined in terms of infinite chains and finite cochains are called *homology and cohomology of the second kind*, respectively.[2] Cohomology groups of the second kind and the cochains that define them are denoted by $H^n_c(K;G), C^n_c(K;G)$, respectively, where the subscript c indicates compact supports. On the other hand for homology groups of the second kind and chains defining them we shall preserve the earlier notation, while the ordinary homology and chains which clearly have compact supports will be denoted by $H^c_n(K;G)$ and $C^c_n(K;G)$ respectively.

The group $C^c_n(K;G)$ is the direct sum $\bigoplus_{\Delta^n} G$ (the summands correspond to the simplices $\Delta^n \in K$) and therefore every simplicial map $f : K \to L$ determines homomorphisms of chains $f_\# : C^c_n(K;G) \to C^c_n(L;G)$ for which, in accordance with (1), we have $f_\#\partial = \partial f_\#$. The existence of this relation means that, by definition, $f_\#$ is a *chain homomorphism* $C^c_*(K;G) \to C^c_*(L;G)$ (here $C^c_*(K;G)$ denotes the sequence of all $C^c_n(K;G)$ connected by the boundary homomorphism ∂). Under a chain homomorphism of chain complexes, cycles are mapped to cycles and boundaries are mapped to boundaries; therefore there are defined the *induced homomorphisms* $f_* : H^c_n(K;G) \to H^c_n(L;G)$. In a dual way (as for functions) there arise the maps of cochains $f^\# : C^*(L;G) \to C^*(K;G)$ and the relationship $f^\#d = d f^\#$. Therefore there are also defined the induced homomorphisms of cohomology groups $f^* : H^n(L;G) \to H^n(K;G)$. Let us point out that for the composition gf we have the following obvious formulas: $(gf)_* = g_*f_*$ and $(gf)^* = f^*g^*$.

Since the inverse image under a simplicial map of a simplex in L may contain, in general, an infinite number of simplices of the same dimension, the map $f_\#$ is not defined for infinite chains. For the same reason, in the general case the map $f^\#$ is not defined for finite cochains — the image of a finite cochain may be infinite. However, the maps $f_\#$ and $f^\#$, and consequently the induced maps f_* and f^* of homology and cohomology of the second kind are defined for *proper maps* of locally finite complexes.

Thus in the category of locally compact polyhedra and their proper maps, in addition to the ordinary groups $H^c_n(K;G)$ and $H^n(K;G)$, there are also defined the homology and cohomology groups of the second kind $H_n(K;G)$ and $H^n_c(K;G)$. A similar situation arises for locally compact Hausdorff spaces (see §4, Chap. 2 and §3, Chap. 5). In spite of the fact that the theories of the second kind have been used extensively for a long time in the literature, until recently they existed only "illegally" in the sense that they were given no treatment in any monographs on algebraic topology. Scant information about these groups for polyhedra is contained in the exercises to Chaps. 6 and 8 of Eilenberg and Steenrod (1952). A short description in terms of singular

---

[2] They were considered for the first time by A.N. Kolmogorov in Kolmogorov (1936e) for infinite (triangulated) manifolds (homology only with compact coefficients in the naturally arising compact topology).

theory (see §2 below) is given in Cartan (1948-49) and a brief definition of H_c^n in terms of Alexander-Spanier cochains is included in Spanier (1966). The first systematic description of the theory in terms of chains and cochains which are used in it was given in Sklyarenko (1969) and in Massey (1978b).

It is possible that this situation could be partially explained by the fact that (since proper maps of locally compact spaces are precisely those that can be extended to their one-point compactifications) homology and cohomology of the second kind can be reduced to the ordinary homology or cohomology by considering the one-point compactification of a space. It will not even be an overstatement to say that they are, in general, the same (see Sect. 1.4, Chap. 2). Nevertheless there is a variety of reasons why theories of the second kind should be considered independently (as it is done below in §4, Chap. 2; §3, Chap. 5; §3, Chap. 7). First of all, very frequently homology and cohomology of the second kind are studied simultaneously with ordinary homology and cohomology. Secondly, the transition to compactifications indicated above is not always natural. For example, in the case of a manifold or a polyhedron, the one-point compactification does not preserve the structure of the space. Thus we have to work in the category of (pointed) compact spaces, and instead of the simplicial or singular homology or cohomology theories we are forced to use more general ones. Let us mention, however, that the orientation determining cycle of a non-compact manifold M belongs to the group $H_n(M; \mathbb{Z})$ and similarly, the fundamental cohomology class is an element of the group $H_c^n(M; \mathbb{Z})$. Finally, the transition to the one-point compactification is simply impossible in the case when the homology (or cohomology) under consideration has nonconstant coefficients.

There are also other connections between the ordinary theories and the theories of the second kind (see §1.5 and §1.7 below). One of them is the exact homology sequence

$$\ldots \to H_n^c(K; G) \to H_n(K; G) \to J_n^\infty(K, G) \to H_{n-1}^c(K; G) \to \ldots \quad (3)$$

corresponding to the following short exact sequence of chain complexes

$$0 \to C_*^c(K; G) \to C_*(K; G) \to C_*^\infty \to 0.$$

The symbol J_n^∞ is used here to denote the homology of the chain complex $C_*^\infty = C_*(K; G)/C_*^c(K; G)$. The following exact cohomology sequence is obtained essentially in the same way.

$$\ldots \to H_c^n(K; G) \to H^n(K; G) \to J_\infty^n(K; G) \to H_c^{n+1}(K; G) \to \ldots \quad (4)$$

The groups J_n^∞ (J_∞^n) are called the *homology (cohomology) of K at infinity*. If K_λ denotes the family of all finite subcomplexes of K and N_λ the family of the closures of $K \setminus K_\lambda$, then it is clear that $C_*^c(K; G)$ is the union of the subcomplexes $C_*(K_\lambda; G)$. Similarly $C_c^*(K_\lambda; G)$ is the union of the subcomplexes $C^*(K, N_\lambda; G)$. Because of this, the complexes C_*^∞ and C_∞^* are the direct limits of $C_*(K, K_\lambda; G)$ and $C^*(N_\lambda; G)$ respectively. Since the $\varprojlim_\lambda$-functor of the direct limit is exact, it commutes with the operations H_n (or H^n), and therefore $J_n^\infty(K;G) = \varinjlim_\lambda H_n(K, K_\lambda; G), J_\infty^n(K;G) = \varinjlim_\lambda H^n(N_\lambda; G)$. We shall see §2, Chap. 8) that $J_n^\infty = H_n^\infty$ and $J_\infty^n = H_\infty^n$ coincide with the *local homology and cohomology* at infinity of the one-point compactification of K.

1.5. Supports. The union of all simplices of the complex K which appear with non-zero coefficients in a chain (on which a cochain is non-zero) is called the *support of the chain (cochain)*. So far we have considered homology and cohomology either with compact or closed supports. As it turns out, there exist homology (or cohomology) with other ("intermediate") *families of supports*. Such groups appear naturally, for example, in the interpretation of the homology (or cohomology) of pairs of spaces (in connection with this see §4, Chap. 3 and §4, Chap. 5).

Let K be a locally finite complex and L be its subcomplex. Recall that the *homology of the pair* (K, L) is determined by the quotient complex of the complex of the chains of K with respect to the subcomplex of the chains of L, and cohomology (dually) as the subcomplex of cochains of K vanishing on L. The short exact sequences of chains (or cochains) which arise naturally in this construction determine the exact homology sequence of L, K and (K, L) (cohomology sequence of (K, L), K and L). The above statement is true for ordinary groups as well as for the groups of the second kind.[3]

It is true that $H_n(K, L; G) = H_n(K \setminus L; G)$. To see this, subdivide each simplex in $K \setminus L$ that is adjacent to L in such way that the subdivision become finer as the simplices get closer to L (an infinite triangulation). Assign to each chain c of the pair (K, L) the chain from $K \setminus L$ in which the simplices of the new triangulation have the same coefficients as the simplices of the original triangulation containing them have in c. It is not difficult to check that the resulting map of chains of (K, L) into the chains of $K \setminus L$ induces the desired isomorphism.

Exactly the same argument guarantees the existence of the isomorphism $H_n^c(K, L; G) = H_n^\varphi(K \setminus L; G)$, where φ is the family of all subcomplexes of $K \setminus L$ (in the new triangulation) whose closures in K are compact and where $H_n^\varphi(K \setminus L; G)$ is the homology group of $K \setminus L$ defined by (infinite) chains whose supports belong to φ. In particular if L contains all but a finite number of simplices of K then $H_n^c(K, L; G) = H_n(K \setminus L; G)$ (in connection with the above see §§3.2, 4.3, Chap. 5).

In a similar way we obtain the fact that $H_c^n(K, L; G) = H_c^n(K \setminus L; G)$. To prove this let us assign to each cochain ξ with compact support of the complex $K \setminus L$ (in the triangulation introduced earlier) the cochain of the pair (K, L) whose value on an arbitrary simplex of K is equal to the sum of the values of ξ on simplices of the triangulation of $K \setminus L$ contained in it. In the same way it is possible to obtain an isomorphism $H^n(K, L,; G) = H_\varphi^n(K \setminus L; G)$ of the

---

[3] Homology groups of a pair were introduced by Lefschetz in 1927. The exact homology sequence of a pair was described for the first time by Hurewicz in 1942.

ordinary cohomology of the pair (K, L) with the cohomology of $K \setminus L$ with supports in the family φ of all subcomplexes of $K \setminus L$ that are closed in K. In particular $H^n(K, L; G) = H^n_c(K \setminus L; G)$ if only finitely many simplices of K are not included in L (in connection with this see §§4.1, 4.2, Chap. 2 and §4, Chap. 3).

1.6. Additivity Properties. Let K_λ denote the family of connected components of the complex K. Obviously $C^c_*(K; G) = \bigoplus_\lambda C^c_*(K_\lambda; G)$ is the direct sum and therefore $H^c_n(K; G) = \bigoplus_\lambda H^c_n(K_\lambda; G)$ is also the direct sum of the homology groups of components. This is the so called $\sum$-*additivity property of the homology* H^c_*. Similarly we have $C^*(K; G) = \prod_\lambda C^*(K_\lambda; G)$ and $H^n(K; G) = \prod_\lambda H^n(K_\lambda; G)$ known as the $\prod$-*additivity property of the cohomology* H^*. On the other hand, in the case of the theories of the second kind, the homology is clearly $\prod$-additive and the cohomology is $\sum$-additive.

The properties of additivity are extremely simple and are customarily mentioned only when one wants to point out that in the study of homology and cohomology it is possible to work only with connected polyhedra. However their significance is much more essential. In a typical situation it is true that the additivity properties, together with the Eilenberg-Steenrod axioms, uniquely characterize a homology or cohomology theory (see Chap. 7). In the category of compact polyhedra $\sum$-additivity (equivalent clearly to $\prod$-additivity) of homology and cohomology is a consequence of Eilenberg-Steenrod axioms (in view of their completeness).

1.7. Universal-Coefficient Formulas. A typical tool in the study related to the applications of homology and cohomology are the universal coefficient formulas which describe the relationship between homology and cohomology groups corresponding to different coefficient groups. In order to give a clear account of the possible directions that might be considered in general situations, we shall include all such relationships for the classical case.

Let us recall that for homology and cohomology it is possible to consider as the coefficients not only ordinary Abelian groups, but also modules over a certain ring R. In this case the homology and cohomology groups also have the structure of R-modules. Most frequently in such situations in the place of R we consider a principal ideal domain. For the validity of the universal coefficient formulas it is enough to require that the ring be *hereditary* (that is, that the functors Ext^n be zero for $n \geq 2$).

Since the modules $C^c_*(K; R)$ are free and $C^c_*(K; G)$ and $C^*(K; G)$ are isomorphic to $C^c_*(K; R) \otimes_R G$ and $\mathrm{Hom}_R(C^c_*(K; R), G)$ respectively, we have the following *universal coefficient formulas* (exact sequences)[4]

$$0 \to H^c_n(K; R) \otimes_R G \to H^c_n(K; G) \to \mathrm{Tor}_R(H^c_{n-1}(K; R), G) \to 0, \quad (5)$$

$$0 \to \mathrm{Ext}_R(H^c_{n-1}(K; R), G) \to H^n(K; G) \to \mathrm{Hom}_R(H^c_n(K; R), G) \to 0. \quad (6)$$

---

[4] All exact sequences in this paragraph are split.

Similar relationships are valid in the case of locally finite polyhedra for the cochains C_c^* (free in the case $G = R$) and the chains C_*, and therefore we also have the following *universal coefficient formulas*:

$$0 \to H_c^n(K;R) \otimes_R G \to H_c^n(K;G) \to \mathrm{Tor}_R(H_c^{n+1}(K,R),G) \to 0, \quad (7)$$

$$0 \to \mathrm{Ext}_R(H_c^{n+1}(K;R),G) \to H_n(K;G) \to \mathrm{Hom}_R(H_c^n(K;R),G) \to 0. \quad (8)$$

For compact K, both (6) and (8) are satisfied. It is clear that (5)-(8) are true not only for a single K, but also for the groups of the pair (K,L). This is also true for (9) and (10) considered below.

In the case of locally finite polyhedra the groups $C_n(K;G)$ are naturally isomorphic to the products $\prod G$ of the coefficient group G (every factor corresponds to some simplex $\Delta^n \in K$). In accordance with Lemma 7 in Sklyarenko (1969), for infinite products we have $\prod G = (\prod R) \otimes_R G$ precisely when the module G is finitely presented. Therefore, for finitely presented G the following *universal coefficient formulas* are true

$$0 \to H_n(K;R) \otimes_R G \to H_n(K;G) \to \mathrm{Tor}_R(H_{n-1}(K;R),G) \to 0. \quad (9)$$

Exactly the same argument with the same restrictions on G guarantees the exactness[5] (for every K) of the sequence

$$0 \to H^n(K;R) \otimes_R G \to H^n(K;G) \to \mathrm{Tor}_R(H^{n+1}(K;R),G) \to 0. \quad (10)$$

The example of an infinite zero-dimensional polyhedron illustrates that (9) and (10) are not valid for modules that are not finitely presented.

§2. Singular Theory

2.1. Introductory Remarks. The origins of the singular theory can be traced back to the works of Veblen who defined chains by means of continuous maps of complexes into a topological space (in 1921). These ideas were later developed by Hurewicz, Dowker and Dugundji. A definition of chains using the concept of a singular simplex appeared in a work of Lefschetz (in 1933). In its final form the theory was introduced by Eilenberg (in 1944).

The singular theory is irreplaceable in problems of homotopic topology, in the study of spaces of continuous maps and in the theory of fibrations. Its importance, however, is not limited to the fact that it is applicable beyond the realm of the category of polyhedra. The singular theory is necessary for the in depth understanding of the homology and cohomology theory of polyhedra themselves. In particular, it is used in the problems of homotopic classification of continuous maps of polyhedra and in the description of homology and

---

[5] The validity of (9) and (10) for the modules of the finite type over a principal ideal domain can be established also by using Theorem 10, §5, Chap. 5 in Spanier (1966).

cohomology of cell complexes. By definition it is topologically invariant and it is not difficult to establish its invariance under homotopy. The fact that the simplicial homology and cohomology share the same properties is proved by showing that they are isomorphic to the singular ones.

The analysis of the singular theory from the point of view of sheaf theory (§5, Chap. 3) reveals the natural limits of its applicability: essentially it is the realm of *(weakly) locally contractible spaces* (see also the beginning of the next chapter).

2.2. Basic Definitions. By a *singular n-dimensional simplex σ^n* in a topological space X we understand a continuous map σ^n of any *ordered simplex* $\Delta^n = (a_0, \ldots, a_n)$ into X (it is assumed that every two ordered simplices are equivalent because there exists an affine map of one onto another that preserves the order of the vertices). Similarly to the case of the simplicial chains which are linear combinations of simplices of a complex, the *singular chains* of X with coefficients in a group G are defined as linear combinations with coefficients in G of its singular simplices. The *boundary operator ∂* : $C_n^s(X; G) \to C_{n-1}^s(X; G)$ in the groups of singular chains is defined for every $\sigma^n \in C_n^s(X; G)$ as the image under σ^n of the boundary (described by formula (1)) of the standard simplex Δ^n. The *singular homology $H_n^s(X; G)$* of the space X is defined as the homology of the chain complex $C_*^s(X; G)$. The *homology of the subspace $A \subset X$* is defined by the subcomplex $C_*^s(A; G)$ of chains with supports in A (the *support of a chain* is the union of the images of all singular simplices in the chain). At the same time the *homology $H_n^s(X, A; G)$ of the pair* (X, A) is defined by the quotient complex $C_*^s(X, A; G) = C_*^s(X; G)/C_*^s(A; G)$. The image of a singular simplex σ^n under a map $f : X \to Y$ is a singular simplex $f\sigma^n : \Delta^n \to Y$. The assignment $\sigma^n \to f\sigma^n$ defines a homomorphism of chain complexes $f_\# : C_*^s(X; G) \to C_*^s(Y; G)$, which induces the map of homology $f_* : H_n^s(X; G) \to H_n^s(Y; G)$.

As in the case of simplicial theory, the *singular cochains* are described as functions defined on singular simplices and assuming values in the coefficient group G (the chains $C_*^s(X; R)$ with coefficients in a ring R are free and therefore the complex of cochains with coefficients in an R-module G can be identified with $\mathrm{Hom}_R(C_*^s(X; R), G))$. The definitions of *singular cohomology groups $H_s^n(X; G)$, $H_s^n(X, A; G)$* are (like the simplicial ones) dual to the definitions of homology groups. In particular, the *cohomology of a pair* (X, A) is defined by the subcomplex of $C_s^*(X; G)$ consisting of all cochains that are equal to zero on the singular simplices in A. The induced homomorphisms $f^\#$ and f^* of cochains and cohomology are also defined by using the same duality.

2.3. Invariance Under Homotopy. In the description of this property the construction of a prism plays a crucial role. By the *prism* over a simplex Δ^n we understand the product $\Delta^n \times I$ of Δ^n with the interval $I = [0, 1]$ considered with the triangulation consisting of the simplices $(a_0, \ldots, a_i, \widetilde{a}_i, \ldots, \widetilde{a}_n)$, where

$(a_0, \ldots, a_n) = \Delta^n = \Delta^n \times 0$ is its "lower" base and $(\widetilde{a}_0, \ldots, \widetilde{a}_n) = \widetilde{\Delta}^n = \Delta^n \times 1$ is the "upper" base. The chain $\prod(\Delta^n) = \sum_i (-1)^i (a_0, \ldots, a_i, \widetilde{a}_i, \ldots, \widetilde{a}_n)$ called an *oriented prism* includes all $(n+1)$-dimensional simplices considered with the same orientation. Its boundary has the form $\partial \prod(\Delta^n) = \widetilde{\Delta}^n - \Delta^n - \prod \partial(\Delta^n)$ (two bases and the sides in the orientation induced by the prism). We have $\prod \partial(\Delta^n) = \sum_i (-1)^i \prod(\Delta_i^{n-1})$, where $\Delta_i^{n-1} = (a_0, \ldots, \hat{a}_i, \ldots, a_n)$ are the sides of Δ^n. Equivalently

$$\widetilde{\Delta}^n - \Delta^n = \partial\prod(\Delta^n) + \prod\partial(\Delta^n). \tag{11}$$

Let $F : X \times I \to Y$ be a *homotopy* between continuous maps $f, g : X \to Y$ $(F(x, 0) = f(x), F(x, 1) = g(x))$. The composition of maps $\Delta^n \times I \to X \times I \to Y$ which assigns to each $\sigma^n : \Delta^n \to X$ the image of the prism-chain $\prod(\Delta^n)$ in $C_{n+1}^s(Y; G)$ defines a homomorphism $D : C_n^s(X; G) \to C_{n+1}^s(Y; G)$ called the *chain homotopy* between the homomorphisms $f_\#, g_\# : C_*^s(X; G) \to C_*^s(Y; G)$ of chain complexes.[6] It follows immediately from (11) that

$$g_\# - f_\# = \partial D + D\partial. \tag{12}$$

Since the above formulas hold for all subspaces $A \subset X, B \subset Y$ (for which $F(A) \subset B$), they remain true also for *homotopic maps of pairs* $f, g : (X, A) \to (Y, B)$. These formulas immediately imply the following *homotopy property*: homotopic maps f, g induce the same homomorphism $f_* = g_* : H_n^s(X, A; G) \to H_n^s(Y, B; G)$.

Similar considerations are valid also for cohomology: by using the functor $\mathrm{Hom}_R(C_*^s(X, R), G)$ in passing to dual constructions we can establish that formula (12) is true for cochains, with the chain homotopy D being a map $D : C_s^{n+1}(Y; G) \to C_s^n(X; G)$.

The spaces X and Y (as well as the pairs (X, A) and (Y, B)) are called *homotopically equivalent* if there exist maps $f : X \to Y$ and $g : Y \to X$ such that the compositions gf and fg are homotopic to the identity homeomorphisms of X and Y respectively (every such map f, g is called a *homotopy equivalence* between X and Y). From the fact that $(gf)_* = g_* f_* = e$ (e denotes the identity isomorphism) and similar formulas for $(fg)_*, (gf)^*$ and $(fg)^*$ it follows that every homotopy equivalence induces an isomorphism of homology and cohomology (this property is called *homotopic invariance* of homology and cohomology). As a typical example of a homotopy equivalence we can consider a deformation of a space to its subspace.

2.4. Homology and Cohomology of Cell Complexes. One of the ways of describing an isomorphism between the simplicial and singular theories consists of demonstrating that the definition of simplicial groups of a complex is a "rule" for computing its singular groups. This is typically done already for arbitrary cell complexes.

---

[6] The concept of a chain homotopy was introduced by Lefschetz (see Lefschetz (1935)).

To this end we consider spaces with a filtration. We say that a topological space X is equipped with a *filtration* if a sequence of closed subspaces $\emptyset = X_{-1} \subset X_0 \subset X_1 \subset \ldots$ is given in X such that $X = \bigcup_i X_i$, and for each subset A of X the set A is closed in X if and only if every $A \cap X_i$ is closed in X_i. It is clear that every compact subset of X is contained in some X_i and therefore (since the supports of chains are compact) $C_*^s(X; G) = \bigcup_i C_*^s(X_i; G)$.

A filtration is called a *cell filtration* if $H_s^q(X_i, X_{i-1}; G) = 0$ for $q \neq i$. Typical examples are provided by filtrations of simplicial or cell complexes K by their i-dimensional skeleta $X_i = K^i$. A cell filtration arises when a piecewise linear n-dimensional manifold M is decomposed into *barycentric stars* (the star of a simplex Δ^k in M is defined as the $(n-k)$-dimensional subcomplex formed by the barycenters of all simplices of M containing Δ^k). This decomposition is used in the classical proof of the Poincaré duality.

For X with a cell filtration the groups $\widetilde{C}_n(X, G) = H_n^s(X_n, X_{n-1}; G)$ form a chain complex with the boundary operator $\widetilde{\partial}$ which is defined as the composition of the *connecting homomorphism* $\delta : H_n^s(X_n, X_{n-1}; G) \to H_{n-1}^s(X_{n-1}; G)$ with $H_{n-1}^s(X_{n-1}; G) \to H_{n-1}^s(X_{n-1}, X_{n-2}; G) = \widetilde{C}_{n-1}(X; G)$ from the exact homology sequence of the pair (X_{n-1}, X_{n-2}) (from its exactness it follows that $\widetilde{\partial}^2 = 0$). A fairly easy diagram-chasing argument shows that $H_n(\widetilde{C}_*^s(X; G)) = H_n^s(X; G)$ and that this isomorphism is compatible with the *filtration-preserving maps* (that is, the maps for which $f(X_i) \subset Y_i$). In the case when $X = K$ is a simplicial (or cell) complex a simple verification demonstrates that $\widetilde{C}_*(X; G)$ coincides with the complex of simplicial (or cell) chains in K and this implies the isomorphism of simplicial (or cell) homology with singular homology. Let us point out that the class of filtration-preserving maps (*cellular maps*) of cell complexes is much broader than the class of piecewise linear maps of polyhedra.

In the case of homology of the pair (X, A) the filtration should start with $X_{-1} = A$. Dual constructions lead to similar results for cohomology.

The construction described above is effective not only for the singular theory, but also for any other axiomatic theory (homology or cohomology) and it plays a fundamental role in proofs of the uniqueness theorems (see §1, Chap. 7). In the case when $X = K$ is a cell complex (and $X_i = K^i$ are the skeleta of K), the homotopy and excision axioms imply, for example, that the groups $\widetilde{C}_n(K; G) = H_n(K^n, K^{n-1}; G)$ are isomorphic with the homology groups of the discret union of pairs (B^n, S^{n-1}), where B^n is the n-dimensional ball and S^{n-1} is its boundary. This is a manifestation of the *additivity property (axiom)*: $\widetilde{C}_n(K; G)$ is equal to $\bigoplus_\lambda G$ or $\prod_\lambda G$ depending on whether the theory is $\sum$- or $\prod$-additive.

2.5. Subdivisions of a Singular Complex. Quite often when working with singular homology it is necessary to consider only sufficiently fine singular simplices and ignore the remaining ("coarse") ones. It turns out that this situation does not lead to any change in the arising homology or cohomology.

Specifically, let $\alpha = \{M_\lambda\}$ be an arbitrary cover of the space X, such that the interiors $\operatorname{Int} M_\lambda$ also cover X (let us agree temporarily to consider only such covers). Let $C_*^\alpha(X;G)$ be the subcomplex of $C_*^s(X;G)$ generated by all *singular simplices* σ^n *of mesh* α (simplices whose images $|\sigma^n| = \sigma^n(\Delta^n)$ are contained in some elements of α). It turns out that the inclusion $i : C_*^\alpha(X;G) \subset C_*^s(X;G)$ induces an isomorphism of homology and thus $H_n^s(X;G) = H_n(C_*^\alpha(X;G))$. If $f : X \to Y$ is a continuous map and if $\alpha \geq f^{-1}\beta$ (that is α is inscribed in the inverse image of β), then there exists a map $C_*^\alpha(X;G) \to C_*^\beta(Y;G)$ which clearly induces $f_* : H_n^s(X;G) \to H_n^s(Y;G)$ introduced earlier.

For simplicity the following constructions are considered only for homology of X. However, the maps of chains appearing in them preserve supports and therefore all definitions and arguments are automatically applicable to arbitrary subsets $A \subset X$ and subsequently to chains and homology of the pair (X, A). By using duality all these constructions can be carried out for cochains and therefore all arguments remain valid also for cohomology. In particular the restriction $j_\alpha : C_s^*(X;G) \to C_\alpha^*(X;G)$ of the singular cochain-functions to the singular simplices of mesh size α induces an isomorphism of cohomology. Therefore, in the definition of cohomology groups it is enough to consider only the values of cochains on sufficiently fine simplices. This remark is essential in the analysis of the singular theory from the point of view of sheaf theory (see §5, Chap. 3).

Let us begin with the more precise definition of the concept of the *barycentric subdivision* $B\Delta^n$ of a simplex. Assume that $B\Delta^0 = \Delta^0$ and define $B\Delta^n$ by induction with respect to the dimension n of Δ^n as the chain $D\Delta^n = cB\partial\Delta^n$, where $\partial\Delta^n$ is the chain defined by formula (1) and the symbol c denotes the addition to each simplex of the chain $B\partial\Delta^n$ (which has already been defined by the inductive assumption) the barycenter c of Δ^n. This formula can also be derived from the description of the prism $\prod \partial(\Delta^n)$ over the boundary of Δ^n (see §2.3) by contracting one of the bases to a point. Therefore, as for a prism, it is true that $\partial B\Delta^n = \partial(cB\partial\Delta^n) = B\partial\Delta^n - c\partial B\partial\Delta^n$. Again using induction with respect to the dimension of a chain we conclude that $c\partial B\partial\Delta^n = cB\partial\partial\Delta^n = 0$, and therefore $\partial B\Delta^n = B\partial\Delta^n$.

For every singular simplex $\sigma^n : \Delta^n \to X$, let $B\sigma^n$ denote the image of the chain $B\Delta^n$ under the map σ^n. Since $\partial B\sigma^n = B\partial\sigma^n$, we can define a map $B : C_*^s(X;G) \to C_*^s(X;G)$ of chain complexes; this map is called the *operator of barycentric subdivision*. There exists a chain homotopy D between B and the identity map e, and therefore the operator B does not have any effect on homology: $B_* = e$ is the identity isomorphism. To make sure that this is true we can compare $B\Delta^n$ with Δ^n by representing their difference in the form $B\Delta^n - \Delta^n = \partial D\Delta^n + D\partial\Delta^n$. Assuming that $D = 0$ for $n = 0$ and that by induction D is determined in dimensions $\leq n - 1$, the unknown term of this formula will be written in the form $\partial D\Delta^n = B\Delta^n - \Delta^n - D\partial\Delta^n$. It is easy to verify that the chain z on the right hand side is a cycle, and since $H_n(\Delta^n) = 0$, we obtain that $z = \partial y$ is the boundary of some $(n+1)$-

dimensional chain y. If we put $D\Delta^n = y$, then by considering in place of $D\sigma^n$ the image of y under the map $\sigma^n : \Delta^n \to X$, we obtain the necessary chain homotopy $D : C_n^s(X;G) \to C_{n+1}^s(X;G)$.

It is clear that the powers B^k of the operator B also do not effect homology. These operators, too, are chain homotopic to the identity: $B^k - e = \partial D_k + D_k \partial$, where $D_k = D(B^{k-1} + \ldots + B + e)$.

It is clear that transition to chains of mesh α using only the powers of B is not possible. For a singular simplex σ let $k(\sigma)$ denote the smallest number such that $B^{k(\sigma)}\sigma \in C_*^\alpha(X;G)$ and let $D_\alpha\sigma = D_{k(\sigma)}\sigma$. Taking into account the fact that for each σ we have $\partial D_\alpha + D_\alpha\partial = \partial D_\alpha + D_{k(\sigma)}\partial - (D_{k(\sigma)}\partial - D_\alpha\partial)$ and $k(\sigma) \geq k(\partial\sigma)$, we find easily that $(\partial D_\alpha + D_\alpha\partial)\sigma = B^{k(\sigma)}\sigma - \sigma - (D_{k(\sigma)} - D_\alpha)\partial\sigma$. In this case the map $B_\alpha\sigma = B^{k(\sigma)}\sigma - (D_{k(\sigma)} - D_\alpha)\partial\sigma$ is a retraction $C_*^s(X;G) \to C_*^\alpha(X;G)$ (that is, $B_\alpha i = e$, where i is the embedding $C_*^\alpha(X;G) \to C_*^s(X;G)$) with D_α being a chain homotopy between B_α and e, and therefore $(B_\alpha)_*$ and i_* are canonical isomorphisms. The necessary condition $\partial B_\alpha = B_\alpha\partial$ is a consequence of the equality $B_\alpha = e + \partial D_\alpha + D_\alpha\partial$. Let us notice that because this condition will be violated we cannot put $B_\alpha\sigma = B^{k(\sigma)}\sigma$.

2.6. Excision Property. Let (X, A) be a pair of topological spaces and let U be an open set, whose closure $\overline{U}$ is contained in $\operatorname{Int} A$. Then the inclusion $(X \setminus U, A \setminus U) \subset (X, A)$ induces an isomorphism of singular homology and cohomology.

To prove the above claim it is enough (using the results of the previous paragraph) to consider the complexes $C_*^\alpha(X;G)$ and $C_\alpha^*(X;G)$. It is easy to see that $C_*^\alpha(X\setminus U, A\setminus U;G) = C_*^\alpha(X, A;G)$ (similar claim is true for cochains), where $\alpha = \{\operatorname{Int} A, X \setminus \overline{U}\}$.

Let us point out that for the simplicial theory and some other theories the excision property is true under the weaker assumption that $\overline{U} \subset A$.

2.7. Sequences of Triads (Mayer-Vietoris Sequences). Let X_1, X_2 be subspaces of X for which $X_1 \cup X_2 = \operatorname{Int} X_1 \cup \operatorname{Int} X_2$ (the operation Int is considered with respect to the subspace $X_1 \cup X_2 \subset X$), $X' = X_1 \cup X_2$, $X'' = X_1 \cap X_2$ and let $\alpha = \{X_1, X_2\}$ be the cover of X'. Then we have the following exact short sequences

$$0 \to C_*^s(X'';G) \to C_*^s(X_1;G) \oplus C_*^s(X_2;G) \to C_*^\alpha(X';G) \to 0,$$

$$0 \to C_*^s(X, X'';G) \to C_*^s(X, X_1;G) \oplus C_*^s(X, X_2;G) \to$$
$$C_*^s(X;G)/C_*^\alpha(X';G) \to 0,$$

in which the first maps are induced by the inclusions $X'' \subset X_i$ (the map induced by the inclusion $X'' \subset X_2$ is considered with the opposite sign). According to §2.5 the inclusion $C_*^\alpha(X';G) = C_*^s(X_1;G) + C_*^s(X_2;G)$ into $C_*^s(X';G)$

induces an isomorphism of homology. Therefore from the exact homology sequence corresponding to this inclusion it follows that $C_*^s(X';G)/C_*^\alpha(X';G)$ is an *acyclic complex* (all its homology groups are equal to zero). Since it is the kernel of the epimorphism $C_*^s(X;G)/C_*^\alpha(X';G) \to C_*^s(X,X';G)$, it follows from the exact homology sequence corresponding to it that this epimorphism induces an isomorphism of homology. Because of this, corresponding to the exact short sequences described above there are the following exact homology sequences: *the sequence of the triad* $(X';X_1,X_2)$

$$\ldots \to H_n^s(X'';G) \to H_n^s(X_1;G) \oplus H_n^s(X_2;G) \to H_n^s(X';G) \to$$
$$H_{n-1}^s(X'';G) \to \ldots$$

and the *additional sequence (of the triad* $(X;X_1,X_2))$

$$\ldots \to H_n^s(X,X'';G) \to H_n^s(X,X_1;G) \oplus H_n^s(X,X_2;G) \to H_n^s(X,X';G) \to$$
$$H_{n-1}^s(X,X'';G) \to \ldots$$

Similar results are true for cohomology. For more general sequences of this type see §5.9 of Chap. 8.

The condition that the inclusion $C_*^s(X_1;G)+C_*^s(X_2;G) \subset C_*^s(X_1 \cup X_2;G)$ induces an isomorphism of homology is in fact one of the definitions of the so called *excisive triads* (for an equivalent definition see Dold (1971)). One of the advantages of the theories considered below is the fact that (at least for closed or open X_1, X_2) for the chains appearing in them it is always true that $C_*(X_1 \cup X_2;G) = C_*(X_1;G) + C_*(X_2;G)$. For the simplicial chains of subcomplexes $K_1, K_2 \subset K$ it is also clear that

$$C_*(K_1 \cup K_2;G) = C_*(K_1;G) + C_*(K_2;G).$$

2.8. The Compact Support Property. Every element $h \in H_n^s(X;G)$ is determined by some singular cycle with a compact support. Consequently, it is possible to find for h a compact set $C \subset X$ such that h belongs to the image of the homomorphism $H_n^s(C;G) \to H_n^s(X;G)$. If h' is an element of the kernel of such a homomorphism, then the cycle z (with support in C) representing it is a boundary: $z = \partial y$. In this case the image of h' is equal to zero already in $H_n^s(C';G)$, where C' is a compact set containing C and the support of y.

This implies that the natural map $\varinjlim_{C \subset X} H_n^s(C;G) \to H_n^s(X;G)$ is an isomorphism. In a similar way it is possible to show that $H_n^s(X,A;G) = \varinjlim_{C' \subset C} H_n^s(C,C';G)$, where $C' \subset C$ are all possible compact pairs contained in (X,A). The above claims follow also from the fact that $C_*^s(X,A;G) = \bigcup_{(C,C')} C_*^s(C,C';G) = \varinjlim_{C' \subset C} C_*^s(C,C';G)$, and that the exact $\varinjlim$-functor commutes with the operator H_n.

Therefore, as with the simplicial homology H_*^c, the singular homology H_*^s has *compact supports*. It is fully determined by the homology of all compact subspaces of X (all finite subcomplexes in the case of a complex K).

2.9. Homology and Cohomology of the Second Kind. If a space is locally compact then in addition to the ordinary (that is, finite) chains it is possible to consider also the infinite *locally finite singular chains* (compare with §1.4). The collection of all singular simplices that appear with nonzero coefficients in such a chain is locally finite (or equivalently, the collection of all singular simplices of a chain that intersect an arbitrary compact set $C \subset X$ is finite). Homology determined by such chains is called the *singular homology of the second kind*.

Analogously, in a complex of singular cochains of a locally compact space X there exists a subcomplex consisting of all *cochains ξ with compact supports*. For every such ξ there exists a compact set $C \subset X$ with the property that ξ is equal to zero on all singular simplices in $X \setminus C$ (or equivalently, ξ is nonzero only on a finite number of simplices of any locally finite collection of singular simplices in X). Cohomology determined by such cochains is called the *singular cohomology with compact supports (cohomology of the second kind)*.

Singular theories of the second kind were introduced by Cartan (1948-49). It is clear that proper maps between locally compact spaces induce natural homomorphisms of the singular groups of the second kind. In the case of locally finite polyhedra the constructions of §2.4 can easily be adapted to show that there exists an isomorphism between the singular homology and cohomology of the second kind and the simplicial theories H_* and H_c^*.

Chapter 2
Cohomology Theories of Čech and Alexander-Spanier

§1. Methods and Goals of Generalizations

1.1. Discussion of the Weaknesses of the Singular Theory. In spite of the fact that the singular homology and cohomology groups are defined for arbitrary topological spaces without any additional restrictions, their application in general categories of topological spaces is not justified. One of the reasons is that they are additive (as are the simplicial theories H_*^c and H^*) not with respect to connected components, but rather with respect to components of path connectedness which are essentially different from them. In particular, zero-dimensional homology (or cohomology) groups of a connected space can be different from the coefficient group if the space is not path connected. A connected compact space can admit obvious essential maps onto a circle, yet at the same time it can have trivial one-dimensional singular homology and cohomology groups. Conversely, in Barrat (1962) and Barrat and Milnor (1962), one can find examples of other "anomalies", such as the existence of compact subsets of the Euclidean space $\mathbb{R}^3$ that have nontrivial singular groups in arbitrarily high dimensions. Some deficiencies of the singular theories were mentioned previously in Sects. 2.6 and 2.7 above.

Some attempts to justify the complications which arise in the consideration of the singular groups of general spaces lead to describing some topological spaces as "pathological". Such a point of view is not justified. For example, the *solenoid* $\sum$ (the inverse limit of a sequence of circles with bonding maps of degree ≥ 2) is a connected compact set for which $H^0(\sum; \mathbb{Z}) = \mathbb{Z}$, and $H^1(\sum; \mathbb{Z})$ is a nontrivial subgroup of the group of rational numbers that is different from $\mathbb{Z}$. At the same time $H^0_s(\sum; \mathbb{Z})$ is the product of a continuum of copies of the group $\mathbb{Z}$, and $H^p_s(\sum; \mathbb{Z}) = 0$, for all $p > 0$, since $\sum$ has a continuum of path components with trivial cohomology groups in positive dimensions. The "anomalies" (every time they rise) simply reflect the fact that the singular theory was used outside of the limits of its applicability within which it still remains a homology or cohomology theory. It is a typical role of the sheaf theory to discover such limits.

1.2. Possible Generalizations and Their Uniqueness. In spite of the existence of a great variety of generalizations of classical theories, corresponding to different categories of topological spaces, the idea that outside of the limits of the category of polyhedra there exist mutually nonequivalent theories of homology and cohomology is, in general, not true. The generalizations, in addition to expanding the realm of application of these theories, were instrumental in the rethinking of the very foundations of homology and cohomology theories and revealed their essentially different and considerably deeper nature, not visible from previous approaches. The language of the sheaf theory, which was created simultaneously with these generalizations, allowed, in particular, a definite answer to the question: what is a cohomology theory? (see Sect. 1.5, Chap. 3). The tendency, that can be noticed at times, to introduce new definitions and develop them separately from others by verifying essentially the same properties expressed in a different language, cannot be justified. In spite of a multitude of different initial descriptions, homology and cohomology theory is essentially unique. The presence of its various descriptions is a result of different approaches which take into account different specific problems and require the application of special constructions corresponding to specific situations. For example, when it is necessary to take into account the specific structure of a space under consideration, it is convenient to apply the approach of Čech which is based on the analysis of special covers. In the study of the behavior of homology (cohomology) with respect to continuous maps it is better to consider chains (cochains) that are functors of the space and of the coefficient group (Alexander–Spanier approach). In problems related to differentiable topology it is natural to use de Rham cochains represented as differential forms.

1.3. Compact Spaces. The uniqueness of the theory can be seen clearly in the category of metric compacta, whose homology and cohomology can be described (even without using any chains or cochains) as a unique natural extension of the simplicial theory. Corresponding constructions were given in Sklyarenko (1971a) (and partially sketched in Milnor (1960)).[7]

According to a well-known theorem of Freudenthal every metrizable compactum X can be represented as the inverse limit of a sequence of finite complexes K_i and their piecewise linear maps $K_i \to K_{i-1}$ (see Sklyarenko (1971a) and for a less precise description Sklyarenko (1969)). By identifying consecutive mapping cylinders of bonding maps along their common base and by "gluing in" X at infinity, we obtain a compactum $\widetilde{X}$ containing X, for which $\widetilde{X} \setminus X$ is a countable, locally finite complex. We can assume that K_1 consists of a single point so that the compactum $\widetilde{X}$ is contractible, and thus acyclic. From the exact sequence of the pair $(\widetilde{X}, X)$ and by using observations from Sect. 1.5, Chap. 1, we deduce that $H_n(X; G) = H_{n+1}(\widetilde{X}, X; G) = H_{n+1}(\widetilde{X} \setminus X; G), H^n(X; G) = H^{n+1}(\widetilde{X}, X; G) = H_c^{n+1}(\widetilde{X} \setminus X; G)$, that is, the homology and cohomology of X are naturally isomorphic with the homology and cohomology of the second kind of the locally finite polyhedron $\widetilde{X} \setminus X$.[8] This is further justified by the fact that for every continuous map $f : X \to Y$, the compacta $\widetilde{X}, \widetilde{Y}$ containing them can be selected in such a way that f can be extended to a map $\widetilde{f} : \widetilde{X} \to \widetilde{Y}$ that defines a proper piecewise linear map of the polyhedra $\widetilde{X} \setminus X \to \widetilde{Y} \setminus Y$. Finally, this construction can be adapted to describe homology and cohomology of pairs of spaces.

Some versions of chains and cochains which determine this theory can be constructed (see Chap. 5 and §4 below) for arbitrary compact Hausdorff (not only metrizable) spaces. In the light of the construction presented above, it is not unexpected that the cochains with coefficients in a ring R are free, while the chains with coefficients in an R-module G can be identified with the complex of homomorphisms of such cochains into G. This construction also provides insight into the axiomatic approach to homology and cohomology of compact spaces that is considered in Chap. 7.

1.4. The Category of Locally Compact Hausdorff Spaces and Their Proper Maps. We shall denote this category by $\mathcal{B}$ and $\mathcal{B}_0$ will be used to denote its subcategory consisting of compact spaces. By taking the one-point compactification of the spaces in $\mathcal{B}$ (and by adding to every object of $\mathcal{B}_0$ an isolated point), we obtain the category $\overline{\mathcal{B}}$ of pointed compacta (whose morphisms always preserve complements of the distinguished point). According to the

---

[7] Independence from chains makes it similar to constructions applicable for extraordinary homology theories (with nontrivial homology groups of a point in nonzero dimensions), see §2, Chap. 7. Compare also with the closing parts of Sect. 5.5, Chap. 8.

[8] In particular, with respect to their variety the groups of the second type of polyhedra, are the same as the homology and cohomology groups of metric compacta.

previous section there exists a unique natural homology and cohomology theory on the subcategory $\mathcal{B}_0 \subset \overline{\mathcal{B}}$. The analogous result for the category $\mathcal{B} \sim \overline{\mathcal{B}}$ follows now from the fact that for every object $\overline{X} \in \overline{\mathcal{B}}$ there exists an object, homeomorphic with it (modulo an isolated point added to $\overline{X}$), in $\mathcal{B}_0 \subset \overline{\mathcal{B}}$.[9]

The uniqueness of a natural extension of simplicial theories of the second kind to the category $\mathcal{B}$ can also be proved directly by making use of the construction, especially adjusted for this case, from the previous section (see Sklyarenko (1971a)). Accordingly, homology and cohomology (of the second kind) in the category $\mathcal{B}$ are indeed equivalent to homology and cohomology of compact spaces belonging to a suitable category $\overline{\mathcal{B}}$ (see Sect. 1.4, Chap. 1). Nevertheless, for the reasons mentioned in this chapter, it is natural to consider them independently.

1.5. The General Case. Internal definitions of homology and cohomology, provided in Chap. 1 and applied to general polyhedra and topological spaces without any restrictions imposed on them, are in no way associated with such a special property as compactness. However, in the case of homology, the so called *compactness factor*, nevertheless manifests itself in an essential way. Every cycle is compact (in the case of homology of the second kind, after passing to the one-point compactification, see Sect. 1.4), and the homology of a space is determined by the homology of its compact subspaces (see Sect. 2.8, Chap. 1). It is true that also in the general case, "good" chains (necessary, say, for a direct description of the theory existing according to Sect. 1.3 above) can be constructed only for compact and locally compact spaces. As we can see, the property of compactness is reflected in the very nature of homology (in contrast with cohomology). Therefore, it appears completely logical to consider first the compact case in the construction of the theory and to define homology of general spaces as *homology with compact supports*: $H_n^c(X, A; G) = \varinjlim_{C' \subset C} H_n(C, C'; G)$, where (C, C') are all possible compact pairs contained in (X, A) (compare with Sect. 2.8, Chap. 1). So far there exist no satisfactory descriptions of homology without a noticeable impact of compactness.

Other observations also speak in favor of the groups H_*^c with compact supports (see §3, Chap. 5). Homology with compact supports is the only acceptable theory for the subspaces of the Euclidean spaces (equivalently, spheres S^n). According to the *Sitnikov duality* (see Sect. 5.6, Chap. 8 and also Sect. 5.5, Chap. 8) it is uniquely determined by the cohomology of their complements in S^n. Incidentally, the homology H_*^c appeared for the first time in Sitnikov (1951), the work concerning the concept of duality. These groups are frequently called the *Steenrod-Sitnikov homology* (see Milnor (1960))[10], and

---

[9] The theory arising in $\mathcal{B}$ is sometimes called *associated* with the theory $\mathcal{B}_0$ which determines it, see §6, Chap. 10 in Eilenberg and Steenrod (1952). The same approach using the language of chains and cochains is presented in §3, Chap. 5 and below in §4.2.

[10] See, in particular Altshuler (1985), Sklyarenko (1980a, 1980b, 1984), Dydak (1986) and other sources.

in Aleksandrov (1959) they are called the *Sitnikov homology*. As homology
of general spaces they became frequently used in the sixties, after the publications of Eilenberg and Steenrod (1952, exercises to Chap. 9), Aleksandrov
(1959), Borel and Moore (1960), Milnor (1960), Raymond (1961) and other
works.

In the case of cohomology the picture is quite different. The impact of compactness is barely noticeable (see Sect. 1.5, Chap. 8). It is natural to consider
arbitrary closed subsets as supports of cochains and cocycles (or polyhedra,
in the case of simplicial theory). Even though it is quite useful to distinguish
cohomology with compact supports in the case of locally compact spaces,
it becomes the ordinary cohomology when we consider the one-point compactifications. Cohomology theory of compact spaces is barely a noteworthy
special case, without any impact on the development of the general theory. In
spite of this, there exist other reasons, due to which cohomology theory (as
pointed out in the introduction) reached a higher degree of completeness than
homology theory. Some of these reasons will become apparent in this chapter.

§2. Čech Cohomology

2.1. Introductory Remarks. Independently of the development of the singular theory, beginning from the 20's, other approaches to constructing homology of general spaces were contemplated. A definition of homology of metric spaces based on the concept of the infinite cycles contained in them and
consisting of infinitely subdivided "abstract" simplices was proposed by Vietoris (1927).[11] A different type of limit procedure, an approximation of a
compactum by its *projection spectrum*, that is, a specific sequence of finite-
dimensional complexes, was proposed by P.S. Aleksandrov (in the years 1925-
1928) who defined in this way the *Betti numbers* of compacta. The key role
in these developments was played by the concept of the *nerve of a cover*,
introduced by P.S. Aleksandrov. A definition of homology groups of metric
spaces as inverse limits of homology groups of complexes forming the projection spectra of P.S. Aleksandrov, was given for the first time by Pontryagin
(1931). Analogous limit process with respect to the family of the nerves of
all finite open covers of arbitrary topological spaces (not only compact) was
proposed by Čech (1932). Later it became clear that for a noncompact space
X (in particular, even for a polyhedron) the consideration of finite covers only
determines (in place of homology of X) quite complex homology of the Čech-
Stone compactification of X. For example, in the case when $X = \mathbb{R}^1$ is the
real line, the one-dimensional homology (and cohomology) groups arising in
this process not only are nontrivial, but they are even uncountable. Dowker

---

[11] The fact that a metrizable compactum itself is an inverse limit of a sequence of
polyhedra and their piecewise linear maps (see Sect. 1.3) was proved by Freudenthal
in 1937.

noticed (in 1950) that this situation can be remedied by considering, together with finite open covers, arbitrary infinite open ones as well. However, this approach takes care completely only of the case of cohomology. For homology the Čech-Dowker construction (ignoring, in particular, the compactness factor, see earlier, Sect. 1.5) needs further improvement (see §§1, 3, Chap. 5).

In spite of the fact, that in addition to Čech, the fundamental ideas and constructions were contributed by P.S. Aleksandrov, L.S. Pontryagin and C.H. Dowker, according to the established tradition, homology (and also cohomology) of general spaces determined by open covers became known as *Čech homology (cohomology)*. Since the operation of the inverse limit does not preserve the property of exactness, Čech homology theory does not satisfy one of the fundamental requirements, namely, the requirement of exactness of the homology sequence of a pair of spaces. For this reason, at present this theory plays only a supporting role in the descriptions of some better behaved theories. On the other hand Čech cohomology theory plays an important role in many applications, not even limited to paracompact spaces (such restriction is sometimes considered).

2.2. Cohomology with Coefficients in Presheaves. As will be seen from the content of the following chapters, the most natural language for describing Čech cohomology is provided by sheaf theory. In connection with the above, the description of major constructions is given, whenever possible, not only for ordinary coefficients, but also for coefficients in presheaves.

Let $\alpha = \{M_\lambda\}, \lambda \in \Lambda_\alpha$ be an open cover of a space X. We shall denote by $|\alpha|$ the *nerve* of α, that is, the complex whose simplices S are formed by all finite sets $\lambda_0, \ldots, \lambda_n \in \Lambda_\alpha$ of indices, for which $M_{\lambda_0 \ldots \lambda_n} = M_{\lambda_0} \cap \ldots \cap M_{\lambda_n} \neq \emptyset$. A simplex S of dimension n is also denoted by $(\lambda_0, \ldots, \lambda_n)$ and the set $M_{\lambda_0 \ldots \lambda_n}$ is denoted by M_S.

Let A be an arbitrary *presheaf* of Abelian groups (or R-modules) on X, that is, a contravariant functor from the category of open subsets of the space X to the category of Abelian groups (or modules). A *cochain* ξ of dimension n on $|\alpha|$ (or simply on α) with coefficients in A is a function assigning to every n-dimensional simplex $S = (\lambda_0, \ldots, \lambda_n)$ some element $\xi(\lambda_0, \ldots, \lambda_n) \in A(M_S)$. We agree to consider only alternating cochains (see Sect. 1.2, Chap. 1) which vanish whenever $\lambda_i = \lambda_j$, for some $i \neq j$.

The *cochains of the cover* α form a group (R-module) $C^n(\alpha; A)$ that is isomorphic to the product $\prod_{\dim S = n} A(M_S)$. The formula

$$d\xi(\lambda_0, \ldots, \lambda_{n+1}) = \sum_i (-1)^i \xi(\lambda_0, \ldots, \widehat{\lambda_i}, \ldots, \lambda_{n+1}), \qquad (1)$$

where it is assumed (compare with formula (2), Chap. 1) that all values of ξ on the right hand side are restricted to the set $M_{\lambda_0 \ldots \lambda_{n+1}} \subset M_{\lambda_0 \ldots \widehat{\lambda_i} \ldots \lambda_{n+1}}$, determines the coboundary operator $d : C^n(\alpha; A) \to C^{n+1}(\alpha; A)$. The cohomology groups of the complex $C^*(\alpha; A)$ obtained in this manner are denoted by $H^n(\alpha; A)$ and are called the *cohomology groups of the cover* α with coefficients in the presheaf A. It is clear that the groups $H^n(\alpha; A)$ are the *cohomology groups of the complex* $|\alpha|$ *with local coefficients* described by the relation $S \to A(M_S)$, where S denotes simplices of $|\alpha|$.

If $\alpha < \beta$, that is, if the cover $\beta = \{N_\mu\}$ is inscribed in α, then by assigning to every $\mu \in M_{\lambda_\beta}$ some $\lambda = \lambda(\mu) \in \Lambda_\alpha$ in such way that $N_\mu \subset M_{\lambda(\mu)}$, we obtain a simplicial map of the nerves $|\beta| \to |\alpha|$. This leads to a homomorphism of cochains $\pi_\beta^\alpha : C^n(\alpha; A) \to C^n(\beta; A)$ which is induced by the restriction homomorphism $A(M_{\lambda_o \ldots \lambda_n}) \to A(N_{\mu_0 \ldots \mu_n})$, where $N_{\mu_0 \ldots \mu_n} = \bigcap_i N_{\mu_i}$ and $\lambda_i = \lambda(\mu_i)$. Clearly $\pi_\beta^\alpha d = d\pi_\beta^\alpha$.

The map π_β^α is not defined uniquely. However, if $\widetilde{\pi}_\beta^\alpha$ is another map defined by the assignment $\mu \to \widetilde{\lambda}(\mu)$, then there exists a chain homotopy (compare with the oriented prism, Sect. 2.3, Chap. 1) $\widetilde{\pi}_\beta^\alpha - \pi_\beta^\alpha - dD + Dd$,

$$(D\xi)(\mu_0, \ldots, \mu_{n-1}) = \sum_i (-1)^i \xi(\lambda_o, \ldots, \lambda_i, \widetilde{\lambda}_i, \ldots, \widetilde{\lambda}_{n-1}). \qquad (2)$$

On the right hand side of the last formula, before summation it is necessary to restrict the values of the cochain $\xi \in C^n(\alpha; A)$ to $N_{\mu_0 \ldots \mu_{n-1}}$.

Thus, for $\alpha < \beta$ there exists a homomorphism $(\pi_\beta^\alpha)_* : H^n(\alpha; A) \to H^n(\beta; A)$ not depending on the selection of the map $|\beta| \to |\alpha|$, and such that $(\pi_\gamma^\alpha)_* = (\pi_\gamma^\beta)_*(\pi_\beta^\alpha)_*$ provided that $\beta < \gamma$. In particular $(\pi_\beta^\alpha)_*$ is an isomorphism if $\alpha < \beta$ and $\beta < \alpha$ simultaneously. It is natural to consider such α and β *equivalent*. The operation of the direct limit with respect to the class of equivalent covers yields the *Čech cohomology* of the space X with coefficients in the presheaf A: $\check{H}^n(X; A) = \varinjlim_\alpha H^n(\alpha; A)$.

The cohomology $\check{H}_*$ is clearly a functor of A. Corresponding to a homomorphism $f : A \to B$ of presheaves there are the maps $f_* : \check{H}^n(X; A) \to \check{H}^n(X; B)$ such that $(gf)_* = g_* f_*$ and $e_* = e$ for the identity map $e : A \to A$. Since the lim-functor is exact, corresponding to each short exact sequence of presheaves $0 \to A' \to A \to A'' \to 0$ (on X) there is an *exact cohomology sequence*

$$\ldots \to \check{H}^n(X; A') \to \check{H}^n(X; A) \to \check{H}^n(X; A'') \to \check{H}^{n+1}(X; A') \to \ldots, \qquad (3)$$

which is the limit of analogous sequences corresponding to the covers α.

2.3. Cohomology of Subspaces and Pairs. Cohomology of a subspace $Y \subset X$ can be defined without any extra effort only in the case when $A = G$ is a *constant presheaf*[12] ($A(U) = G$, where G is the same group for all $U \subset X$), as cohomology determined by open covers of Y independently of the embedding of $Y \subset X$. In the case when Y is a closed subset of a paracompact Hausdorff space X, $\check{H}^*(Y : G)$ can de defined by using covers of X.

---

[12] Such cohomology (see the following chapters) are always defined in a cohomology theory with coefficients in a sheaf.

For an open cover α of the space X let $\widetilde{\alpha}$ denote all $M_\lambda \in \alpha$ for which $M_\lambda \cap Y \neq \emptyset$. This leads to the following short exact sequence of cochain complexes

$$0 \to C^*(\alpha, \widetilde{\alpha}; A) \to C^*(\alpha; A) \to C^*(\widetilde{\alpha}; A) \to 0.$$

Since every open cover α' of a closed $Y \subset X$ is a restriction to Y of an open cover α of X, and for a paracompact X, in addition, there exist arbitrarily fine covers α for which $|\widetilde{\alpha}| = |\alpha'|$ (α' is the restriction of α to Y), we have $\check{H}^*(Y; G) = \varinjlim_\alpha H^n(\widetilde{\alpha}; A)$. Therefore also in the general case we put $\check{H}^n(Y; A) = \varinjlim_\alpha H^n(\widetilde{\alpha}; A)$. The exact cohomology sequences obtained by passing to the direct limit with respect to α from the short exact sequences described above determine the *exact Čech cohomology sequence of the pair* (X, A) with coefficients in the presheaf A:

$$\ldots \to \check{H}^n(X, Y; A) \to \check{H}^n(X; A) \to \check{H}^n(Y; A) \to \check{H}^{n+1}(X, Y; A) \to \ldots \quad (4)$$

2.4. Some Properties. The groups $\check{H}^*$ satisfy all usual properties expected of cohomology. Homotopic continuous maps of paracompact spaces induce the same homomorphism of these groups, the *excision property* for a pair (X, Y) and open $U \subset Y$ holds under the assumption that $\bar{U} \subset Y$ (compare Sect. 2.6, Chap. 1). Standard proofs of these and other statements can be found, for example, in Eilenberg and Steenrod (1952). Deeper results, among them statements relating to paracompact spaces (about the place of $\check{H}^*$ among other theories and the connections between them, about the connections between the cohomology of a space and the cohomology of a cover, about the role of Čech cohomology in the theory of algebraic varieties, etc.) are obtained by using the tools of sheaf theory (see Chap. 4 below).

The well known fact that the cohomology groups $H^n(X; G)$ of polyhedra can be identified in a natural way with the groups of homotopy classes of continuous maps of X into the *Eilenberg-MacLane spaces* $K(G, n)$ (*homotopic representability of cohomology*) remains true without any changes also for Čech cohomology of paracompact spaces (see Bartik (1968)).

Let us mention that for every topological space, the group $\check{H}^0(X; G)$ is the group of all locally constant functions on X with values in G (in particular, $\check{H}^0(X; G) = G$ for a connected space X). Indeed, the *zero-dimensional cocycles* ξ *of a cover* α assign to the sets $M_\lambda \in \alpha$ the elements $\xi(M_\lambda) \in G$ in such way that $\xi(M_\lambda) = \xi(M_\mu)$ if $M_\lambda \cap M_\mu \neq \emptyset$, and therefore the group $H^0(\alpha; G)$ consisting of all such cocycles is the same as the group of all functions on X (with values in G) that are constant on M_λ. For $\alpha < \beta$ there exists a monomorphism $H^0(\alpha; G) \to H^0(\beta; G)$ and therefore it is enough to make sure that by passing to the direct limit all *locally constant functions* fall into $\varinjlim_\alpha H^0(\alpha; G)$. This follows from the fact that every locally constant function on X is constant on the elements of some open cover.

§3. Alexander-Spanier Cohomology

3.1. Basic Constructions. Of great importance in the development of cohomology theory was the approach based on the ideas of Alexander (in 1935). Analogous ideas were expressed by Kolmogorov (in 1936) for locally compact spaces (see Kolmogorov (1936c)). In its final form the theory was formulated by Spanier (in 1948).

Similarly to the approach proposed by Vietoris (see Sect. 2.1), in place of singular simplices which play the fundamental role in the singular theory, we consider arbitrary ordered collections of points $(a_0, \ldots, a_n)$ of a space X. The cochains are not different from the singular cochains; they are functions on such ordered collections. As usual, the coboundary is defined by formula (2) of Chap. 1.

The fundamental difference by comparison with the singular theory is the fact that with this definition of cochains every cocycle ξ is a coboundary: $\xi = d\eta$, where $\eta(a_0, \ldots, a_{n-1}) = \xi(x, a_0, \ldots, a_{n-1})$ and x is an arbitrary fixed point of the space X. This can be remedied if we consider the so called *localized Alexander-Spanier cochains* which are defined as equivalence classes of the cochains considered above. Here, the cochains ξ_1 and ξ_2 are considered *equivalent* if the values of ξ_1 and ξ_2 coincide on all "simplices" $(a_0, \ldots, a_n)$ of size smaller than some open cover α of the space X (compare Sect. 2.5, Chap. 1). It is clear that localized cochains form a cochain complex that determines the *Alexander-Spanier cohomology* of the space X.

In an obvious way one can define cohomology of an arbitrary subspace $Y \subset X$. Every (localized) cochain on Y can be extended to a cochain on the whole space X. Accordingly, the restriction of cochains on X to an arbitrary subspace Y is always an epimorphism. The kernel of such an epimorphism determines the *cohomology of the pair* (X, Y). For every pair (X, Y) there exists an exact Alexander-Spanier cohomology sequence that is similar to (4) presented above. An exact cohomology sequence corresponds also to every short exact sequence of the coefficient groups (compare with (3)).

Let us point out that the *zero-dimensional cochains* are functions ξ of the points of X with values in the coefficient group. In the language of localized cochains the condition that ξ is a cocycle means that $d\xi$ vanishes for all pairs of points belonging to some open cover of X, that is, the function ξ is locally constant. Accordingly, as is also the case for Čech cohomology, the *zero-dimensional cohomology group* $H^0(X; G)$ is the *group of locally constant functions* on X with values in G.

3.2. Remarks. There exist also special types of Alexander-Spanier cochains which are frequently defined for more specific purposes. For example, it is quite natural to define cohomology by using the subcomplex consisting of all locally finite-valued cochains, see Massey (1978b). Cohomology can also be defined by using continuous Alexander-Spanier cochains including the ones that locally can be represented as finite sums of products of functions of one variable (see Korobov (1969)). In any case the *Alexander-Spanier cochains* form sheaves and therefore the theories defined by them are easily identified with the sheaf theories.

For Alexander-Spanier cohomology the typical properties (excision, invariance under homotopy) can be established by considerations similar to those used in the singular theory. However, they follow also from the fact that for paracompact spaces (see below) Alexander-Spanier cohomology is equivalent to Čech cohomology. The Alexander-Spanier cochains are convenient because they are not only a functor of the coefficient group, but also of a topological space on which they are defined, and this makes them valuable in the study of continuous maps. It is easy to define the *support* $|\xi|$ *of a cochain* ξ (and also of the representative cocycle of any element in a cohomology group) as the closed set such that the restriction of ξ to the complement of this set is equal to zero (compare with the definition of the support of a singular cochain in Sect. 2.9, Chap. 1).

§4. Cohomology with Compact Supports

4.1. Alexander-Spanier Cochains. In this section we construct *cochains with compact supports* which define the cohomology H_c^* (of the second kind) on the category $\mathcal{B}$ of locally compact spaces and their proper maps.[13] Since the theory obtained in this process (with any description of cochains) satisfies all the usual requirements, it coincides (in the subcategory of the first countable spaces) with the one that was briefly mentioned in §1.3 and §1.4 above.

The subcomplex of the complex of the (localized) Alexander–Spanier cochains formed by all cochains with compact supports determines the cohomology $H_c^*(X;G)$ of the second kind. The groups $H_c^n(X;G)$ are clearly compatible with proper maps. They are invariant with respect to proper homotopies. Those and other standard properties are established, for example, in Spanier (1966). In particular, it is known (see also §4, Chap. 3) that the groups $H_c^n(X;G)$ of a noncompact space X are isomorphic to the *reduced cohomology groups* $\widetilde{H}^n(\overline{X};G)$ of the one-point compactification $\overline{X}$ of the space X ($\widetilde{H}^n(\overline{X};G) = H^n(\overline{X},\infty;G)$, where $\infty = \overline{X} \setminus X$; in particular $\widetilde{H}^n = H^n$, for $n > 0$).

It is true that in order to define $H_c^n(X;G)$ we can consider also the *finite-valued Alexander-Spanier cochains* with compact supports, the *Massey cochains* (see Massey (1978b)). The fact that the cohomology groups determined by them are isomorphic to the ones defined previously can be established by using sheaf theory (see §2.3, Chap. 4). For every closed subset $Y \subset X$ the restriction $C_c^*(X;G) \to C_c^*(Y;G)$ of cochains with compact supports to Y is clearly an epimorphism. Its kernel $C_c^*(X,Y;G)$ defines the cohomology $H_c^n(X,Y;G)$ of the pair (at the same time for H_c^* we obtain

---

[13] The groups H_c^n were introduced in Kolmogorov (1936c).

the exact sequence of a pair similar to (4)). It is also true that the inclusion $C_c^*(X \setminus Y; G) \to C_c^*(X, Y; G)$ of Massey cochains induces the isomorphism $H_c^*(X, Y; G) = H_c^*(X \setminus Y; G)$ (the *excision property*, see Sect. 1.5, Chap. 1). The proof (see Massey (1978b)) is not completely trivial.[14] As an obvious corollary we obtain the isomorphism $H_c^*(X; G) = H^*(\overline{X}, \infty; G)$ which was mentioned earlier. Another simple corollary is the fact that if Y is compact, then $H_c^*(X, Y; G) = H_c^*(X/Y, y; G)$ (where y is the image of Y under the quotient map $X \to X/Y$).

Massey cochains are noteworthy because of the fact that the complexes $C_c^*(X; \mathbb{Z})$ are *free* ($\mathbb{Z}$ is the group of integers). This fact is established in Massey (1978b) with the aid of a theorem of Nöbeling related to the theory of Abelian groups (whose proof is not constructive and uses the transfinite induction). It plays a crucial role in the development of homology theory (see also Chap. 5).

4.2. Čech-Type Cochains. The free cochains for the theory H_c^* slightly earlier than in Massey (1978b) were constructed in Sklyarenko (1969) (for spaces without the second countability axiom, see the foot-note to §1 in Sklyarenko (1971b)[15]).

For this aim it is necessary to consider locally finite covers α consisting of *"canonical"* compact sets F (coinciding with the closure of Int F) and such that Int $F \cap$ Int $F' = \emptyset$ for all $F, F' \in \alpha$. If α, β are such covers then $\alpha \wedge \beta$ denotes the cover consisting of the closures of the sets Int $F \cap$ Int $F', F \in \alpha, F' \in \beta$. Let ω be any system of covers of this type satisfying the properties: a) for each $\alpha, \beta \in \omega$ there exists $\gamma \in \omega$ such that $\gamma > \alpha \wedge \beta$; b) for each $x \in X$ and each neighborhood U of x there exists $\alpha \in \omega$ such that $F \subset U$ if $x \in F, F \in \alpha$. The projections of nerves of covers (in the case when one cover is inscribed in another) are generally not uniquely defined and therefore in the definition of Čech cohomology the operation of the direct limit with respect to covers is applied to cohomology, but not to cochains (see §2.2). In the case considered here such projections are defined uniquely and they are proper maps (even surjections) and thus we can define a cochain complex $\check{C}_{\omega c}^*(X; G) = \varinjlim_{\alpha \in \omega} C_c^*(\alpha; G)$. It is true that $H^n(\check{C}_{\omega c}^*(X; G)) = H_c^n(X; G)$. Since the $\varinjlim$-functor is exact, it commutes with the operation of cohomology and we have also $H^n(\check{C}_{\omega c}^*(X; G)) = \check{H}_c^n(X; G)$.

It is not always convenient to consider infinite covers (particularly in homology theory, see Chap. 5). In such cases it is possible to consider pairs $(\alpha, \overline{\alpha})$, where $\alpha \in \omega$ and $\overline{\alpha}$ is any part of α containing all but a finite number of elements of α. We assume that $(\alpha, \overline{\alpha}) < (\beta, \overline{\beta})$ if $\alpha < \beta$ and the union of all the sets in $\overline{\beta}$ is contained in the union of all the sets in $\overline{\alpha}$. Since $C_c^*(\alpha; G) = \varinjlim_{\overline{\alpha}} C^*(\alpha, \overline{\alpha}; G)$, we obtain $\check{C}_{\omega c}^*(X; G) = \varinjlim_{(\alpha, \overline{\alpha})} C^*(\alpha, \overline{\alpha}; G)$.

---

[14] See also Kolmogorov (1936d).
[15] See also the remarks following Theorem 3.1 in Sklyarenko (1984).

The system of pairs $(\alpha, \overline{\alpha})$ can be considered as the system $\overline{\omega}$ of canonical covers of the one-point compactification $\overline{X}$ (for non-compact X), and therefore $\check{C}^*_{\omega c}(X; G) = \check{C}^*_{\overline{\omega}}(\overline{X}, \infty; G)$. In particular $H^*_c(X; G) = \widetilde{H}^*(\overline{X}; G)$.

The condition that ω be a system of covers of X with canonical sets is not so essential. It is important that there exist unique (and surjective) projections of the nerves of such covers. In particular, in order to determine the cohomology of a closed subspace Y it is convenient to consider the system ω' consisting of the restrictions α' of the covers $\alpha \in \omega$ to Y (the elements of α' can be, and in general are, repeated; the projections of the nerves in ω' are determined by the projections in ω). One can equally well consider the system $\widetilde{\omega}$ consisting of the families $\widetilde{\alpha}$ of these elements of α that intersect Y. In this way the system ω determines the cohomologies $H^*_c(Y; G)$ and $H^*_c(X, Y; G)$ (and also the exact cohomology sequence of the pair (X, Y)).

In the case of a compact Y the proper map $X \to X/Y$ clearly induces an isomorphism $\check{C}^*_{\omega c}(X, Y; G) = \check{C}^*_{\overline{\omega} c}(X/Y; G)$ (here $\overline{\omega}$ is obtained by taking the union, in each $\alpha \in \omega$, of a finite set of the elements that intersect Y). In particular, therefore, $H^*_c(X, Y; G) = H^*_c(X/Y; G)$. Since, as in the case of non-compact spaces X, $H^*_c(X, Y; G) = H^*(\overline{X}; Y \cup \infty; G)$, as one obvious corollary we obtain the *excision property*, that is the isomorphism $H^*_c(X, Y; G) = H^*_c(X \setminus Y; G)$.[16]

The cochains $\check{C}^*_{\omega c}$ are convenient in the sense that they allow us, if possible, to obtain additional information by a suitable choice of a system ω of covers of a space. In particular, in the metric case, in place of ω one can take a fundamental sequence of covers, such that each cover is inscribed in the previous one. In such case, for countable G the groups $\check{C}^*_{\omega c}(X; G)$ and thus $H^*_c(X; G)$ are also countable. By selecting covers of specific order we can take into account the dimension of the space (Massey cochains are different from zero in all positive dimensions and in connection with this, the fact that the cohomology groups of a manifold M in dimensions $p > \dim M$ vanish requires a separate proof, see Massey (1978b)). The advantage of Massey cochains (reflected primarily in consideration of homomorphisms of cohomology groups induced by continuous maps of spaces) is due to the fact that for a fixed coefficient group (as with arbitrary Alexander-Spanier cochains) they are uniquely determined by a space and thus they are a functor of a space.

For every principal ideal domain R the cochains $\check{C}^*_{\omega c}(X; R)$ (and for closed $Y \subset X$ both $\check{C}^*_{\omega c}(Y; R)$ and $\check{C}^*_{\omega c}(X, Y; R)$) are *free*.[17] This can be established more easily than for Massey cochains by taking the direct limit of free com-

---

[16] In connection with this topic see also Sect. 4.1, Chap. 5. The theories H^*_c and H_* are therefore sometimes called *absolute* (or the *theories of a "single space"*), see §7, Chap. 10 in Eilenberg and Steenrod (1952)).

[17] Before the publication of Sklyarenko (1969, 1971b) free cochains for Čech cohomology $H^*(X, \mathbb{Z})$ for compact spaces were constructed in Milnor (1960). As with Massey cochains they are functorial. Their deficiency, however, is due to the fact that they are different from zero in all dimensions (including negative). In addition, they are not sections of the sheaves defined by them.

plexes of the form $C_c^*(\alpha, R)$ (surjectivity of projections between the nerves of covers is essential). It follows that (for every coefficient group) in each dimension the short exact sequences of cochains with compact supports of a pair (X, Y) are split.

The independence of cohomology determined by the complexes $\check{C}_{\omega c}^*$ of the choice of the system ω of covers can be established easily. The comparison of such complexes for different systems ω (by using the operation $\omega \wedge \omega'$ with $G = \mathbb{R}$, see Sklyarenko (1969)) shows that they are all homotopically equivalent. Independence follows also (see Sklyarenko (1969)) from the fact (compare the following chapter) that the cochains $\check{C}_{\omega c}^*(X, Y; G)$ are the sections with compact supports (over $U = X \backslash Y$) of the soft resolutions of the constant sheaf G.[18] By using similar methods the analogous fact can be established also for Massey cochains (from which we obtain the equality $H_c^*(X; G) = \check{H}_c^*(X; G)$ mentioned earlier).

4.3. Remarks. In the case of compact spaces the covers with canonical sets are clearly finite and the symbol "c" in the notation for cochains and cohomology can be deleted. This leads to the chains $\check{C}_\omega^*(X; G)$ (or $C^*(X; G)$) which define cohomology in the category of compact Hausdorff spaces. A brief description of it for the case of compact metric spaces, based on its connection with the simplicial cohomology in the category of polyhedra, was given in Sect. 1.3 above.

Chapter 3
Cohomology and Sheaf Theory

Sheaf theory has its origin in algebraic topology. Its fundamental ideas (as well as the theory of spectral sequences) are due to Leray (in the 40's). The first reasonably complete presentation was given in Cartan (1950-51). Cohomology theory, by serving as a catalyst for the creation of sheaf theory, even presently remains the area in which the main ideas of sheaf theory appear in their simplest and most clear form.

The interpretation of cohomology as cohomology of complexes of sections of acyclic resolutions of the coefficient sheaf, which was the result of the extensive and long development of the theory in general categories of topological spaces, radically changed our ideas about the nature of cohomology. According to the new approach, cohomology is regarded as a derived functor of the functor of the zero-dimensional cohomology, that is, the group of sections of the coefficient sheaf. It is constructed for arbitrary topological spaces. Cohomology

---

[18] See Sect. 4.1, Chap. 5.

groups of polyhedra defined in Sect. 1.2, Chap. 1 by using the coboundary operator (2) are one of the natural reflections of the general construction. Other concrete theories arising in practice (considered, in particular, in Chap. 1 and 2), defined generally in sufficiently broad categories of topological spaces, are frequently applied only under certain additional assumptions.

It is remarkable that the "anomalous" behavior of specific cohomology theories accompanies, as a rule, their diversion from the sheaf cohomology theories, and conversely, when they coincide with sheaf theories, no problems seem to arise. It is true that every cohomology theory admits a natural comparison with a sheaf theory. In such comparison the conditions that guarantee the isomorphism of theories (in most cases such conditions are easily formulated in the process of comparison) typically allow us to identify the category of topological spaces that is most natural for a specific cohomology theory.

However, it is worth pointing out that resolutions frequently appear independently of specific definitions of cohomology (in such areas as differential calculus, functional analysis, complex analysis, algebraic geometry and others). The sheaves included in them are typically generated by objects which are apparently not related to any type of cochains, but which determine (as a result of acyclicity of resolutions formed by them) exactly the same cohomology (some examples will be given below). In all such cases cohomology reflects in a natural way the influence of the topological nature of an object on the solution of a problem that is superficially not related to topology.

§1. Constructions of Sheaves and Resolutions

1.1. Presheaves of Cochains. Quite frequently cochains are functions defined on some objects located in a topological space X or in its subsets (like singular simplices, ordered collections of points, intersections of elements of covers, collections of tangent vectors, etc.) with values in a coefficient group G. In such cases they provide the most typical examples of presheaves[19] (Sect. 2.2, Chap. 2).

In the case when the values $A(U)$ of a presheaf A on X are ordinary functions of points $x \in X$ (continuous, smooth, or of any other kind), they clearly satisfy the following conditions:

(S1) If $U = \bigcup_\lambda U_\lambda$ and if the restriction of $\xi \in A(U)$ to each U_λ is equal to zero, then $\xi = 0$.

(S2) If $\xi_\lambda \in A(U_\lambda)$ are such that the restrictions of every pair ξ_λ, ξ_μ to $U_\lambda \cap U_\mu$ are the same, then for $U = \bigcup_\lambda U_\lambda$ there exists an element $\xi \in A(U)$, whose restriction to U_λ coincides with ξ_λ, for every λ.

Even though cochains are always essentially different from ordinary functions, the presheaves determined by them obviously satisfy condition (S2).

---

[19] For this purpose it is convenient to consider only cochains without any restrictions on their supports. For example, the restriction of a cochain with compact support to a smaller open set does not have to be a cochain with compact support.

(The condition (S2) is satisfied for non localized Alexander-Spanier cochains.) The condition (S1), in general, is not satisfied. For example, a singular cochain ξ can be different from zero only on "large" singular simplices (not contained in U_λ).

The significance of conditions (S1), (S2) will be explained in the following section.

1.2. Sheaves. An arbitrary presheaf A leads to the following construction on X, called a *sheaf*. For each point $x \in X$ let $\mathcal{A}_x$ (called the *stalk of the sheaf*) be the direct limit $\varinjlim_{x \in U} A(U)$ with respect to all neighborhoods U of this point, and then $\mathcal{A} = \bigcup_{x \in X} \mathcal{A}_x$ is called a sheaf. Let us mention that in the case of a presheaf of a family of functions, the stalk $\mathcal{A}_x$ is the group of all germs of such functions in x (two functions define the same *germ* at a point if they agree in some neighborhood of this point) and then $\mathcal{A}$ is a *sheaf of germs* at all points.

For each $a \in A(U)$, let a_x be the image of a under the homomorphism $A(U) \to \mathcal{A}_x$ defined for $x \in U$, and let $U_a \subset \mathcal{A}$ be the set of all such a_x. The sets of the form U_a (clearly covering all of $\mathcal{A}$) are considered open in $\mathcal{A}$. The topology on $\mathcal{A}$ arising in this manner is, in general, not Hausdorff. The natural projection $p : \mathcal{A} \to X$ is clearly a local homeomorphism in this topology, and all stalks $\mathcal{A}_x$ are discrete. From the definition of the stalks it follows that algebraic operations in them are continuous in the topology on $\mathcal{A}$. We say that the *sheaf $\mathcal{A}$ is generated by the presheaf A.*[20]

A *section s of a sheaf $\mathcal{A}$* over a set $Y \subset X$ is defined as an arbitrary continuous map $s : Y \to \mathcal{A}$ for which $s(x) \in \mathcal{A}_x$ for each $x \in Y$. Examples of sections are provided by the maps $s_a : x \to a_x$, for $a \in A(U)$. In particular, $a = 0$ determines the *zero section* (whose values are zero elements in $\mathcal{A}_x$), and therefore the sets $\mathcal{A}(U)$ of sections over U are nonempty. Moreover, the $\mathcal{A}(U)$ define a presheaf on X (of the same type as A, that is a presheaf of groups, modules, etc.) called the *presheaf of sections* of the sheaf $\mathcal{A}$. The assignment $a \mapsto s_a$ defines maps $r : A(U) \to \mathcal{A}(U)$ that are presheaf homomorphisms and are compatible with the operations of the restriction to $V \subset U$. In this context it is natural to ask when is the homomorphism r an isomorphism.

Like any presheaf of functions, the presheaf $\mathcal{A}(U)$ of sections clearly satisfies the conditions (S1) and (S2). It turns out that it is precisely these two conditions which guarantee that r is an isomorphism establishing the equivalence of the presheaf A with the presheaf of sections. Indeed, if $s_a = 0$ for $a \in A(U)$, then due to the fact that the zero element in $\mathcal{A}_x$ is determined by the zero element in $A(U_x)$ for some neighborhood $U_x \subset U$ of a point $x \in U$, the map r is a monomorphism if condition (S1) is satisfied. Furthermore, every section s over U is locally determined by some $a_\lambda \in A(U_\lambda)$, $U_\lambda \subset U$, and $s_{a_\lambda} = s_{a_\mu}$ on $U_\lambda \cap U_\mu$. If A satisfies condition (S1), then the restrictions of a_λ

---

[20] It is not correct to assume that typical sheaves are generated only by presheaves of cochains or functions (for a suitable example see Sect. 4.4, Chap. 5).

and a_μ to $U_\lambda \cap U_\mu$ coincide and therefore condition (S2) implies that $s = s_\alpha$, for some $a \in A(U)$. This proves that r is an epimorphism.

Since condition (S1) is not always satisfied for presheaves of cochains, we turn our attention to the fact that condition (S2) alone guarantees that $A(U) \to \mathcal{A}(U)$ is an epimorphism for an arbitrary paracompact open $U \subset X$ (the proof is simple, see §3.9, Chap. 2 in Godement (1958). The same argument proves that if X is paracompact and condition (S2) is satisfied, then for every section s over a closed set $Y \subset X$ (and if X is hereditarily paracompact then also for every Y) there exists a neighborhood U of Y and an element $a \in A(U)$ such that s is the restriction to Y of a section $s_a \in \mathcal{A}(U)$.

1.3. Flabby and Soft Sheaves. A sheaf $\mathcal{A}$ on X is called *flabby* (respectively *soft*) if every section of $\mathcal{A}$ over any open (closed) subset of X can be extended to a section over all of X.

The requirements included in these definitions can be too strong, particularly if we think of sheaves as analogues of ordinary covering spaces. However, for the moment we are interested primarily in sheaves generated by presheaves of cochains. The fact that an arbitrary cochain can be extended from an arbitrary subset to the whole space is obvious. Thus, according to the previous section, for a paracompact (respectively, hereditarily paracompact) space X the sheaves of cochains are always soft (respectively, flabby). A similar property is also true for sheaves of germs of functions that are similar in nature to cochains.

In the case when X is paracompact, by using local extensions of a section on a closed set $Y \subset X$ and by taking into account local finiteness, it is easy to show (see §4.11, Chap. 2 in Godement (1958)) that every section of a sheaf $\mathcal{A}$ over Y can be extended to some neighborhood of Y. Therefore, if a space X is paracompact, then every flabby sheaf is soft. The same argument shows (see the end of Sect. 1.2) that for hereditarily paracompact spaces X the restriction $\mathcal{A}|_Y$ of a flabby sheaf $\mathcal{A}$ to an arbitrary subspace Y of X is also flabby. It is clear that the restriction of a flabby (soft) sheaf $\mathcal{A}$ to an open (respectively, closed) $Y \subset X$ is a flabby (soft) sheaf.

1.4. Resolutions. It is natural to expect that for $p > 0$ every nontrivial cocycle $\xi \in A^p(U)$ of a presheaf A^* of cochain complexes, occupies some "volume" in a space in the sense that for every point $x \in U$ there exists a neighborhood U_x small enough for the restriction of ξ to U_x to be a coboundary. This means that for each point $x \in X$ the condition $\varinjlim_{x \in U} H^p(U) = H^p(x)$ is satisfied. Also, since the exact $\varinjlim$-functor commutes with the operation H_n of the nth homology, at each point x the cochain complex $\mathcal{A}_x^*$ consisting of the stalks of the sheaf $\mathcal{A}^*$ generated by the presheaf A^* is acyclic. Thus, the sheaves $\mathcal{A}^n$ together with the coboundary operators should form an exact sequence of the form

$$0 \to \mathcal{G} \to \mathcal{A}^0 \to \mathcal{A}^1 \to \ldots \to \mathcal{A}^n \to \ldots \tag{1}$$

In the general case the sheaf $\mathcal{G}$ in the above formula is simply the kernel of the map $\mathcal{A}^0 \to \mathcal{A}^1$, and is called the *coefficient sheaf*. In the case of ordinary cochains, $\mathcal{G} = G$ is the *constant sheaf* corresponding to the coefficient group G (that is the sheaf of the form $X \times G$). This follows from the fact that it is always true that $A^0(U) = \mathcal{A}^0(U)$ and the kernel of $\mathcal{A}^0(U) \to \mathcal{A}^1(U)$ consists of locally constant functions from U to G, that is of sections of the sheaf G (see Sect. 2.4 and 3.1 in Chap. 2).

The exact sequence (1) is called the *resolution of the sheaf $\mathcal{G}$*. Note that the condition which guarantees its existence (that every cocycle is locally cohomologous to zero) is also necessary. If a cocycle $\xi \in A^p(U)$ defines a germ $a_x^p \in \mathcal{A}_x^p$ at a point $x \in U$ and if $da_x^p = 0$, then $a_x^p = da_x^{p-1}$ and in a sufficiently small neighborhood V of this point we have $\xi|_V = d\eta$, where $\eta \in A^{p-1}(V)$ determines the germ a_x^{p-1}.

1.5. A Fundamental Idea. So far we have determined that either the cochain complexes $A^*(U)$ satisfy condition (S1) (condition (S2) for such complexes is satisfied automatically) and they coincide with sections of the sheaves $\mathcal{A}^*$ defined by them, or there exist epimorphisms $A^*(X) \to \mathcal{A}^*(X)$ leading to the interpretation of cohomology of a space as the cohomology of the complex $\mathcal{A}^*(X)$ (see §5). Accordingly, the cohomology is typically obtained as the cohomology of the complex of sections over X of some resolution $\mathcal{A}^*$ of the "coefficient" sheaf $\mathcal{G}$. In particular, $H^0(X; \mathcal{G})$ is the group of the sections of this sheaf (which in the case of $\mathcal{G} = G$ is simply the group of locally constant functions). From this we easily obtain the $\prod$-*additivity* of cohomology in the general case, and the $\sum$-*additivity* of cohomology with compact supports.

However, the role of sheaves is not just restricted to the fact that they provide a convenient language for a description of cochains. Of fundamental importance is the fact that for all resolutions of a sheaf $\mathcal{G}$, independently of the way in which they are constructed, the cohomology groups of the complex of sections $\mathcal{A}^*$ are always the same provided that the sheaves $\mathcal{A}^P$ forming $\mathcal{A}^*$ are either flabby or soft (or more generally *acyclic*, see below). As a result, it is correct to define the *cohomology groups $H^n(X; \mathcal{G})$ of a space X with coefficients in a sheaf $\mathcal{G}$* as the cohomology groups of the complex of sections of any resolution $\mathcal{A}^*$ of the type indicated earlier of the sheaf $\mathcal{G}$. Since among flabby resolutions there are injective ones, in the category of sheaves on X the "sheaf" cohomology groups $H^n(X; \mathcal{G})$ are the *derived functors* of the left exact functor $\mathcal{G}(X) = H^0(X; \mathcal{G})$ (*left exactness* means that the sequence $0 \to \mathcal{G}'(X) \to \mathcal{G}(X) \to \mathcal{G}''(X)$, corresponding to the exact sequence $0 \to \mathcal{G}' \to \mathcal{G} \to \mathcal{G}''$, is exact, obviously, in the first two terms). The groups $H^n(X; \mathcal{G})$ represent one of the most typical examples of such functors. In no other place can the fundamental ideas of derived functors be studied with such ease and clarity as for $H^n(X; \mathcal{G})$.

One of the consequences of the approach described above is the possibility of computing the *cohomology of a subspace $Y \subset X$* by using the complex $\mathcal{A}^*$ of sections over Y in the cases when the *restrictions of sheaves $\mathcal{A}^p$ to Y are* flabby or soft (see the end of Sect. 1.3). In particular the stalks $\mathcal{A}^*_x$, being sections of the sheaves $\mathcal{A}^*$ over the points x, should yield the cohomology of a point, which is another indication in favor of the fact that sheaves of cochains should form an exact sequence (a resolution). Moreover, the possibility to extend sections over Y, for $Y \subset X$, to some neighborhoods of Y in X (for example in the case of closed subspaces of a paracompact spaces or, in the case of arbitrary subspaces, of a hereditarily paracompact space, see the end of Sect. 1.2), means that the groups of sections over such Y, and consequently the cohomology groups $H^n(Y; \mathcal{G})$, coincide with the direct limit of corresponding groups of neighborhoods U of Y in X. This is called the *tautness property* for cohomology (in the case of constant coefficient this property is a natural result of the homotopic representability of cohomology, see Sect. 2.4, Chap. 2).

The reason why it is necessary to consider cohomologies with non constant coefficients is related to the fact that such cohomologies have many applications and that they are necessary for the internal development of the theory. Such applications exist in topology as well as in other areas of mathematics, and in several areas (like algebraic geometry, functional and complex analysis among others) they have fundamental importance (cohomology of algebraic manifolds with coefficients in coherent algebraic sheaves, cohomology of complex analytic manifolds, in particular Stein manifolds, with coefficients in coherent analytic sheaves, etc.)

In the internal development of the theory, cohomology groups with non-constant coefficients are necessary, in particular, in the proof of the independence of the groups $H^n(\mathcal{A}^*(X))$ of the choice of the acyclic resolution $\mathcal{A}^*$ of a constant sheaf G. While, if for example, for algebraic manifolds over the field of complex numbers the proofs of fundamental facts related to cohomology groups in most cases can still be left "topological" in nature, then for cohomology of manifolds over arbitrary fields and in other more general situations the proofs are formal in nature and are based primarily on the methods of homological algebra.

Accordingly, sheaves provide a natural language, the coefficients for cohomology and finally a powerful medium for the development of the theory and for its applications. Similarly to the situation arising in the category of simplicial complexes where the only acceptable definition of cohomology is the one based on the coboundary formula (2) of Chap. 1, the only natural definition of cohomology groups of arbitrary spaces is the definition in terms of sheaf theory that was described above. Regardless of the description of cochains, the cohomology groups that arise by using sheaf theory are always unique, the only distinction being caused by using different families of supports (see below).

To verify that a specific approach leads to a sheaf cohomology, it is usually necessary to verify two conditions: a) is it true that the cochain sheaves arising in this approach form a resolution of the coefficient sheaf (*"local" conditions*); b) are those sheaves flabby, soft or just acyclic (*"global" conditions*). These two conditions determine, in a natural way, the limits of applicability of any concrete cohomology theory that is the result of a specific approach.

The *sheaf complexes* $\mathcal{A}^*$ on a topological space X can arise not only as sheaves of cochains, but they can appear in different specific situations, sometimes not even superficially connected with cohomology. The coboundary operator d in such a case is frequently called a *differential* (the name originates from the theory of differential forms in analysis), and the complexes $\mathcal{A}^*$ are called *differential sheaves*. Sometimes the sheaves forming $\mathcal{A}^*$ are not acyclic (in particular, not flabby or soft) or they do not form a resolution of a sheaf. However, in all cases there exists a mechanism of "comparison" of the cohomology of the complex $\mathcal{A}^*(X)$ of sections with the cohomology of X. Typically, this is accomplished by using the concept of spectral sequences. Spectral sequences of fibrations and continuous maps, covering spaces, as well as a spectral sequence of a cover and many others arise in this way. In particular, in a similar way we can even compare the homology and cohomology of a space (see the end of Chap. 5). In such situations, of great interest are always those particular cases when the spectral sequences become "degenerate" which leads to the isomorphism between $H^p(\mathcal{A}^*(X))$ and cohomology of X. The existence of such isomorphisms, together with the conditions that guarantee their existence, are usually the assertions of separate theorems (see below).

§2. Cohomology with Coefficients in a Sheaf

2.1. Grothendieck's Definition. Previous observations indicate that the *cohomology groups $H^n(X, \mathcal{G})$ of a space X with coefficients in a sheaf $\mathcal{A}$* should be defined as right *derived functors* of the left exact *functor of sections $\mathcal{G}(X)$* of the sheaf $\mathcal{G}$ over X, that is as $H^n(\mathcal{J}^*(X))$, where $\mathcal{J}^*$ is an arbitrary *injective* (consisting of injective sheaves) *resolution* of $\mathcal{G}$. In the category of sheaves over an arbitrary topological space there exist sufficiently many injective objects. Every sheaf can be embedded into an injective one (see Bredon (1967), Godement (1958), Grothendieck (1957)) and therefore it is clear that injective resolutions exist (as $\mathcal{J}^n$ we can take an arbitrary injective sheaf containing the *quotient sheaf*[21] $\mathcal{J}^{n-1}/\operatorname{Im} d$).

If $\mathcal{A}^*$ is any other resolution of $\mathcal{G}$ then (in virtue of the injectivity of $\mathcal{J}^*$) there exists a homomorphism of resolutions $\mathcal{A}^* \to \mathcal{J}^*$ which is the identity on $\mathcal{G}$, and any two such homomorphisms are chain homotopic. From the above observation it follows that, first of all, any two injective resolutions of $\mathcal{G}$ are homotopically equivalent (independence of the cohomology of the choice of $\mathcal{J}^*$), and secondly, that there exists a *comparison homomorphism* $\gamma : H^n(\mathcal{A}^*(X)) \to H^n(X; \mathcal{G})$.

It is clear that $H^0(X; \mathcal{G}) = \mathcal{G}(X)$ is the group of sections over X.

---

[21] For a subsheaf $\mathcal{A}' \subset \mathcal{A}$ the *quotient sheaf* $\mathcal{A}/\mathcal{A}'$ is the sheaf generated by the presheaf $U \to \mathcal{A}(U)/\mathcal{A}'(U)$.

2.2. Cohomology with Supports. There exist a variety of situations when in addition to ordinary cohomology groups it is necessary to consider cohomology groups with special types of supports. Examples of such situations were presented in Sect. 1.4 of Chap. 1 and in §4 of Chap. 2 (compact supports) as well as in Sect. 1.5 oc Chap. 1. The role of cohomology with supports will be explained much more extensively in §4 below.

The *support of a section s* of a sheaf $\mathcal{A}$ is defined as the set $|s|$ of all points x for which $s(x) \neq 0$. The projection $p : \mathcal{A} \to X$ is a local homeomorphism, and all zeros of the sheaf $\mathcal{A}$ form a section of $\mathcal{A}$. Therefore a section s equal to zero at some point is also equal to zero in some neighborhood of this point. Consequently, the set $|s|$ is always closed (in the domain of s). A *family of supports* is defined as an arbitrary family φ of closed subsets of X satisfying the following conditions: a) if $F_1, F_2 \in \varphi$, then $F_1 \cup F_2 \in \varphi$; b) if $F' \subset F$ and $F \in \varphi$, then $F' \in \varphi$. A family φ is called *paracompactyfying* if all $F \in \varphi$ are paracompact and for each $F \in \varphi$ there exists an $F' \in \varphi$, such that $F \subset \operatorname{Int} F'$.

The subfunctor $\Gamma_\varphi(\mathcal{A})$ of the functor $\Gamma(\mathcal{A}) = \mathcal{A}(X)$ consisting of all sections of the sheaf $\mathcal{A}$ with supports in φ is also left exact. In the same way as $H^n(X; \mathcal{G})$ we can define its right derived functors $H^n_\varphi(X; \mathcal{G})$. They are called the *cohomology groups of X with coefficients in the sheaf $\mathcal{G}$ and supports in the family φ*. As before there exists the *comparison homomorphism* $\gamma : H^n(\Gamma_\varphi(\mathcal{A}^*)) \to H^n_\varphi(X; \mathcal{G})$, and it is also true that $H^0_\varphi(X; \mathcal{G}) = \Gamma_\varphi(\mathcal{G})$.

2.3. Acyclic Resolutions. The sheaves of cochains which arise in concrete situations are essentially never injective (even though they determine the desired cohomology groups). In connection with the above a natural question arises as to which other resolutions (in addition to the injective ones) can be used to define cohomology groups.

A sheaf $\mathcal{A}$ is called *acyclic (φ-acyclic)* if $H^n(X; \mathcal{A})) = 0$ (respectively, $H^n_\varphi(X; \mathcal{A}) = 0$) for all $n > 0$. One of the fundamental results of the theory of sheaf cohomology is the statement that for any *φ-acyclic resolution $\mathcal{A}^*$* (consisting of φ-acyclic sheaves) of the sheaf $\mathcal{G}$, the groups $H^n(\Gamma_\varphi(\mathcal{A}^*))$ coincide with $H^n_\varphi(X; \mathcal{G})$. This statement is usually obtained as a consequence of a spectral sequence corresponding to the resolution $\mathcal{A}^*$ (see §6 below) but it is easy to obtain it directly. This can be accomplished most efficiently by using the construction of the *iterated connecting homomorphism Δ*. This construction, better than any other, reveals the nature of the requirement of acyclicity of a resolution.

Let $\mathcal{A}^*$ be an arbitrary resolution of $\mathcal{G}$, and let $\mathcal{Z}^n \subset \mathcal{A}^n$ denote the kernel (and the image) of the differential d. From the initial terms of the *exact cohomology sequence* (as the derived functors) corresponding to the short exact sequence $0 \to \mathcal{Z}^{n-1} \to \mathcal{A}^{n-1} \to \mathcal{Z}^n \to 0$ we obtain the monomorphism $H^n(\Gamma_\varphi(\mathcal{A}^*)) \to H^1_\varphi(X; \mathcal{Z}^{n-1})$. Its composition with the *connecting homomorphisms δ* $: H^{n-k-1}_\varphi(X; \mathcal{Z}^{k+1}) \to H^{n-k}_\varphi(X; \mathcal{Z}^k)$ that correspond to the short exact sequences of sheaves $0 \to \mathcal{Z}^k \to \mathcal{A}^k \to \mathcal{Z}^{k+1} \to 0$ defines $\Delta : H^n(\Gamma_\varphi(\mathcal{A}^*)) \to H^n_\varphi(X; \mathcal{G})$.

Frequently the map Δ is mistakenly identified with the comparison homomorphism γ (see Godement (1958), Bredon (1967)). In reality $\Delta = (-1)^{\frac{n(n+1)}{2}}\gamma$ (this can be established by induction with respect to n by using the fact that $\delta\Delta = (-1)^{n+1}\Delta\delta$; compare with §7 of Chap. 5 in Cartan and Eilenberg (1956))) provided that for the triple of resolutions the sequence $0 \to \Gamma_\varphi(\mathcal{A}_1^*) \to \Gamma_\varphi(\mathcal{A}^*) \to \Gamma_\varphi(\mathcal{A}_2^*) \to 0$ is exact.

Acyclicity of all the $\mathcal{A}^k$ is clearly equivalent to the property that all δ are isomorphisms and therefore for acyclic resolutions the map Δ, and consequently γ, is an isomorphism.

All flabby sheaves are acyclic (and even φ-acyclic sheaves for every family φ). This is an obvious consequence of the fact that the functor Γ_φ preserves exactness of short exact sequences that begin with a flabby sheaf (see §3.1 of Chap. 2 in Godement (1958)). Accordingly the quotient sheaf $\mathcal{A}/\mathcal{A}'$ is flabby provided that $\mathcal{A}$ and $\mathcal{A}'$ are flabby sheaves. In a similar way we can establish acyclicity of soft sheaves on paracompact (Hausdorff) spaces (see Godement (1958)).

2.4. Godement's Approach. Almost at the same time as Grothendieck, Godement gave another description of sheaf cohomology (see Godement (1958)). For an arbitrary sheaf $\mathcal{G}$, let $\mathcal{C}(\mathcal{G}) = \mathcal{C}^0(\mathcal{G})$ be the sheaf of germs of all (not only continuous) sections of the sheaf $\mathcal{G}$ (the sections in $\mathcal{C}(\mathcal{G})$ over an open set $U \subset X$ are defined as arbitrary maps $s : U \to \mathcal{G}$ for which $ps(x) = x$). It is clear that $\mathcal{C}(\mathcal{G})$ is flabby. Since $\mathcal{G} \subset \mathcal{C}^0(\mathcal{G})$, there exists a sheaf $\mathcal{C}^1(\mathcal{G}) = \mathcal{C}(\mathcal{C}^0(\mathcal{G})/\mathcal{G})$, and by obvious induction with respect to n we can also define the sheaves $\mathcal{C}^n(\mathcal{G})$ that form a resolution of $\mathcal{G}$. This resolution is called the *canonical flabby resolution* (or *Godement's resolution*). By definition $H_\varphi^n(X;\mathcal{G}) = H^n(\Gamma_\varphi(\mathcal{C}^*(\mathcal{G})))$. The equivalence of this definition with the one given in Sect. 2.1 follows from the fact that the resolution $\mathcal{C}^*(\mathcal{G}))$ is flabby.

The Godement resolution is convenient because not only it, but also (by the flabbiness of the sheaves $\mathcal{C}^n(\mathcal{G})$) the complexes of sections $\Gamma_\varphi(\mathcal{C}^*(\mathcal{G}))$ are an exact functor of $\mathcal{G}$. The usual properties of cohomology (such as functoriality with respect to $\mathcal{G}$, exactness of the cohomology sequence corresponding to an exact triple of coefficient sheaves, etc.) follow easily from the previous observation.

2.5. One More Description of the Comparison Homomorphism. Since the Godement resolution is a functor of $\mathcal{G}$, for each differential sheaf $\mathcal{A}^*$ we can define the *bigraded differential sheaf* $\mathcal{C}^*(\mathcal{A}^*) = \{\mathcal{C}^p(\mathcal{A}^q)\}$. Let us denote by d' and d'' its differentials induced by the differentials of $\mathcal{C}^*$ and $\mathcal{A}^*$, and by $d = d' + (-1)^p d''$ the *full differential* (the addition of the sign $(-1)^p$ assures that $d^2 = 0$). A standard diagram chasing proves that in the case when $\mathcal{A}^*$ is a resolution of $\mathcal{G}$ (see, for example, Lemma B.32 and Theorem B.32 in Schapira (1970)) the inclusion of $\mathcal{C}^*(\mathcal{G})$ into $\mathcal{C}^*(\mathcal{A}^0) \subset \mathcal{C}^*(\mathcal{A}^*)$ allows us to identify cohomology of X (with supports in φ) with cohomology of the *double* *complex* $\Gamma_\varphi(C^*(\mathcal{A}^*))$, considered with *full gradation* (in which the groups of n-dimensional cochains are the direct sums $\bigoplus_{p+q=n} \Gamma_\varphi(C^p(\mathcal{A}^q))$). In connection with the previous remarks, the inclusion $\mathcal{A}^*$ into $C^0(\mathcal{A}^*) \subset C^*(\mathcal{A}^*)$ defines the homomorphisms $\gamma' : H^n(\Gamma_\varphi(\mathcal{A}^*)) \to H^n_\varphi(X; \mathcal{G})$ (which are isomorphisms in the case when the sheaves $\mathcal{A}^q$ are acyclic).

Since the map $\mathcal{A}^* \to \mathcal{J}^*$ which defines γ can be extended to a map of bigraded sheaves $C^*(\mathcal{A}^*) \to C^*(\mathcal{J}^*)$ that is the identity on the differential subsheaf $C^*(\mathcal{G})$, it is true that $\gamma' = \gamma$ (see also the following paragraph).

We note that a similar argument can be carried out by considering instead of $C^*(\mathcal{A}^*))$ the injective bigraded differential sheaf $\mathcal{J}^*(\mathcal{A}^*)$. This last sheaf, however, is more difficult to construct.

§3. Axiomatic Characterization of the Comparison Homomorphism and Cohomology

3.1. Introductory Remarks. Above we illustrated several ways of comparing cohomology of a space X with cohomology groups $H^n(\Gamma_\varphi(\mathcal{A}^*))$ defined by a resolution. There exist also other constructions (for example, based on the concept of a spectral sequence, see §6 below). It is not always easy to establish directly whether two specific methods of comparison are equivalent (particularly that, as we have seen earlier, they can be different). Therefore it is natural to consider this question from a more general point of view. Similar remarks are also true with respect to cohomological functors.

By taking into account the fact that H^n_φ are derived functors of the zero-dimensional cohomology Γ_φ, we shall present the material of this paragraph in the language of derived functors F^n of a functor F (that is, we shall change the notation by replacing H^n_φ and Γ_φ with F^n and F, respectively).

3.2. Uniqueness of the Comparison Homomorphism. Since the functor F is left exact, it is true that $F^0 = F$ and $H^0(F(\mathcal{A}^*)) = F(\mathcal{G})$ for each resolution $\mathcal{A}^*$ of $\mathcal{G}$. The comparison homomorphism $\gamma : H^n(F(\mathcal{A}^*)) \to F^n(\mathcal{G})$ has the following properties: a) it is the identity in dimension zero; b) if a morphism $f : \mathcal{G} \to \mathcal{G}_1$ can be extended to a map $\tilde{f} : \mathcal{A}^* \to \mathcal{A}_1^*$ of resolutions, then $\gamma \tilde{f}_* = f_* \gamma$; c) for each short exact sequence $0 \to \mathcal{G}' \to \mathcal{G} \to \mathcal{G}'' \to 0$ of resolutions of objects that remains exact after the application of the functor F, it is true that $\gamma \tilde{\delta} = \delta \gamma$ ($\tilde{\delta}$ denotes the connecting homomorphism for cohomology of resolutions). These properties can be established with the aid of maps of the given resolutions into specially selected injective resolutions of these objects (in the case of cohomology groups H^n_φ one can apply the construction from Sect. 2.5 based on a consideration of Godement's resolutions).

There exists only one map γ with the above properties. To prove this, let $\mathcal{Z}^1$ be the kernel of the morphism $d : \mathcal{A}^1 \to \mathcal{A}^2$, and let $\mathcal{A}_1^*$ be the resolution of $\mathcal{Z}^1$ consisting of the terms of $\mathcal{A}^*$ beginning with the object $\mathcal{A}^1$. The exact triple of objects $0 \to \mathcal{G} \to \mathcal{A}^0 \to \mathcal{Z}^1 \to 0$ can be embedded into the exact sequence of resolutions $0 \to \mathcal{A}^* \to \mathcal{M}^* \to \mathcal{A}_1^* \to 0$. In the last sequence $\mathcal{M}^n = \mathcal{A}^n \oplus \mathcal{A}_1^n$ (where $\mathcal{A}_1^n = \mathcal{A}^{n+1}$), the embedding $\mathcal{A}^0 \to \mathcal{M}^0$ is defined by $a^0 \mapsto (a^0, da^0) \in \mathcal{M}^0$, and the maps $d : \mathcal{M}^n \to \mathcal{M}^{n+1}$ are described by the formula $d(a^n, a^{n+1}) = (da^n + (-1)^{n+1}a^{n+1}, da^{n+1})$. Clearly, the exactness is preserved by the functor F. By means of the differential d the resolution $\mathcal{M}^*$ is split at all its terms into direct sums, and therefore the complex $F(\mathcal{M}^*)$ is acyclic in positive dimensions and $H^0(F(\mathcal{M}^*)) = F(\mathcal{A}^0)$. Accordingly, we obtain the following commutative diagram with exact rows:

$$
\begin{array}{ccccccccc}
& & & & H^1(F(A^*)) & \to & 0 & & \\
& & & \nearrow & & & & & \\
\cdots \to & F(A^0) & \to & F(Z^1) & & & \downarrow \gamma & & \\
& & & \searrow & & & & & \\
& & & & F^1(\mathcal{G}) & \to & F^1(A^0) & \to \cdots & \\
& 0 \to & H^{n-1}(F(A_1^*)) & \to & H^n(F(A^*)) & \to 0 & & & \\
& & \downarrow \gamma & & \downarrow \gamma & & & & \\
\cdots \to F^{n-1}(A^0) & \to & F^{n-1}(Z^1) & \xrightarrow{\;\delta\;} & F^n(\mathcal{G}) & \to F^n(A^0) & \to \cdots & &
\end{array}
$$

from which the uniqueness of γ is easily established by induction with respect to the dimension n.

3.3. Universality Property. The method of comparison is not the only one applicable in the process of identifying some specific cohomology with sheaf cohomology. The following approach is also useful (see, for example, §2, Chap. 4 below).

Let Φ^n be an arbitrary *connected sequence of (cohomological) functors* (this means that, as in the case of $\{F^n\}$, to every exact short sequence $0 \to \mathcal{G}' \to \mathcal{G} \to \mathcal{G}'' \to 0$ there corresponds the exact sequence

$$
\cdots \to \Phi^n(\mathcal{G}') \to \Phi^n(\mathcal{G}) \to \Phi^n(\mathcal{G}'') \xrightarrow{\delta\Phi} \Phi^{n+1}(\mathcal{G}') \to \cdots,
$$

and for every transformation f of the first short exact sequence into the second one $0 \to \mathcal{F}' \to \mathcal{F} \to \mathcal{F}'' \to 0$ we have the identity $f_*\delta_\Phi = \delta_\Phi f_*$). It is natural to ask to what extent is the sequence $\{\Phi^n\}$ different from the sequence of derived functors $\{F^n\}$. Obviously, we shall assume that $\Phi^n = 0$, for $n < 0$ and that $\Phi^0 = F$.

It is true that there exists a unique natural transformation of connected sequences of functors $\mu : F^n(\mathcal{G}) \to \Phi^n(\mathcal{G})$ that is the identity in dimension zero (the *property of universality of derived functors F^n*). The transformation μ is an isomorphism if and only if the functors Φ^n, for $n > 0$, vanish on F-acyclic objects.

Recall that *F-acyclicity* of an object $\mathcal{C}$ means that $F^n(\mathcal{C}) = 0$ for $n > 0$. It is clear that the condition $\Phi^n(\mathcal{C}) = 0$, for $n > 0$ on F-acyclic objects is necessary. The claim remains valid also in the case when $\Phi^n(\mathcal{C}) = 0$ for $n > 0$ on injective objects only.

In order to prove this, consider the exact sequence $0 \to \mathcal{G} \to \mathcal{C} \to \mathcal{G}' \to 0$ in which $\mathcal{C}$ is F-acyclic (for example, injective), and compare the resulting exact sequences of cohomological functors F^n and Φ^n. The first consists of the isomorphisms $\delta : F^n(\mathcal{G}') \to F^{n+1}(\mathcal{G})$, for $n \geq 1$, and from its initial terms we obtain the description of F^1 as the subfunctor of the functor Φ^1, since $\Phi^0 = F$ and $F^1(\mathcal{G})$ is the quotient of $F(\mathcal{G}')$ with respect to the image of $F(\mathcal{C})$. This implies (by induction with respect to n) the uniqueness of μ and also the fact that μ is an isomorphism when $\Phi^n(\mathcal{C}) = 0$ for $n > 0$ on F-acyclic objects $\mathcal{C}$.

The morphism $f : \mathcal{G} \to \mathcal{F}$ typically can be extended to the morphism $\mathcal{C}(\mathcal{G}) \to \mathcal{C}(\mathcal{F})$ of F-acyclic objects (for example, injective, canonical flabby Godement sheaves in sheaf category, etc.) containing the objects $\mathcal{G}$ and $\mathcal{F}$. Accordingly, the transformation μ constructed earlier commutes with the induced transformations f_* of cohomological functors. For that reason the definition of μ does not depend on the choice of the embedding of $\mathcal{G}$ into an F-acyclic object. For an embedding of $\mathcal{G}$ into $\mathcal{C}'$ and into $\mathcal{C}''$ there exists an embedding of $\mathcal{G}$ into $\mathcal{C}' \oplus \mathcal{C}''$ extending the first two embeddings and the problem is reduced to a simple verification of the claim in the case when the identity morphism of $\mathcal{G}$ can be extended to a morphism $\mathcal{C}' \to \mathcal{C}''$.

It remains to show that μ commutes with the connecting homomorphisms δ and δ_Φ. Let $0 \to \mathcal{G}' \to \mathcal{G} \to \mathcal{G}'' \to 0$ be an arbitrary short exact sequence of objects. After embedding $\mathcal{G}$ into an F-acyclic object $\mathcal{C}$, consider its map f to a short exact sequence $0 \to \mathcal{G}' \to \mathcal{C} \to \mathcal{F} \to 0$. For every n we obtain the following diagram

$$
\begin{array}{ccccc}
F^{n-1}(\mathcal{G}'') & & \xrightarrow{\ \ \delta\ \ } & & F^n(\mathcal{G}') \\
& \searrow^{f_*} & & \nearrow^{\delta} & \\
& & F^{n-1}(\mathcal{F}) & & \\
\Big\downarrow{\mu} & & \Big\downarrow{\mu} & & \Big\downarrow{\mu} \\
& & \Phi^{n-1}(\mathcal{F}) & & \\
& \nearrow^{f_*} & & \searrow^{\delta_\Phi} & \\
\Phi^{n-1}(\mathcal{G}'') & & \xrightarrow{\ \delta_\Phi\ } & & \Phi^n(\mathcal{G}')
\end{array}
$$

Commutativity of the outside square in this diagram is a consequence of the commutativity of two trapezoids and the two triangles inside it, which in turn is a result of the definition of μ for an embedding of $\mathcal{G}'$ into an F-acyclic object $\mathcal{C}$, the properties of μ established earlier and other simple considerations.

§4. Cohomology of Subspaces and Pairs.
The Role of Supports

4.1. Cohomology of Subspaces. Let Y be a subspace of X and let $\mathcal{A}^*$ be an acyclic resolution of a sheaf $\mathcal{G}$ on X. Its restriction $\mathcal{A}^*|_Y$ to Y is a resolution of the sheaf $\mathcal{G}|_Y$. The comparison homomorphisms γ for $\mathcal{A}^*$ and $\mathcal{A}^*|_Y$ (represented, for example, as iterated connecting homomorphism Δ) are compatible with the operation of *restriction of sheaves* (and *sections*) to Y. Accordingly the restriction of sections of $\mathcal{A}^*$ to Y together with the transformation γ for $\mathcal{A}^*|_Y$ define a homomorphism of cohomology $i^* : H^n(X; \mathcal{G}) \to H^n(Y; \mathcal{G})$ which is induced by the inclusion $i : Y \subset X$ (here $H^n(Y; \mathcal{G}) = H^n(Y; \mathcal{G}|_Y)$). The independence of i^* of the choice of $\mathcal{A}^*$ follows, for example, from considering homomorphisms of $\mathcal{A}^*$ into injective resolutions $\mathcal{J}^*$ (for which the analogous claim can be checked routinely). When $\mathcal{A}^* = \mathcal{C}^*(\mathcal{G})$ is the Godement resolution, the homomorphism γ (as is easily seen from its naturality, see Sect. 3.2 above) is induced by the map $v : \mathcal{C}^*(\mathcal{G})|_Y \to \mathcal{C}^*(\mathcal{G}|_Y)$ of sheaves on Y, which can easily be defined by induction in each dimension (without any restrictions on $X, Y, \mathcal{G}$, see Sect. 4.9, Chap. 2 in Godement (1958).

In order to define cohomology of Y it is useful to know when $\mathcal{A}^*|_Y$ is an acyclic resolution of $\mathcal{G}|_Y$ (or when γ is an isomorphism for $\mathcal{A}^*|_Y$). It is clear that a space X with trivial cohomology can contain subspaces with nontrivial cohomology and therefore the restriction of an acyclic sheaf to a subspace does not have to be acyclic. This depends on the properties of X and Y, as well as on the nature of the sheaf itself. Restrictions of flabby sheaves (and resolutions) to Y are flabby for open Y (obvious) and for every Y if X is hereditarily paracompact (see Sect. 3.3, Chap. 2 in Godement (1958). Restrictions of soft sheaves (in particular, also flabby sheaves, if X is paracompact) are soft for every closed Y (obvious), and if X is hereditarily paracompact also (as follows, for example, from Sect. 3.4, Chap. 2 in Godement (1958) for arbitrary locally closed subspaces Y (in particular, open subspaces), even though the restriction of sections to Y in this case is not an epimorphism (contrary to the previously considered situations).

4.2. Cohomology of Pairs of Spaces. It can be shown that the construction of $v : \mathcal{C}^*(\mathcal{G})|_Y \to \mathcal{C}^*(\mathcal{G}|_Y)$ presented earlier insures that $\kappa : \Gamma(\mathcal{C}^*(\mathcal{G})) \to \Gamma(\mathcal{C}^*(\mathcal{G}|_Y))$ is an epimorphism (moreover, every section of $\mathcal{C}^*(\mathcal{G}|_Y)$ with support in $F \subset Y$ is the image of a section of the sheaf $\mathcal{C}^*(\mathcal{G})$ with support in the closure of F in X). It is natural to define the *cohomology groups of the pair* $H^n(X, Y; \mathcal{G})$ as $H^n(\mathrm{Ker}\,\kappa)$. Accordingly, in every case we obtain the following *exact cohomology sequence of a pair.*

$$\ldots \to H^n(X, Y; \mathcal{G}) \to H^n(X; \mathcal{G}) \to H^n(Y; \mathcal{G}) \to H^{n+1}(X, Y; \mathcal{G}) \to \ldots \quad (2)$$

In the cases when the resolutions $\mathcal{A}^*$ and $\mathcal{A}^*|_Y$ are acyclic and the restriction of sections of $\mathcal{A}^*$ to Y is an epimorphism (see Sect. 4.1), the exact sequence (2) can be defined by considering the short exact sequence

$0 \to \Gamma_\psi(\mathcal{A}^*) \to \Gamma(\mathcal{A}^*) \to \Gamma(\mathcal{A}^*|_Y) \to 0$ (ψ is a family of closed subsets of X contained in $X \setminus Y$). Typically, $\mathcal{A}^*$ is a flabby resolution and since flabby sheaves are φ-acyclic for every family φ (see Sect. 2.3), by the very definition of cohomology of a pair we obtain $H^n(X, Y; \mathcal{G}) = H^n_\psi(X; \mathcal{G})$. In the case when Y is closed in X the same argument can be made also by considering soft resolutions $\mathcal{A}^*$ (see Sect. 4.4 below). It is obvious that in this case $H^n(X, Y; \mathcal{G}) = H^n_\psi(X \setminus Y; \mathcal{G})$. The conclusion agrees fully with the observations we made for simplicial cohomology (the end of Sect. 1.5, Chap. 1).

It is clear that the groups $H^n(X, Y; \mathcal{G})$, interpreted as cohomology groups of the space $X \setminus Y$ (or X) with supports in the family ψ, are different from the ordinary cohomology groups of $X \setminus Y$ (or X) and depend on the topology of $X \setminus Y$, as well as on the embedding $X \setminus Y \subset X$ (which determines the family ψ). For example, in the case of the n-dimensional ball B^n and its boundary S^{n-1}, we have $H^n(B^n, S^{n-1}) = H^n_c(B^n \setminus S^{n-1}) \neq 0 = H^n(B^n \setminus S^{n-1})$.

4.3. Cohomology with Supports. It is necessary to consider cohomology of a subspace also in the case when the cohomology of X is determined by supports in some family φ (frequently with $\varphi = c$). Since the map κ described above carries the sections in $\Gamma_\varphi(\mathcal{C}^*(\mathcal{G}))$ to $\Gamma_{\varphi \cap Y}(\mathcal{C}^*(\mathcal{G}|_Y))$, where $\varphi \cap Y$ is the family of all $F \cap Y$ with $F \in \varphi$, it is natural to compare $H^*_\varphi(X; \mathcal{G})$ with $H^*_{\varphi \cap Y}(Y; \, sg)$. Exactly in the same way as at the beginning of Sect. 4.2 we obtain the following *exact cohomology sequence of the pair* (X, A).

$$\ldots \to H^n_\varphi(X, Y; \mathcal{G}) \to H^n_\varphi(X; \mathcal{G}) \to H^n_{\varphi \cap Y}(Y; \mathcal{G}) \to H^{n+1}_\varphi(X, Y; \mathcal{G}) \to \ldots \tag{3}$$

If $\mathcal{A}^*$ and $\mathcal{A}^*|_Y$ are flabby (for example, in the case of a hereditarily paracompact space X) and if the family φ is paracompactifying, then the restriction $\Gamma_\varphi(\mathcal{A}^*) \to \Gamma_{\varphi \cap Y}(\mathcal{A}^*|Y)$ is an epimorphism, the groups $H^n_\varphi(X, Y; \mathcal{G})$ are determined by its kernel (see Sect. 4.2) and are equal to $H^n_\psi(X; \mathcal{G})$, where $\psi = \varphi|_{X \setminus Y}$ consists of all $F \in \varphi$ for which $F \subset X \setminus Y$.[22] A similar conclusion is true for closed $Y \subset X$ and for soft resolutions of $\mathcal{A}^*$, see Sect. 4.4 below, and also Sect. 5.8 of Chap. 8. In this case $H^n_\varphi(X, Y; \mathcal{G}) = H^n_\psi(X \setminus Y; \mathcal{G})$.

In the case when Y is closed and $\varphi = c$, the symbols $\varphi \cap Y$ and $\varphi|_{X \setminus Y}$ can be replaced by c in the groups described above. In particular, for locally compact spaces we obtain still another confirmation of the results of §4 of Chap. 2 (about the isomorphism $H^n_c(X, Y; G) = H^n_c(X \setminus Y; G)$, about the identity of the groups $H^n_c(X; G)$ with the cohomology groups of the one-point compactification of X, etc.). If Y is an open subset of a locally compact space X, then certainly $H^n_{c \cap Y}(Y; \mathcal{G}) \neq H^n_c(Y; \mathcal{G})$. If φ is the family of all closed subsets of X, then for every Y the symbols φ and $\varphi \cap Y$ in the sequence (3) are omitted (see sequence (2)).

---

[22] For homology analogous operations lead to the groups of the pair $(X, X \setminus Y)$ and of the subspace $X \setminus Y$, respectively, see Sect. 4.3, Chap. 5 and Sect. 5.4, Chap. 8.

4.4. The Case of Closed Subspaces. We have seen that the cohomology of a pair (X, Y) can be treated as the cohomology of a single space $X \setminus Y$ with a special family of supports. We can accomplish more, namely, all terms in the sequence (3) can be interpreted as the cohomology of the space X itself with the same family of supports, but with special coefficient sheaves. To this end, let $\mathcal{A}_{X \setminus Y}$ be the subsheaf of the sheaf $\mathcal{A}$ that is the union of the restriction of $\mathcal{A}$ to $X \setminus Y$ with the zero section of $\mathcal{A}$ over X, and $\mathcal{A}_Y = \mathcal{A}/\mathcal{A}_{X \setminus Y}$. Consider the following short exact sequence of resolutions

$$0 \to \mathcal{A}_{X \setminus Y}^* \to \mathcal{A}^* \to \mathcal{A}_Y^* \to 0.$$

Recall that a sheaf $\mathcal{A}$ is called φ-*soft* if the sections of $\mathcal{A}$ with supports in φ, defined over any closed subset $Y \subset X$, can be extended to sections with supports in φ over all of X. In the case of a paracompactifying family this is equivalent to the requirement that the restriction homomorphism $\Gamma_\varphi(\mathcal{A}) \to \Gamma(\mathcal{A}|_F)$ (equivalently $\Gamma(\mathcal{A}) \to \Gamma(\mathcal{A}|_F)$) be an epimorphism for every $F \in \varphi$ (and consequently, to the requirement that the sheaves $\mathcal{A}|_F$ be soft for all $F \in \varphi$, see §3.5, Chap. 2 in Godement (1958)). For every family of supports we have $\Gamma_\varphi(\mathcal{A}_{X \setminus Y}) = \Gamma_{\varphi|_{X \setminus Y}}(\mathcal{A}|_{X \setminus Y})$, and therefore φ-softness of $\mathcal{A}$ implies φ-softness of $\mathcal{A}_{X \setminus Y}$. We note also that all sections of $\mathcal{A}_Y$ over any open set U (among them also sections with supports in φ) are completely determined by the sections of $\mathcal{A}|_Y$ over the set $U \cap Y$ (by the sections with supports in φ, respectively), and therefore φ-softness of $\mathcal{A}$ implies φ-softness of both $\mathcal{A}|_Y$ and $\mathcal{A}_Y$ as well as the exactness of the sequence of sections $0 \to \Gamma_\varphi(\mathcal{A}_{X \setminus Y}) \to \Gamma_\varphi(\mathcal{A}) \to \Gamma_\varphi(\mathcal{A}_Y) \to 0$. Exactly in the same way as in Sects. 2.3 and 1.3 above, we can establish φ-acyclicity of any φ-soft sheaves and φ-softness of flabby (and soft) sheaves for paracompactifying families φ. For every family φ the exact sequence (3) coincides with the cohomology sequence of the space X

$$\to H_\varphi^n(X; \mathcal{G}_{X \setminus Y}) \to H_\varphi^n(X; \mathcal{G}) \to H_\varphi^n(X; \mathcal{G}_Y) \to H_\varphi^{n+1}(X; \mathcal{G}_{X \setminus Y}) \to \dots \quad (4)$$

corresponding to the short exact sequence of coefficient sheaves

$$0 \to \mathcal{G}_{X \setminus Y} \to \mathcal{G} \to \mathcal{G}_Y \to 0.$$

For paracompactifying φ (according to the remarks made above) we have
$$H_\varphi^n(X; \mathcal{G}_{X \setminus Y}) = H_{\varphi|_{X \setminus Y}}^n(X; \mathcal{G}) = H_{\varphi|_{X \setminus Y}}^n(X \setminus Y; \mathcal{G}).$$

§5. Comparison with Concrete Theories. Typical Examples of Resolutions

5.1. General Observations. Assume that a presheaf of cochains $U \to A^*(U)$ satisfying condition (S2) induces an acyclic resolution $\mathcal{A}^*$ of some sheaf $\mathcal{G}$, K^* is the kernel of the epimorphism $j : A(X) \to \mathcal{A}^*(X)$ and $\bar{A}^*$ is the cochain complex for which $\bar{A}^n = A^n(X) \oplus K^{n+1}$ and whose coboundary d is defined as $d(a, k) = (da + k, -dk)$. In the general case, $H^n(X; \mathcal{G}) = H^n(\mathcal{A}^*(X)) = H^n(\bar{A}^*)$. Indeed, the kernel $\bar{K}^*$ of the natural epimorphism $\bar{A}^* \to \mathcal{A}^*(X)$ formed by the groups $\bar{K}^n = K^n \oplus K^{n+1}$ is an acyclic complex because the coboundary operator in $\bar{K}^*$ induced by the inclusion $\bar{K}^* \subset \bar{A}^*$ is of the form $d(k^n, k^{n+1}) = (dk^n + k^{n+1}, -dk^{n+1})$ and splits $\bar{K}^*$ (compare with the resolution $\mathcal{M}^*$ in Sect. 3.2). Accordingly, $H^n(X; \mathcal{G}) = H^n(A^*(X))$ only in the case when the complex K^* is acyclic (compare with Sect. 5.2), and for acyclic $A^*(U)$ (compare Sect. 5.3) we have $H^p(X; \mathcal{G}) = H^{p+1}(K^*)$.

5.2. Singular Theory. Acyclicity of differential *sheaves of singular cochains* $\mathcal{S}^*$ follows from the "global" conditions (see Sect. 1.5) of paracompactness of X. In accordance with Sect. 1.3, the sheaves $\mathcal{S}^*$ are soft if X is paracompact and flabby if X is hereditarily paracompact. For a paracompact X there exists an epimorphism $j : C_s^*(X; G) \to \Gamma(\mathcal{S}^*)$ (see the end of Sect. 1.2). Note that $j(\xi) = 0$ if and only if ξ vanishes on singular simplices inscribed in some open cover α, and therefore the complex $\mathrm{Ker}\, j$ is the union with respect to all α of acyclic complexes $\mathrm{Ker}\, j_\alpha$ (see Sect. 2.5 Chap. 1) and consequently it is acyclic. As a matter of fact $\Gamma(\mathcal{S}^*)$ are the *localized singular cochains*.

This, however, does not imply yet that $H_s^n(X; G) = H^n(X; G)$[23], since the differential sheaves $\mathcal{S}^*$ in general do not form a resolution of G. The local conditions found in Sect. 1.4 should be satisfied. Namely, for every neighborhood U of every point $x \in X$, and every $h \in H_s^n(U; G)$ there should exist a neighborhood $V \subset U$ of this point such that the image of h in $H_s^n(V; G)$ is equal to zero. As always, for $n = 0$ the cohomology group is assumed to be reduced (this condition means that every point in V can be joined with x by a path in U). These conditions are satisfied when X is *weakly locally contractible* (for every neighborhood U of a point x there exists a neighborhood $V \subset U$ which is contractible in U to x) or homologically locally connected in the sense of the singular theory (see Chap. 8).

Thus if X is paracompact and if it satisfies the local conditions formulated above, then the groups $H_s^n(X; G)$ are isomorphic to sheaf cohomology. For the same reason it is also true for every paracompact open $Y \subset X$ as well as for $H_s^n(X, Y; G)$. The same assertion is also true for the cohomology of a pair (X, Y) for closed subsets Y that satisfy (together with X) the local conditions formulated above. Let $C_l^*(Y; G)$ be the complex of localized singular cochains of Y. Since Y has a fundamental system of paracompact neighborhoods W, we have $\varinjlim_W C_l^*(W; G) = \varinjlim_W \Gamma(\mathcal{S}^*|W) = \Gamma(\mathcal{S}^*|_Y)$, and the maps $C_l^*(W; G) \to C_l^*(Y; G)$ define a homomorphism of cochain complexes $\Gamma(\mathcal{S}^*|_Y) \to C_l^*(X; G)$ inducing the isomorphism of cohomology of

---

[23] Sheaf cohomology, together with Čech cohomology (Chap. 4) is free from the deficiences mentioned in Sect. 1.1, Chap. 2 (in particular, $H^n(X; \mathcal{G}) = 0$, for $n > \dim X$ and therefore examples of the type presented in Barrat (1962), Barrat and Milnor (1962) in this case are impossible).

Y (because of the conditions imposed on Y). The singular cohomology of the pair (X, Y) is determined by the kernel $C_l^*(X, Y; G)$ of the epimorphism $\Gamma(\mathcal{S}^*) = C_l^*(X; G) \to C_l^*(Y; G)$. By the so called "five lemma" the inclusion $\Gamma(\mathcal{S}_{X\backslash Y}^*) \to C_l^*(X, Y; G)$ induces an isomorphism of cohomology and this proves the earlier assertion.

5.3. Alexander-Spanier Cohomology. For the presheaf $U \to A^*(U)$ of Alexander-Spanier cochains (non localized) we have $H^n(A^*(U)) = 0$ for $n > 0$ and $H^0(A^*(U))$ is the group of locally constant functions from U to the coefficient group G (see Sect. 3.1, Chap. 2). Therefore, in accordance with Sect. 1.4, the *Alexander-Spanier sheaves* $\mathcal{A}^*$ generated by them form a resolution of G (for every topological space X).

The remaining observations are similar to the ones made in the previous section. A cochain $\xi \in A^n(U)$ is equivalent to the zero cochain (that is, equal to zero as a localized cochain) if its restrictions to the elements of some open cover of U are equal to zero. Therefore, in accordance with Sects. 1.1-1.3, for a paracompact X the sheaves $\mathcal{A}^*$ are always soft, and in the case of a hereditarily paracompact space they are flabby. As in the previous section the sections of $\mathcal{A}^*$ are *localized cochains* (of Alexander-Spanier). Thus Alexander-Spanier cohomology of a paracompact space X is isomorphic with sheaf cohomology.

5.4. De Rham Theorem. Quite frequently (in particular, in the areas of differential geometry, functional and complex analysis, etc.) cohomology groups find their natural representation in terms of objects closely related to a particular area of study but having very little in common (at least superficially) with cochains in the usual sense. However, as with cochains, these objects (primarily the differential forms and their generalizations) typically generate resolutions, so that their connection with cohomology should not seem to be so unusual.

Let X be a smooth n-dimensional manifold. By assigning to each open subset $U \subset X$ the collection of all differential forms of degree k on U we define a presheaf which is the *sheaf Ω^k of germs of differential forms*. The operation d of exterior differentiation introduces the structure of a differential sheaf in Ω^* formed by all Ω^k. According to Poincaré lemma, for a differential form ω of degree k, it follows from the equality $d\omega = 0$ that locally $\omega = d\bar{\omega}$ for $k > 0$ and that ω is a locally constant function for $k = 0$. Accordingly, Ω^* is a resolution of the constant sheaf $\mathbb{R}$ (the symbol $\mathbb{R}$ denotes the field of real numbers).

In order to establish acyclicity of Ω^* we shall consider the concept of a fine sheaf (such sheaves are useful in questions related to analysis). A sheaf $\mathcal{A}$ is called *fine* if for any two disjoint closed subsets A and B there exists an automorphism of $\mathcal{A}$ that is the identity on A and equal to zero on B. A sheaf $\mathcal{A}$ is called *φ-fine* if for every $F \in \varphi$ the sheaf $\mathcal{A}|_F$ is fine. Every φ-fine sheaf is clearly φ-soft. One of the useful properties of φ-fine sheaves is the following: if $\mathcal{A}$ is φ-fine and $\mathcal{B}$ is arbitrary, then the sheaf $\mathcal{A} \otimes \mathcal{B}$ is φ-fine.

Since for any two disjoint closed subsets A, B there exists a smooth function equal to 1 on A and vanishing on B and since this function can be multiplied by any other function, the sheaf Ω^0 of germs of smooth functions is fine. Since the sheaves Ω^k are Ω^0-modules, they are also fine (and moreover, φ-fine for every family φ). Accordingly, for every paracompactifying family φ the resolution Ω^* is φ-acyclic and therefore $H^k(\Gamma_{varphi}(\Omega^*)) = H_\varphi^k(X; \mathbb{R})$ (*de Rham's theorem*). Therefore the answer to the question whether for a given k-form ω the equation $\omega = d\tilde{\omega}$ is satisfied (with the necessary condition $d\omega = 0$) depends only on the topological structure of X (or the domain of ω). Let us point out that it is possible to express in terms of differential forms some specific tensors arising in geometry, mechanics and physics, while some differential equations can be interpreted in terms the operation of exterior differentiation.

5.5. Dolbeault's Cohomology. An analogue of De Rham's theorem holds also for complex-valued differential forms on complex manifolds. Let $\mathcal{E}^*(X) = \bigoplus_r \mathcal{E}^r(X)$ be the algebra of exterior complex-valued differential forms on X (with partial derivatives of all orders with respect to all $2n$ real variables). Since each such form of degree r can be locally expressed as a linear combination of the forms of the type $dz^{i_1} \wedge \ldots \wedge dz^{i_p} \wedge d\bar{z}^{j_1} \wedge \ldots \wedge d\bar{z}^{j_q}, p + q = r$, the space $\mathcal{E}^r(X)$ of such forms can be represented as the direct sum $\mathcal{E}^r(X) = \bigoplus_{p+q=r} \mathcal{E}^{p,q}(X)$ of differential forms of bidegree (p, q). The exterior derivative $d : \mathcal{E}^r(X) \to \mathcal{E}^{r+1}(X)$ of each homogeneous term can be represented as the sum $d = \partial \oplus \bar{\partial}$ of the derivatives $\partial : \mathcal{E}^{p,q}(X) \to \mathcal{E}^{p+1,q}(X), \bar{\partial} : \mathcal{E}^{p,q}(X) \to \mathcal{E}^{p,q+1}(X)$, and it follows from $d^2 = 0$ that $\partial^2 = \bar{\partial}^2 = \partial\bar{\partial} + \bar{\partial}\partial = 0$.

The space of (p, q)-forms $\mathcal{E}^{p,q}(X)$ on an n-dimensional complex manifold X coincides with the space of sections of sheaves of germs of (p, q)-forms $\mathcal{E}^{p,q}$. For a fixed $p \geq 0$ we obtain a differential sheaf

$$0 \to \Omega^p \to \mathcal{E}^{p,0} \xrightarrow{\partial} \mathcal{E}^{p,1} \to \ldots \xrightarrow{\partial} \mathcal{E}^{p,n} \to 0,$$

where the kernel Ω^p of the operator $\bar{\partial}$ is the sheaf of holomorphic differential forms of the type $(p, 0)$ which are traditionally called the holomorphic forms of degree p ($\omega \in \Omega^p(U)$ if and only if ω in local coordinates can be expressed in terms of the differentials $dz^{i_1} \wedge \ldots \wedge dz^{i_p}$). According to an analogue of the Poincaré lemma for the operator $\bar{\partial}$, proved by Grothendieck, the differential sheaf $\mathcal{E}^{p,*}$ is the (*Dolbeault-Grothendieck*) *resolution* of the sheaf Ω^p. Since it consists of *fine* sheaves, we have $H^q(\Gamma_\varphi(\mathcal{E}^{p,*})) = H_\varphi^q(X; \Omega^p)$ for every paracompactifying family φ (*Dolbeault's theorem*, see Wells (1973)).

There exists analogues of the theorems of Dolbeault and Grothendieck for differential forms with values in vector bundles and consequently in holomorphic vector bundles over manifolds.

5.6. Other Examples. It is true that the Dolbeault-Grothendieck resolutions of the sheaves Ω^p exist also in the case of forms (of bidegree (p, q)) in whose representations the coefficients are analytic functions on $\mathbb{R}^{2n} = \mathbb{C}^n$, distributions in the sense of Schwartz (see Hörmander (1973)), and also hyperfunctions in the sense of Sato (see Schapira (1970)). Such resolutions, because they posses many unique properties, find many applications in analysis. The last type plays an important role in the proof of *Sato's theorem* which provides one of their fundamental interpretations: if U is an open subset in $\mathbb{R}^n \subset \mathbb{C}^n$, then the cohomology $H_U^p(\mathbb{C}^n; \mathcal{O})$ (with supports in U) with coefficients in the sheaf of germs of holomorphic functions on $\mathbb{C}^n$ (complexification of $\mathbb{R}^n$) are equal to zero for $p \neq n$, and for $p = n$ determine the presheaf $U \to H_U^n(\mathbb{C}^n, \mathcal{O})$ that coincides with the sheaf of hyperfunctions on $\mathbb{R}^n$ (see Schapira (1970)). This sheaf is flabby.

Whenever a topological space X is equipped with additional structure (when it is a manifold or a polyhedron, or when it is locally compact, etc.) it is possible , in most cases, to construct special resolutions of cochains taking into account such structure. Therefore, on a manifold, in addition to the standard singular cochains, it is possible to consider cochains that are defined only on smooth singular simplices (on simplices that are homeomorphic or diffeomorphic to a standard simplex, etc.) and such cochains (after localization) form acyclic resolutions. As was mentioned earlier (§4, Chap. 2), on a locally compact space there exist free cochains with compact supports (with coefficients in the ring of integers or some other rings). They turn out to be the sections (with compact supports) of soft resolutions (see Sklyarenko (1969)). There exist many different versions of Alexander-Spanier cochains (compare with Sect. 3.2, Chap. 2) that also form acyclic resolutions.[24]

§6. Basic Spectral Sequences. Behavior of Cohomology with Respect to Continuous Maps

6.1. Main Types of Spectral Sequences. The application of the section functor Γ_φ to the bigraded differential sheaf $\mathcal{C}^*(\mathcal{A}^*)$ or equivalently to the injective bigraded sheaf $\mathcal{J}^*(\mathcal{A}^*)$ (see Sect. 2.5) yields a double complex, which in turn (as every double complex of such type) determines two spectral sequences, each convergent to the cohomology of this complex (see Sect.4.5, Chap. 2 in Godement (1958)). The second term has the form $E_2^{pq} = H^p(H_\varphi^q(X; \mathcal{A}^*))$ for the first sequence and $E_2^{pq} = H_\varphi^p(X; \mathcal{H}^q(\mathcal{A}^*))$ for the second (here $\mathcal{H}^q(\mathcal{A}^*)$ denotes the cohomology of the complex $\mathcal{A}^*$ in the category of sheaves). In terms and notation of the theory of derived functors (see Sect. 3.1 above) the second term of these spectral sequences has the form $E_2^{pq} = H^p(F^q(\mathcal{A}^*))$, and $E_2^{pq} = F^p(H^q(\mathcal{A}^*))$, respectively.

Two situations are most typical. In the first, $\mathcal{A}^*$ is a resolution of an object $\mathcal{G}$. It is clear that in this case we have $H^q(\mathcal{A}^*) = 0$ for $q > 0$ and therefore the first spectral sequence is convergent to $F^{p+q}(\mathcal{G})$ (to $H_\varphi^{p+q}(X; \mathcal{G})$ in

---

[24] "Standard" resolutions of different nature, both acyclic and not, are a typical phenomenon in homological algebra.

the case of cohomology). In the second case $\mathcal{A}^*$ consists of acyclic objects, $F^q(\mathcal{A}^*) = 0$ for $q > 0$ and therefore the second spectral sequence is convergent to $H^{p+q}(F(A^*))$ (to $H^{p+q}(\Gamma_\varphi(\mathcal{A}^*))$ in the case of cohomology). In both situations the nature of the term E_2 is typically more precisely determined by taking into consideration the additional conditions assumed in the actual problems.

We shall present here some examples of the most typical resolutions. The resolution of the constant sheaf $\mathbb{C}$ ($\mathbb{C}$ is the field of complex numbers) consists of the sheaves Ω^p, considered in the previous paragraph, of germs of holomorphic forms of degree p (see §2, Chap. 2 in Wells (1973)). Below (Sect. 3.2, Chap. 4) it will be shown, that resolutions of a coefficient sheaf $\mathcal{G}$ and the associated spectral sequences (of Leray) correspond to covers of a space. See also the spectral sequence in §4, Chap. 4. Resolutions and associated spectral sequences arise from certain continuous maps and contain information about the behavior of cohomology under those maps as well as about the structure of such maps.

The most general resolutions of such type were constructed in Zarelua (1969) for closed, finite-to-one maps $f : X \to Y$, later for arbitrary closed maps by G.S. Skordev, together with simplifications and a description of some of its typical properties and cohomological operations (see Skordev (1970a, 1971)), and finally by A.V. Zarelua (see Zarelua (1977)) for arbitrary continuous maps. This technique is of greatest interest in the case of zero-dimensional maps (that is, in the situation when the Leray spectral sequence considered below, being degenerate, does not in itself contain any information). In this class it allows a characterization of proper maps and open proper maps of spaces (see Zarelua (1983, 1979)) and yields theorems about dimension raising maps (see Zarelua (1969)) (the proofs not making use of spectral sequences were obtained later in Skordev (1970b). In the class of proper open maps it is possible to obtain characterizations of zero-dimensional and monotone maps (see Zarelua (1979)). The general constructions of the *resolution in the sense of Zarelua* was given by G.S. Skordev (see Skordev (1982)). Skordev used this construction (see Skordev (1983)) to give a thorough analysis of the spectral sequence corresponding to such a resolution. Special cases of this construction include the spectral sequences of a cover mentioned above (see Zarelua (1977)), the *Cartan's spectral sequence of a regular cover* (Zarelua (1972)) and its generalization, the *Borel spectral sequence of a semifree action* of a finite group on a topological space (Skordev (1983)) and spectral sequences (of Leray) corresponding to covers of the space (Zarelua (1977)). By using the same approach in considering maps of classifying spaces corresponding to pairs of groups $H \subset G$, we can obtain a natural description of the *Snapper spectral sequence* which establishes a connection between the cohomology of a finite group G and cohomology of its arbitrary subgroup H. It also reveals its connection with the *Hochschild-Serre spectral sequence* arising in the case when H is a normal subgroup of G (for a survey of results of this sort see Skordev (1983)). All spectral sequences corresponding to resolutions give rise to standard maps $H^p(F(\mathcal{A}^*)) = E_2^{p0} \to F^p(\mathcal{G})$ leading to a new description of the comparison homomorphism.

Both in topology and in homological algebra there exists a variety of examples of $\mathcal{A}^*$ in which the acyclicity of $\mathcal{A}^n$ is never in doubt, but for which the sequence $\mathcal{A}^*$ is not exact (the second typical situation). In this way we obtain the Leray spectral sequence (see below) and its generalizations, the *Farey spectral sequence* (Bredon (1967), the *Borel spectral sequence in the theory of G-spaces* (Bredon (1967), the *Cartan spectral sequence corresponding to a differential sheaf of chains* on a topological space ($\S 4$, Chap. 5). See also the examples in Sect. 1.5, Chap. 8. In such spectral sequences $E_2^{p0} = F^p(\mathcal{G})$ ($\mathcal{G}$ is the kernel of the morphism $\mathcal{A}^0 \to \mathcal{A}^1$), and therefore there exist maps $\nu : F^p(\mathcal{G}) \to H^p(F(\mathcal{A}^*)))$ (in topology $\nu : H^p_\varphi(X; \mathcal{G}) \to H^p(\Gamma_\varphi(\mathcal{A}^*)))$ analogous to the comparison homomorphism γ. Like γ, the homomorphisms ν have numerous different descriptions (which, however, are not well represented in the literature).

6.2. Leray Spectral Sequence. Let $f : X \to Y$ be a continuous map and let $\mathcal{A}$ be a sheaf on X. The presheaf $f\mathcal{A}$ on Y described by the relation $f\mathcal{A}(U) = \mathcal{A}(f^{-1}U)$ satisfies conditions (S1) and (S2) of Sect. 1.1 and accordingly, it is a sheaf. It is called the *direct image of the sheaf* $\mathcal{A}$ under the map f. Since the sections of $f\mathcal{A}$ over U are the same as the sections of $\mathcal{A}$ over $f^{-1}U$, the direct image of a flabby sheaf is flabby.

Assume, furthermore, that $0 \to \mathcal{G} \to \mathcal{C}^*$ is an arbitrary flabby resolution of $\mathcal{G}$ on X. Then the differential sheaf $f\mathcal{C}^*$ consists of flabby sheaves on Y. Consequently, there exists the *(Leray) spectral sequence* for which $E_2^{p,q} = H^p(Y; \mathcal{H}^q(f\mathcal{C}^*))$ and which is convergent to $H^{p+q}(\Gamma(f\mathcal{C}^*)) = H^{p+q}(X; \mathcal{G})$, since $\Gamma(f\mathcal{C}^*) = \Gamma(\mathcal{C}^*)$.

The sheaves $\mathcal{H}^q(f\mathcal{C}^*)$ are generated (according to the very definition of the quotient sheaf) by the presheaves

$$U \to H^q(f\mathcal{C}^*(U)) = H^q(\mathcal{C}^*(f^{-1}U)) = H^q(f^{-1}U; \mathcal{G}).$$

Accordingly, $\mathcal{H}^0(f\mathcal{C}^*) = f(\mathcal{G})$ and the stalks of $\mathcal{H}^q(f\mathcal{C}^*)$ at the points $y \in Y$ are the direct limits $\varinjlim H^q(f^{-1}U; \mathcal{G})$ (with respect to the neighborhoods U of y). This leads to the natural question whether these direct limits coincide with the cohomology of the stalks $f^{-1}y$. It is clear that this is the case when Y is locally contractible and f is a locally trivial fibrationsof X over Y. In this case, for constant coefficients and for small U, it is true that $H^q(f^{-1}U; G) = H^q(f^{-1}y; G)$, the sheaves $\mathcal{H}^q(f\mathcal{C}^*)$ are locally constant (with the stalk $H^q(f^{-1}y; G)$) and there exists the *spectral sequence of a fibration*. The stalks of sheaves $\mathcal{H}^*(f\mathcal{C}^*)$ coincide with $H^q(f^{-1}y; \mathcal{G})$ also in all those cases when the sets $f^{-1}U$ form the fundamental system of neighborhoods of the stalks $f^{-1}y$. This happens, in particular, for every closed map (for example, for maps of compact spaces and for proper maps of locally compact spaces).

The condition $E_2^{p0} = H^p(Y; f\mathcal{G})$ implies the existence of the map $v :$ $H^p(Y; f\mathcal{G}) \to H^p(X; \mathcal{G})$ induced by f. However, in the general case, the $f\mathcal{G}$ form quite specific coefficients and since in reality it is necessary to work with coefficients prescribed a priori, we shall consider a useful construction of the *inverse image* of a sheaf.

For a map $f : X \to Y$ and a sheaf $\mathcal{G}$ on Y let $f^*\mathcal{G}$ denote the presheaf assigning to each open subset $U \subset X$ the group of all continuous maps $\sigma :$ $U \to \mathcal{G}$, for which $p\sigma = f(x), x \in U$ (p is the projection $\mathcal{G} \to Y$). Again, the conditions (S1) and (S2) are satisfied and therefore $f^*\mathcal{G}$ is a sheaf (the *inverse image of $\mathcal{G}$ under f*). It is clear that $(f^*\mathcal{G})_x = \mathcal{G}_{f(x)}$ and that $f^*\mathcal{G}$ is constant on the inverse images of points $y \in Y$. It is clear that there exists a map $\tilde{f} :$ $f^*\mathcal{G} \to \mathcal{G}$ covering the map f that is an isomorphism on corresponding stalks. It is easy to verify the inclusion $\mathcal{G} \subset ff^*\mathcal{G}$, and therefore there exist also the homomorphisms $H^p(Y; \mathcal{G}) \to H^p(Y; ff^*\mathcal{G}) \xrightarrow{\nu} H^p(X; f^*\mathcal{G})$. Since $f^*G = G$ for the constant sheaf G, it is possible to "filter" through the Leray spectral sequence the map of cohomology $f^* : H^p(Y; G) \to H^p(X; G)$ (induced by f). It is useful to keep in mind, that if f is monotone and $f(X) = Y$, then there exists the obvious isomorphism $\mathcal{G} = ff^*\mathcal{G}$.

The independence of the spectral sequence from the choice of the resolution $\mathcal{C}^*$ follows from taking into consideration a map of $\mathcal{C}^*$ into any other injective resolution of $\mathcal{G}$.

6.3. Theorems Related to the Vietoris-Begle Theorem. The Leray spectral sequence has many applications and is an indispensable tool in the study of specific maps. This is due to the fact that it reveals the homological structure of continuous maps (among them fibrations), in particular the impact of the structure of the inverse images of points (fibers) on the global properties of the map. An obvious corollary of this is the following *Vietoris-Begle theorem:* for a closed map $f : X \to Y$ of paracompact spaces such that $f(X) = Y$, the *acyclicity* (that is the vanishing of the reduced cohomology) of the inverse images $f^{-1}y, y \in Y$ up to dimension n implies that the induced homomorphism $f^* : H^p(Y; G) \to H^p(X; G)$ is an isomorphism for $p \leq n$, and a monomorphism for $p = n + 1$.[25]

There are many generalizations of this theorem. In particular, for maps that are not acyclic in low dimensions and on sets of points $y \in Y$ of small dimension, typically there exists a number m, $m < n$, such that f^* is an isomorphism for $m \leq p \leq n$, while for $p = m - 1$ ($p = n + 1$) the map f^* is an epimorphism (respectively, a monomorphism). In such form the theorem has even a partial converse. Analogous results hold also for maps $f : X \to Y$ of spaces over some third (base) space B (in particular, for fibrations over B), but instead of acyclicity of the inverse images of points it is necessary to assume that f induces an isomorphism in corresponding dimensions of the fibers over B. Theorems related to the Vietoris-Begle result have numerous

---

[25] For an analogous result for homology see Sect. 6.2, Chap. 8.

applications (the behavior of dimension under zero-dimensional maps, the statement that acyclic maps do not rise dimension, the structure of certain multi-valued maps, etc.). The indicated questions are discussed in Sklyarenko (1964).

Let us point out that for each inclusion $i : X \to Y$ the stalk of the Leray sheaf $\mathcal{H}^q(i\mathcal{C}^*)$ at the point $y \in Y$ is the direct limit with respect to the neighborhoods U of y of the cohomology groups $H^q(U \cap X; \mathcal{G})$, and since for $y \in X$ these limits, being identical with the cohomology groups of a point, are trivial, the stalks of $\mathcal{H}^q(i\mathcal{C}^*)$ are concentrated at the points of the closure of X in Y not belonging to X. Thus we obtain a tool for comparing cohomology of X and Y in the cases when X is dense in Y (in particular, for compactifications of X). It is easy to formulate conditions which insure that f^* is an isomorphism, monomorphism or epimorphism in dimensions 0 and 1, and with more careful analysis this can be accomplished also in higher dimensions (with local conditions at the points of $Y \setminus X$). This provides a foundation for the study of compactifications of X of special types. For inclusions there exist analogues of the Vietoris-Begle theorem, their generalizations presented above and their converses (in place of the requirement that the inverse images of points are acyclic it is necessary to take into account the appropriate behavior of the cohomology of X "at infinity"). In connection with this subject see the survey in Sklyarenko (1964) and in some later papers.

6.4. Homomorphism of Cohomology Induced by Continuous Maps. Such homomorphism can be defined directly, without using Leray spectral sequences, as the composition of the comparison homomorphism over X with the homomorphism induced by the natural map of complexes of sections $\Gamma(\mathcal{A}^*) \to \Gamma(f^*\mathcal{A}^*)$ corresponding to an arbitrary acyclic resolution $\mathcal{A}^*$ of the sheaf $\mathcal{G}$ on Y. It is clear that such a homomorphism $f^* : H_\varphi^p(Y; \mathcal{G}) \to H_\varphi^p(X; f^*\mathcal{G})$ can be defined also for cohomology with supports in the families φ, ψ for which $f^{-1}\varphi \subset \psi$. As usual, the independence of the choice of $\mathcal{A}^*$ is established by comparing f^* for different resolutions through a map $\mathcal{A}^* \to \mathcal{J}^*$ into the injective resolution $\mathcal{J}^*$ of the sheaf $\mathcal{G}$.

If $f^*\mathcal{A}^* \to \mathcal{J}^*$ is a map into an injective resolution of the sheaf $f^*\mathcal{G}$, then our homomorphism f^* is induced by the maps $\Gamma(\mathcal{A}^*) \to \Gamma(f^*\mathcal{A}^*) \to \Gamma(\mathcal{J}^*)$ of complexes of sections which are obviously identified with the maps $\Gamma(\mathcal{A}^*) \to \Gamma(ff^*\mathcal{A}^*) \to \Gamma(f\mathcal{J}^*)$, which, in turn, are defined by the homomorphisms $\mathcal{A}^* \to ff^*\mathcal{A}^* \to f\mathcal{J}^*$ extending the homomorphism $\mathcal{G} \to ff^*\mathcal{G}$. Since the homomorphism f^* from Sect. 6.2 is defined by the spectral sequence of the double complex $\Gamma(\mathcal{C}^*(f\mathcal{J}^*))$, the statement that it coincides with the homomorphism f^* defined above by considering the map $\Gamma(\mathcal{A}^*) \to \Gamma(f\mathcal{J}^*)$, follows from the fact that the latter coincides with the homomorphism induced by the map $\Gamma(\mathcal{C}^*(\mathcal{A}^*)) \to \Gamma(\mathcal{C}^*(f\mathcal{J}^*))$ of double complexes (compare with the similar considerations in Sect. 2.5).

Chapter 4
Čech Cohomology in the Light of Sheaf Theory

Sheaf theory unifies different approaches to defining cohomology. However, in solving concrete problems it is customary to apply a specific cohomology theory which is suitable in each case. In many situations the most natural choice is Čech cohomology. First of all, this is true in the case when it is natural to take into account the "concept of dimension". A collection of covers of required order, reflecting the dimension of a space or its subsets, allows us to disregard cohomology in higher dimensions (which typically vanish together with the corresponding cochains). In addition, since a space X can always be approximated by the nerves of suitable covers, it is natural to compare its cohomology with the cohomology of such nerves and there exist natural conditions which determine whether the cohomology of X coincides with the cohomology of covers of X. In such situations a cover plays, to a certain extent, the role of a triangulation. The computation of its cohomology is a way of computing the cohomology of the space itself and cohomology of a space is realized on such covers (similarly to the situation when the cohomology of a polyhedron is realized by some of its triangulation). This explains why, among other factors, Čech cohomology, which originated in topology, found other applications in different areas of mathematics (such as algebraic geometry, functional analysis, the theory of analytic manifolds, etc.).

In all cases when the applicability of Čech cohomology $\check{H}^*$ is beyond doubt, it coincides with sheaf cohomology. For paracompact spaces this is a consequence of the fact that $\check{H}^*$ is an example of a cohomological functor (see §3, Chap. 3) vanishing on flabby sheaves. In other situations this follows from the study of the connection between $\check{H}^*$ and H^* realised originally through a resolution of a cover (open or locally finite closed). Without restrictions on the topological space X one can prove the *Leray theorem* on the isomorphism of $H^*(X; \mathcal{G})$ with the cohomology of a cover $\alpha = \{M_\lambda\}$ under the assumption that cohomology of the finite intersections of elements of α is trivial in positive dimensions (such covers are called *acyclic*[26]). In the general case spectral sequences correspond to resolutions whose limit with respect to all open covers of X turns out to be the spectral sequence associated with Čech cohomology. With its aid the theorem of Leray mentioned above leads to the result of Cartan on the isomorphism $\check{H}^*(X; \mathcal{G}) = H^*(X; \mathcal{G})$ in the case when there exists sufficiently many open subsets U in X for which $\check{H}^q(U; \mathcal{G}) = 0$ when $q > 0$. Such a situation, which was studied by Serre, arises in the Čech cohomology of algebraic manifolds M with coefficients in coherent algebraic

---

[26] For example, covers with regions of holomorphy in complex analysis are acyclic, and the fact that the cohomology can be realized through the nerves of such covers plays a crucial role in obtaining one of the representations of hyperfunctions, see Schapira (1970).

sheaves $\mathcal{F}$, and therefore in such cases we have $\check{H}^*(M;\mathcal{F}) = H^*(M;\mathcal{F})$. In particular, the covers of M with open affine subsets are acyclic and thus the statement that $\check{H}^*(M;\mathcal{F})$ can be realized on such covers is in fact a special case of Leray's theorem (there exist generalizations of this result to the case of the Noetherian separable schemes).

§1. Čech Cohomology with Coefficients in a Sheaf. Cohomology of Subspaces and Pairs

1.1. The Role of Paracompactness. Čech cohomology with coefficients in a presheaf A on X was defined in §2, Chap. 2. By the cohomology of X with coefficients in a sheaf $\mathcal{A}$ we shall understand the cohomology of X with coefficients in the presheaf of sections $U \to \mathcal{A}(U) = \Gamma(\mathcal{A}|_U)$. Let $\mathcal{A}$ be the sheaf generated by the presheaf A. It turns out that for a paracompact space X the map $r : A(U) \to \mathcal{A}(U)$ (see §1, Chap. 3) induces the isomorphism of cohomology $\check{H}^*(X;A) = \check{H}^*(X;\mathcal{A})$.

A fundamental step in the proof of this result is the following statement: if a presheaf A on a paracompact space X induces a zero sheaf, then $\check{H}^*(X;A) = 0$. It is proved by a straightforward verification which reduces to the observation that for every cochain $\xi \in C^*(\alpha;A)$ of a cover α there exists a cover $\beta > \alpha$ and a projection $|\beta| \to |\alpha|$ under which the image of ξ in $C^k(\beta;A)$ is equal to zero. Paracompactness is used in an essential way, namely to show that the cover α can be considered locally finite and that it admits a *"combinatorial" shrinking* $\tilde{\alpha} = \{\tilde{M}_\lambda\}$ with the same nerve (the closure of the set $\tilde{M}_\lambda$ is contained in $M_\lambda \in \alpha$ for each λ). The existence of an isomorphism follows now from the exact cohomology sequences corresponding to short exact sequences of presheaves $0 \to A'(U) \to A(U) \xrightarrow{f} B'(U) \to 0$ and $0 \to B'(U) \xrightarrow{g} A(U) \to B''(U) \to 0$, where $A' = \operatorname{Ker} r, B' = \operatorname{Im} r, r = gf$. Indeed, since the $\varinjlim$-functor is exact, the presheaves A' and B'' induce the zero sheaf and at the same time B' induces $\mathcal{A}$.

1.2. Cohomological Sequence. As we have seen, for paracompact spaces it is possible to consider sheaves rather than presheaves as coefficients in the cohomology theory $\check{H}^*$. In particular, Čech cohomology with coefficients in a group G is the same as the cohomology with coefficients in the constant sheaf G.

In spite of the fact that transition to presheaves of sections in the short exact sequences of sheaves $0 \to \mathcal{A}' \to \mathcal{A} \to \mathcal{A}'' \to 0$ does not preserve exactness, in the case of a paracompact X, it leads nevertheless to the following *exact cohomology sequence*

$$\ldots \to \check{H}^n(X;\mathcal{A}') \to \check{H}^n(X;\mathcal{A}) \to \check{H}^n(X;\mathcal{A}'') \to \check{H}^{n+1}(X;\mathcal{A}') \to \ldots \quad (1)$$

Since the presheaf $A(U) = \mathcal{A}(U)/\mathcal{A}'(U)$, as well as the presheaf $\mathcal{A}''(U)$ containing it, generate the sheaf $\mathcal{A}''$, the above sequence coincides with the sequence (3) from Sect. 2.2, Chap. 2 corresponding to the short exact sequence of presheaves $0 \to \mathcal{A}'(U) \to \mathcal{A}(U) \to A(U) \to 0$.

1.3. The Case of Pairs of Spaces. Let us return to the study of the cohomology of subspaces and pairs. The exact sequence of the cohomology of a pair (X, Y) presented in Sect. 2.3, Chap. 2 obviously corresponds to the short exact sequence of presheaves $0 \to A' \to A \to A'' \to 0$ in which the presheaf A' coincides with A on all $U \subset X \setminus Y$ and is equal to zero on the remaining open subsets, and $A'' = A/A'$. For a closed $Y \subset X$ these presheaves generate the short exact sequence of sheaves $0 \to \mathcal{A}_{X \setminus Y} \to \mathcal{A} \to \mathcal{A}_Y \to 0$. In the case of a paracompact space X, the *exact cohomology sequence*

$$\to \check{H}^n(X; \mathcal{A}_{X \setminus Y}) \to \check{H}^n(X; \mathcal{A}) \to \check{H}^n(X; \mathcal{A}_Y) \to \check{H}^{n+1}(X; \mathcal{A}_{X \setminus Y}) \to \dots$$
$$(2)$$

corresponding to it coincides with the cohomology sequence of the pair (X, Y). Accordingly, it is enough to establish the isomorphism of the Čech cohomology with the sheaf cohomology only for one space X.

Let us point out that in agreement with the previous section, the cohomology sequence of a pair (2) can also be obtained (see Sect. 2.3, Chap. 2) by taking the direct limit with respect to open covers α from the cohomology sequences corresponding to the short exact sequences of cochain complexes

$$0 \to C^*(\alpha; \mathcal{A}_{X \setminus Y}) \to C^*(\alpha; \mathcal{A}) \to \tilde{C}^*(\alpha) \to 0,$$

in which the $\tilde{C}^*(\alpha) \subset C^*(\alpha; \mathcal{A}_Y)$ are the images of $C^*(\alpha; \mathcal{A})$ under the map of sheaves $\mathcal{A} \to \mathcal{A}_Y$.

§2. The Isomorphism $\check{H}^* = H^*$ for Paracompact Spaces. Paracompactifying Families of Supports

2.1. Resolution of a Cover. As was established in §1, for a fixed paracompact space X the groups $\check{H}^k(X; \mathcal{G})$ form a connected sequence of cohomological functors, therefore to prove the existence of the isomorphism of $\check{H}^k$ with sheaf cohomology it is enough to establish (see Sect. 3.3., Chap. 3) that $\check{H}^k(X; \mathcal{A}) = 0$ for $k > 0$ and for an arbitrary flabby sheaf $\mathcal{A}$, and that $\check{H}^0(X; \mathcal{G}) = \Gamma(\mathcal{G})$ (for arbitrary $\mathcal{G}$). Both these facts are true for arbitrary topological spaces. They are a result of the existence of a resolution of the coefficient sheaf determined by a cover. Its description is provided below.

Let α be an open, or locally finite, closed cover of a space X. Let us consider the differential presheaf $\mathcal{C}^*(\alpha; \mathcal{G})$ determined by the assignment $U \to C^*(\alpha \cap U; \mathcal{G}|_U)$. In every dimension we have $\mathcal{C}^k(\alpha; \mathcal{G})(U) = \prod_{\dim S = k} \mathcal{G}(M_S \cap U)$. It is true that the presheaf $\mathcal{C}^*(\alpha; \mathcal{G})$ is always a sheaf (that is, it coincides with the presheaf of sections of the sheaf generated by it). Indeed, the assignment $U \to \mathcal{G}(M \cap U) = \Gamma(\mathcal{G}|_{M \cap U})$ for every set M (in particular for M_S) determines, in an obvious way, the direct image $\mathcal{G}|_M$ under the inclusion $M \subset X$, that is a sheaf on X. Consequently, this assignment (and thus the presheaf $C^*(\alpha; \mathcal{G})$) satisfies conditions (S1) and (S2), which yields the desired conclusion (see Sect. 1.2, Chap. 3).

A straightforward verification demonstrates (see Sect. 5.2 Chap. 2 in Godement (1958)), that for a cover α described above, $C^*(\alpha; \mathcal{G})$ is a resolution of $\mathcal{G}$. From this fundamental fact it follows that, first of all, $H^0(\alpha; \mathcal{G}) = \Gamma(\mathcal{G})$ and after passing to the direct limit, that $\check{H}^0(G; \mathcal{G}) = \Gamma(\mathcal{G}) = H^0(X; \mathcal{G})$. Moreover, since the direct images of flabby sheaves are flabby, it follows from the definition of $C^k(\alpha; \mathcal{A})$ that in the case of an open cover α these sheaves are flabby when $\mathcal{A}$ are flabby. Consequently, for flabby $\mathcal{A}$ we always have $H^k(\alpha; \mathcal{A}) = 0 = \check{H}^k(X; \mathcal{A})$ for $k > 0$. Similarly one can show (and this fact is useful in some applications), that $\check{H}^k(X; \mathcal{A}) = 0$ for $k > 0$ in the case when $\mathcal{A}$ is a soft sheaf on a paracompact space X and the cover α is closed and locally finite (see §5.2, Chap. 2 in Godement (1958)).

2.2. Cohomology with Supports. We shall define the complex $C_\varphi^*(\alpha; \mathcal{G})$ of cochains of an open cover α with supports in a family φ as the complex of all sections with supports in φ of the resolution $C^*(\alpha; \mathcal{G})$. The *support Φ of a cochain ξ* (even for a presheaf A of coefficients) can also be defined independently: for every simplex $S \in |\alpha|$ the restriction to $(X \setminus \Phi) \cap M_S$ of the value of ξ on M_S is equal to zero. The supports of cochains are preserved by the projections π_β^α of cochains defined for $\alpha < \beta$ (see Sect. 2.2, Chap. 2). The direct limit with respect to α of the groups $H_\varphi^k(\alpha; \mathcal{G})$ determined by such cochains coincides with the Čech cohomology $\check{H}_\varphi^k(X; \mathcal{G})$ with supports in φ. For paracompactifying families φ these groups are isomorphic with the sheaf cohomology $H_\varphi^k(X; \mathcal{G})$.

To prove this it is necessary, as before to verify that $\check{H}_\varphi^k(X; \mathcal{G})$ is a connected sequence of functors in which $\check{H}_\varphi^0(X; \mathcal{G}) = \Gamma_\varphi(\mathcal{G})$ and $\check{H}_\varphi^k(X; \mathcal{A}) = 0$ for $k > 0$ and for flabby $\mathcal{A}$. The last two claims are valid for any family φ and are both obvious since the application of the functor Γ_φ to the resolution $C^*(\alpha; \mathcal{G})$ yields both claims already for the cover α. The first one follows from the fact that for the cohomology $\check{H}_\varphi^*$ with paracompactifying family φ the construction presented in Sects. 1.1 and 1.2 remain valid (in Sect. 1.1 we use the fact that the support of a cochain ξ is contained in the interior of a closed paracompact subset of X).

2.3. Sheaves of Čech Cochains. The obvious difficulty in consideration of the limits of cochains of covers (not the cohomologies) is the non-uniqueness of projections of the nerves of covers for $\alpha < \beta$. This difficulty can be avoided, for example, in the following way. Consider open covers of a space X of the form $\{U_x\}, x \in U_x$. We assume that $\{V_x\} > \{U_x\}$ if $V_x \subset U_x$ for all $x \in X$. For any two covers of such type there exists a canonical projection of nerves (preserving the indices $x \in X$), therefore we have $\check{C}^*_\varphi(X; \mathcal{G}) = \varinjlim_\alpha \check{C}^*_\varphi(\alpha; \mathcal{G})$, where the α are the covers of the indicated type. Since the $\varinjlim_\alpha$-functor is exact, it follows that $H^*(\check{C}^*_\varphi(X; \mathcal{G})) = \check{H}^*_\varphi(X; \mathcal{G})$.

It is possible to apply the operation of the direct limit in the case of resolutions of covers, $\check{C}^*(X; \mathcal{G}) = \varinjlim_\alpha C^*(\alpha; \mathcal{G})$ is a resolution of the sheaf $\mathcal{G}$ (as the direct limit of resolutions). It is easy to show (see §5.10, Chap. 2 in Godement (1958)) that for every paracompactifying family φ this resolution is φ-fine (see Sect. 5.3, Chap. 3) and therefore $H^n(\Gamma_\varphi(\check{C}^*(X; \mathcal{G}))) = H^n_\varphi(X; \mathcal{G})$. In the general case, the sections $\check{C}^*(X; \mathcal{G})$ do not have to coincide with the Čech cochains $\check{C}^*(X; \mathcal{G})$. They do coincide, however, when X is a Zariskii space, and also for cochains with supports in a paracompactifying family. This approach gives another method of proving the isomorphism $\check{H}^*_\varphi(X; \mathcal{G}) = H^*_\varphi(X; \mathcal{G})$ (for paracompactifying families φ).

The indicated approach is by no means the best, since the covers used in it are too restricted, they do not reflect the "concept of dimension", etc. In the particular case of locally compact spaces X, the construction of cochains considered in Sect. 4.2, Chap. 2, is quite useful. It leads to soft resolutions whose sections with compact supports are the same as cochains with compact supports[27] defined in Sect. 4.2 Chap. 2. Since the families of locally finite covers which are considered there are not cofinal in the system of all covers, the direst limits of cochains of covers yield the cochains of Čech type only in the case of cochains with compact supports. Conversely, the complexes of sections of the resulting soft resolutions that do not coincide with these direct limits (yielding $H^*_\varphi(X)$ for arbitrary paracompactifying families φ) can be interpreted as Čech type cochains. Recall that in the case of a locally constant sheaf $\mathcal{G}$ with the stalk isomorphic with a ring R, the cochains (that is sections of resolutions) with compact supports form a complex of free R-modules (of countable rank in the case of metrizable spaces); see Sklyarenko (1969, 1971b).

§3. Comparison of Cohomology of a Cover with Cohomology of a Space

3.1. Comparison Homomorphism. With the definition of $\check{H}^n_\varphi(X : \mathcal{G})$ as the direct limit of $H^n_\varphi(\alpha; \mathcal{A})$, for every open cover α there exists a natural limit homomorphism $\pi_\alpha : H^n_\varphi(\alpha; \mathcal{G}) \to \check{H}_\varphi(X; \mathcal{G})$. On the other hand, since $H^n_\varphi(\alpha; \mathcal{G})$ is determined by the sections of the resolution $C^*(\alpha; \mathcal{G})$, there exists a comparison homomorphism $\gamma^\alpha : H^n_\varphi(\alpha; \mathcal{G}) \to H^n_\varphi(X; \mathcal{G})$ (see Sect. 2.1, Chap. 3). For paracompactifying families of supports $\check{H}^n_\varphi(X; \mathcal{G}) = H^n_\varphi(X; \mathcal{G})$ and it is natural to ask whether π^α and γ^α coincide.

---

[27] See Sect. 4.1, Chap. 5.

It is possible to verify that this indeed is the case in the following way (see also §4). By passing to the direct limit (with respect to α), the system $\{\gamma^\alpha\}$ defines a natural map $\check{\gamma} : \check{H}^*_\varphi(X;\mathcal{G}) \to H^*_\varphi(X;\mathcal{G})$ for which $\gamma^\alpha = \check{\gamma}\pi^\alpha$. It is enough to verify that $\check{\gamma}$ is the identity.

It follows from the definition that $\check{\gamma}f_* = f_*\check{\gamma}$ for every map $f : \mathcal{G} \to \mathcal{G}'$ of coefficients. A direct verification reveals that $\check{\gamma}\delta = \delta\check{\gamma}$ for connecting homomorphisms δ corresponding to short exact sequences of coefficients. Since $\check{\gamma}$ is the indentity in dimension zero, $\check{\gamma}$ is an automorphism of a connected sequence of functors $\{H^n_\varphi\}$ and therefore the fact that it is the identity is a consequence of the universality property in the special case $\Phi^n = F^n = H^n_\varphi$ (see Sect. 3.3 Chap. 3).

3.2. Spectral Sequence of a Cover. As we know (see Sect. 6.1 Chap. 3), to every resolution (consisting of nonacyclic objects) there corresponds a certain spectral sequence. To describe the spectral sequence corresponding to the resolution $\mathcal{C}^*(\alpha;\mathcal{G})$ of a cover (the map γ^α can be "factored" through this sequence) we introduce the coefficient systems $\mathcal{H}^q(\mathcal{G}), q = 0, 1, 2, \ldots$ on $|\alpha|$ in which the groups $H^q(M_S;\mathcal{G})$ are assigned to the simplices $S \in |\alpha|$. It is clear that when S' is a face of S, the inclusion $M_S \subset M_{S'}$ induces a natural homomorphism $H^q(M_{S'};\mathcal{G}) \to H^q(M_S;\mathcal{G})$. In the case of an open cover α the system $\mathcal{H}^0(\mathcal{G})$ is clearly determined by the presheaf of sections of $\mathcal{G}$ (considered earlier in the definition of Čech cohomology). If α is either an open or a locally finite closed cover of an arbitrary space X, then the resolution of the cover yields a spectral sequence in which $E_2^{pq} = H^p(\alpha;\mathcal{H}^q(\mathcal{G}))$ and the term E_∞ is a bigraded group associated with the filtration of $H^*(X;\mathcal{G})$ For an open cover α a simple verification of this fact is obtained by considering the sections of the double complex $C^*(\alpha;\mathcal{C}^*(\mathcal{G}))$ where $\mathcal{C}^*(\mathcal{G})$ is the canonical flabby Godement resolution of the sheaf $\mathcal{G}$ (see Sect. 5.4, Chap. 2 in Godement (1958)). Analogous considerations can be applied also in the second case, if one considers the double complex $C^*(X;\mathcal{C}^*(\alpha;\mathcal{G}))$ of sections over X of the canonical flabby resolution of the differential sheaf $\mathcal{C}^*(\alpha;\mathcal{G})$ (see Sect. 5.2, Chap. 2 in Godement (1958)). An important step in this approach is the observation that for locally finite systems of sheaves on X the direct products from Sect. 2.1 above coincide (for small U) with direct sums (in particular the fibers $\mathcal{C}^p(\alpha;\mathcal{G})$ in this case are isomorphic with direct sums of fibers of direct images of $\mathcal{G}|_{M_S}$ under the inclusions $M_S \subset X$).

3.3. Leray's Acyclic Cover Theorem. If in the situation described above a cover α is *acyclic*, in the sense that $H^q(M_S;\mathcal{G}) = 0$ for $q > 0$ and for every simplex $S \in |\alpha|$, then $H^*(\alpha;\mathcal{G}) = H^*(X;\mathcal{G})$. This is a consequence of the spectral sequence of a cover. The proof can be obtained also directly without using the theory of spectral sequences (see Schapira (1970)). This theorem is of importance not only in topology, but also in functional and complex analysis as well as in algebraic geometry (see also the following section).

§4. Spectral Sequence Associated With Čech Cohomology

4.1. A Spectral Sequence. In general, the maps π^α and γ^α of the cohomology $H^*(\alpha;\mathcal{G})$ of a cover into $\check{H}^*(X;\mathcal{G})$ and $H^*(X;\mathcal{G})$, respectively (Sect. 3.1), define, on passing to the direct limit with respect to α, a map $\check{\gamma} : \check{H}^*(X;\mathcal{G}) \to H^*(X;\mathcal{G})$ for which $\gamma^\alpha = \check{\gamma}\pi^\alpha$. It "factors" through the spectral sequence which is the direct limit of spectral sequences of the cover α. We note that the cohomology of a cover α with coefficients in the system $\mathcal{H}^q(\mathcal{G})$ on $|\alpha|$ considered above is simply the cohomology of α with coefficients in the presheaf $\mathcal{H}^q(\mathcal{G})$ for which $\mathcal{H}(\mathcal{G})(U) = H^q(U;\mathcal{G})$. Therefore after passing to the direct limit with respect to α in the spectral sequence associated with Čech cohomology we conclude that $E_2^{pq} = \check{H}^p(X;\mathcal{H}^q(\mathcal{G}))$ and the spectral sequence converges to the appropriately filtrated cohomology $H^*(X;\mathcal{G})$.

4.2. Corollaries. For every presheaf $\mathcal{H}$ which induces the zero sheaf, every zero-dimensional cochain of a cover can be anihilated by passing to a finer cover, and therefore $\check{H}^0(X,\mathcal{H}) = 0$. Consequently, in the resulting spectral sequence $E_2^{0q} = 0$, for $q > 0$. Accordingly, $\check{\gamma}$ is an isomorphism in dimensions 0 and 1 and a monomorphism in dimension 2 (since $\mathcal{H}^0(\mathcal{G})$ is the presheaf of sections of $\mathcal{G}$, $E_2^{p0} = \check{H}^p(X;\mathcal{G})$ which provides still another definition of $\check{\gamma}$). Since for $q > 0$ the presheaves $\check{H}(\mathcal{G})$ induce the zero sheaf, $E_2^{pq} = 0$ for $q > 0$ for a paracompact space, which once more yields the isomorphism $\check{\gamma} : \check{H}^* \to H^*$ (and also the relation $\gamma^\alpha = \pi^\alpha$) for such spaces.

4.3. Cartan's Theorem. Let us assume that for a sheaf $\mathcal{G}$ on X there exists a base of open sets $U \subset X$ which is closed with respect to finite intersections and such that $\check{H}^q(U;\mathcal{G}) = 0$ for $q > 0$ and for all elements U. In the case when such a base exists we have $\check{H}^*(X;\mathcal{G}) = H^*(X;\mathcal{G})$ and this statement is known as *Cartan's theorem* (see the beginning of this chapter for the connections of this theorem with other results). Moreover, for every cover α consisting of the elements of this base already $\pi^\alpha : H^*(\alpha;\mathcal{G}) \to H^*(X;\mathcal{G})$ is an isomorphism. Indeed, since $\check{H}^1 = H^1$, for the sets U in the base we have $H^1(U;\mathcal{G}) = 0$. This means that for spectral sequences of such covers and the sequence associated with Čech cohomology, $E_2^{p1} = 0$. But then $\check{H}^2 = H^2$, and for the elements U of the base we have $H^2(U;\mathcal{G}) = 0$. Consequently, for these spectral sequences $E_2^{pq} = 0$ also for $q = 2$. Continuation of this process proves that $E_2^{pq} = 0$ for all $q > 0$, from which it follows that $\check{H}^p = E_2^{p0} = H^p$.

Chapter 5
Homology Theory

§1. Introductory Remarks

1.1. Fundamental Difficulties in the Development of the Theory. From the content of the previous chapters is clear that cohomology exhibits some form of "stability" with respect to a variety of different natural versions of its basic definition. Many apparently diverse descriptions lead to the same theory, at least under natural restrictions on the topological spaces for which they are introduced. Such restrictions arise from the specific nature of these constructions or are otherwise easily formulated in each case. The essence of this phenomenon was disclosed by us in §1, Chap. 3 with the aid of the methods of sheaf theory and is a result of the fact that cohomology is a derived functor of the functor of the zero-dimensional cohomology.

By contrast, homology typically is sensitive to any logically possible changes in its original definition. For example, since the functor of the inverse limit $\varprojlim$ is not exact, the *Čech homology groups* $\check{H}_n(X;G)$ (Sect. 2.1, Chap. 2) defined as $\varprojlim_{\alpha} H_n(\alpha;\mathcal{G})$ (compare with Sect. 2.2, Chap. 2), are typically different from the groups $H_n(\check{C}_*(X;G)))$, where $\check{C}_*(X;G)$ is the inverse limit (if it exists) of chain complexes of nerves of some fundamental sequence of open covers α of X. For a non-compact space X the groups $\check{H}_n(X;G)$ are typically different from $\check{H}_n^c(X;G) = \varinjlim_{C \subset X} \check{H}_n(C;\mathcal{G})$ (C belongs to the family of all compact subsets of X). In addition, homology of the complex $\check{C}_*(X;G)$ (in the case when it is defined) is different from homology of the complex $\check{C}_*^c(X;G) = \varinjlim \check{C}_*(C;\mathcal{G})$.

The result can also be different when we use chains of type $\check{C}_*$ by taking into account the topology of the inverse limit in $\check{C}_*$. Of course not every version of a definition leads to a new theory but sometimes good intuition is necessary to select from different approaches the best one. It was assumed that the fundamental difficulty in the development of a homology theory was the non-exactness of the $\varprojlim$-functor. The earlier examples indicate that another difficulty, no less fundamental, is the problem whether in the construction of a theory it is necessary to take into account (and if so, in what way) the *compactness factor*. The role of this factor for homology was already indicated earlier (see Sect. 1.5, Chap. 2) and we shall return to it again (the end of §3, Chap. 5). There are also certain other important features resulting in differences between homology and cohomology theories (see, for example, Sect. 2.3, Chap. 7).

The pressures of the various difficulties which arose in an attempt to create a homology theory of general spaces resulted in the creation of many theories (that not only are not mutually equivalent, but sometimes are even different from the simplicial theory when restricted to the category of infinite

polyhcdra)[28]. This also meant that the homologu theory lagged far behind
the cohomology theory. As a result the correct theory was finally accepted in
topology only in the early 60's. Its development was stimulated by the existence of many problems which involve naturally the concept of homology,
but to which the existing theories were not applicable at all, or were poorly
suited for applications. For example, in the topological theory of transformation groups, "even if one is interested mainly in a statement involving only
cohomology, one has to use in the proof groups which play the role of homology groups, and therefore this presupposes some homology theory" (p. 23 in
Borel (1960)). Similar situation arises in the fixed point theory and in the theory of coincidence points of maps, in homological dimension theory and other
areas. In individual situations (for example, in duality theory, see Sect(s). 5.5,
5.6, Chap. 8) such issues can even serve as a distinctive criteria for identifying
a suitable homology theory among others.

1.2. The Contemporary State of the Theory. In spite of the intensive study
of the theory in numerous publications (the subject under consideration included not only homology with coefficients in a group, but also with coefficients in sheaves, presheaves, cosheaves, copresheaves, etc.), the problem of
creating a homology theory with non constant coefficients, dual to a cohomology theory with coefficients in a sheaf (that is, a theory in which the homology
groups H_n would be the derived functors of the functor H_0) is still awaiting a
solution. At present, a completely satisfactory theory without flaws has been
constructed only for locally constant systems of coefficients. It originated in
the work of Steenrod (see Steenrod (1940)) and K.A. Sitnikov (see Sitnikov
(1951)) and in general categories of topological spaces it is known as the *homology theory with compact supports.*

§2. Compact Spaces

In this paragraph we shall present two types of chains which determine the
homology of compact spaces. Since the resulting homology has all the standard
properties,[29] in the category of metrizable compacta we obtain a theory whose
presence (together with the corresponding cohomology) was established in
Sect. 1.3, Chap. 2 independently from any specific way of describing chains.

2.1. Čech-Type Chains. Let ω be any fundamental system of finite covers of
a compactum X by closed canonical sets (see Sect. 4.2, Chap. 2).[30] In considering homology with locally constant coefficients $\mathcal{G}$ (that is, with coefficients

---

[28] Which are, in essence, no longer homology theories, since the homologies of polyhedra are determined by the boundary formula (1), see Chap. 1.
[29] In connection with this see also §3.
[30] See also the following Sect. 2.3.

in a locally constant sheaf) we require that the covers $\alpha \in \omega$ be sufficiently fine (that is, the restrictions of $\mathcal{G}$ to the elements of α are constant). The uniqueness of projections of nerves in the inverse limit of the chains of covers allows us to define the chain complex $C_*^\omega(X; \mathcal{G}) = \varprojlim_{\alpha \in \omega} C_*(\alpha; \mathcal{G})$. It turns out that the homology of $C_*^\omega(X; \mathcal{G})$ does not depend on the selection of the system of covers ω defining it and that it is the desired homology $H_*(X; \mathcal{G})$. In particular, for metrizable X in place of the system ω we can consider sequences covers of the indicated type inscribed in each other.

Even though the chains $C_*^\omega(X; \mathcal{G})$ are defined independently of cochains, in the case of constant coefficients $\mathcal{G} = G$ there exists a natural connection between them, namely $C_*^\omega(X; G) = \mathrm{Hom}_R(\check{C}_\omega^*(X; R), G)$ ($\check{C}_\omega^*(X; R)$ is the free complex $\varinjlim_{\alpha} C^*(\alpha; R)$, see Sect. 4.2, Chap. 2). This relationship is obtained, by passing to the limit with respect to $\alpha \in \omega$, from the relations $C_*(\alpha; G) = \mathrm{Hom}_R(C^*(\alpha; R), G)$ (we recall that always

$$\mathrm{Hom}(\varinjlim \ldots, G) = \varprojlim \mathrm{Hom}(\ldots, G)). \tag{1}$$

The connection mentioned above (used, for example, in the following section) leads to the universal coefficient formulas that express $H_p(X; G)$ in terms of cohomology $H^*(X; R)$ (see §4, Chap. 8). For the same reason that the cochain complexes $\check{C}_\omega^*(X; R)$ are free (and $\check{C}_\omega^p(X; G)$ have the structure of direct sums $\bigoplus G$), the chains $C_p^\omega(X; G)$ in every dimension are isomorphic with the products $\prod G$ of the module G of coefficients (this is clear also from the relation between chains and cochains that was considered above). In particular,

$$C_*^\omega(X; G) = C_*^\omega(X; R) \otimes_R G \tag{2}$$

only for finitely presented R-modules G (see Sect. 1.7, Chap. 1).

2.2. Independence of H_* of the Choice of ω . If ρ is another system of covers of X, then we define $\sigma = \omega \wedge \rho$ to be the family of all intersections $\alpha \wedge \beta, \alpha \in \omega, \beta \in \rho$. We have also the homomorphisms $C_*^\sigma \to C_*^\omega, C_*^\sigma \to C_*^\rho$ of chain complexes that are dual to the homomorphisms of cochain complexes $\check{C}_\omega^* \to \check{C}_\sigma^*, \check{C}_\rho^* \to \check{C}_\sigma^*$ which induce the identity isomorphism of cohomology $\check{H}^*(X; R) = H^*(X; R)$ (see Sect. 4.2, Chap. 2). Since these cochain complexes are free, the indicated homomorphisms, and consequently the dual homomorphisms of chain complexes, are homotopy equivalences (§4 in Sklyarenko (1969)). This establishes the independence in the case of constant coefficients.

In the case of non-constant coefficients $\mathcal{G}$ we argue differently. The maps $C_*^\omega(X; \mathcal{G}) \to C_*(\alpha; \mathcal{G})$ induce, in conjunction with the operation of the inverse limit with respect to α, a natural map $H_*(X; \mathcal{G}) \to \check{H}_*(X; \mathcal{G})$ into Čech *homology*, and accordingly we obtain (see §1, Chap. 8) the following exact sequences

$$0 \to \varprojlim_{\alpha \in \omega}{}^1 H_{k+1}(\alpha; \mathcal{G}) \to H_k(X; \mathcal{G}) \to \check{H}_k(X; \mathcal{G}) \to 0. \tag{3}$$

The last term in these sequences does not depend on ω. A technique applicable
for defining the derived functors of the $\varprojlim$-functor allows us to show (Theorem
3.1 in Sklyarenko (1984)) that the first term also does not depend on ω.
The independence of the middle term follows from the comparison of exact
sequences (3) for ω, ρ and $\omega \wedge \rho$ and the five lemma.

2.3. Exact Homology Sequences. To every short exact sequence $0 \to \mathcal{G}' \to$
$\mathcal{G} \to \mathcal{G}'' \to 0$ of locally constant coefficients there corresponds a standard exact
sequence of homology groups of X. Indeed, the sequence $0 \to C_*^\omega(X; \mathcal{G}') \to$
$C_*^\omega(X; \mathcal{G}) \to C_*^\omega(X; \mathcal{G}'') \to 0$ of chain complexes is exact. The exactness at
the first two terms is a consequence of the left exactness of the $\varprojlim$-functor.
For the constant coefficients the exactness at the third term is a consequence
of the fact that it is obtained by applying the functor Hom to the modules of
coefficients of the free complex of cochains of X. The chains in $C_*^\omega(X; \mathcal{G})$ are
sections of certain flabby sheaves of chains (see §4), which implies that every
chain $\xi'' \in C_*^\omega(X; \mathcal{G}'')$ can be represented as a sum of chains ξ_i'' over whose
supports the sheaf $\mathcal{G}''$ is constant. Therefore ξ'' is the image of $\xi \in C_*^\omega(X; \mathcal{G})$,
where $\xi = \sum_i \xi_i$ and ξ_i cover ξ_i''.

In order to define the homology sequences of a pair (X, Y), where Y is a
closed subset of X, we note that the only essential assumption in the selection
of the system ω of covers was that the projections of nerves of covers in ω
were unique and surjective.[31]. The first condition allows us to obtain the
inverse limits of chains, and the second guarantees the $\varprojlim$-*acyclicity* of the
projective systems of chains. Indeed, since an arbitrary system of coefficients
$\mathcal{G}$ can be embedded into a system $\mathcal{S}$ with compact stalks (by using a standard
embedding of G into a compact group), the projective system $\{C_*(\alpha; \mathcal{G})\}$ can
be embedded into a $\varprojlim$-acyclic system $\{G_*(\alpha; \mathcal{S})\}$. From this and from the
previous result it follows easily that $\varprojlim^p\{C_*(\alpha; \mathcal{G})\} = 0$, for $p \geq 1$, see §3 in
Sklyarenko (1984).

Accordingly, in defining the homology of $Y \subset X$ we can use the restriction
ω' to Y of the system ω (see Sect. 4.2, Chap. 2). The inverse limit with respect
to $\alpha \in \omega$ of the subcomplexes $C_*(\alpha'; \mathcal{G}) \subset C_*(\alpha; \mathcal{G})$ determines the exact
sequence of chain complexes $0 \to C_*^\omega(Y; \mathcal{G}) \to C_*^\omega(X; \mathcal{G}) \to C_*^\omega(X, Y; \mathcal{G}) \to 0$
which results in the *exact homology sequence of the pair* (X, Y). The fact that
the third map is an epimorphism follows from the relation $\varprojlim_\alpha^1\{C_*(\alpha'; \mathcal{G})\} =$
0.[32] The independence of $H_*(X, Y; \mathcal{G})$ from the choice of the system ω follows
from the independence of the homology of X and Y from ω, from the exact
homology sequence of a pair and from the five lemma (in comparing the
homologies for ω, ρ and $\omega \wedge \rho$).

---

[31] In particular, it is possible to use the intersections of X with covers with canonical
sets (not necessarily compact) of any topological space containing X.

[32] For a constant $\mathcal{G} = G$ this follows also from Sect. 4.2, Chap. 2 by applying the
functor Hom into G.

It is easy to verify that $C_*^\omega(Y;\mathcal{G})$ is the subcomplex of $C_*^\omega(X;\mathcal{G})$ consisting of all chains with supports in Y (by definition, a point $x \in X$ is not contained in the *support of a chain* ξ if for some neighborhood U of x there exists an $\alpha \in \omega$ such that the image of ξ in $C_*(\alpha;\mathcal{G})$ has zero coefficients for those intersections of elements of α (simplices of $|\alpha|$), that intersect U).

2.4. Remarks. The construction of chains described above for constant coefficients was given in Sklyarenko (1969) (without the assumption of metrizability of a space, in §1 of (Sklyarenko (1971b)) in the description of homology of the "second kind" for locally compact spaces. For locally constant coefficients this construction was given in §3 of Sklyarenko (1971b). The most complete description that is independent of cohomology, was given in §3 of Sklyarenko (1984).

The idea of using limits of chains with respect to covers with unique projections in a description of homology is not new. It was already known in the 30's (see the survey in Sklyarenko (1969)), but since then it was applied primarily in different descriptions of Čech homology determined by finite covers. It was gradually eluciated that such definitions of homology were not suitable, in particular outside of the category of compact spaces (not only because of the lack of exactness in the homology sequence of a pair, see Chap. 9 in Eilenberg and Steenrod (1952), but also for certain other reasons, see Sect. 2.1, Chap. 2 and Sect. 1.1, Chap. 5). However, it was customary to consider compact coefficient groups G, since in this case there exists a duality, in the sense of Pontryagin's character theory, between homology and cohomology with discrete coefficients dual to G. The equivalence of several types of such groups (for finite covers and partitions) with Čech homology was established in Chogoshvili (1940, 1951) (in one of these papers, the resulting chains, when applied to the case of compact spaces, are the same as those obtained in Sect. 2.1 above and corresponding to the system ω of all finite partitions of X).

According to the established tradition, the chain complexes with non-compact coefficients which arise as a result of limit operations with respect to covers (as well as the groups $\check{H}_*$ themselves) were also considered for a long time in the topology of the inverse limit. Since it was assumed unnatural to consider quotient groups with respect to a nonclosed subgroup (see, for example, §(40.6), Chap. 3 in Lefschetz (1942)), the subgroups of cycles, which are always closed, were factored with respect to the closure of the subgroup of the boundaries. The homologies arising in this way were called *spectral* (Sect. 16, §4, Chap. 6 in Lefschetz (1942)). In the case when G is a compact group or a field, the boundaries subgroup is always closed and the spectral homology coincides with Čech homology (§4, Chap. 6 in Lefschetz (1942)); for compact G see also (Chogoshvili (1940)). Accordingly, for such G the operation of the inverse limit of chains with respect to finite covers commutes with the homology functor,[33] and consequently Čech theory $\check{H}_*$, as being the same as H_*, is

---

[33] The fact that the same commutativity with direct limits also takes place was first noticed in Chogoshvili (1940).

also exact (Chap. 8 in Eilenberg and Steenrod (1952)). This explains its wide application for these coefficient groups.

Clearly, there are other approaches which do not use covers (or partitions). As an independent, separate theory (without restrictions on the coefficient groups) the homology H_* appeared in Steenrod (1940) and Sitnikov (1951, 1954) (see also Sect. 3.1) devoted to duality for subsets of spheres, in the situation in which no other homology theory was applicable (see Sect. 5.6, Chap. 8). However, the definitions based on an application of modified constructions of Vietoris cycles were difficult to adapt for other purposes and therefore the significance of *Steenrod-Sitnikov homology* was not immediately apparent. In particular, even though they were mentioned in Eilenberg and MacLane (1942), in the well known monograph Eilenberg and Steenrod (1952), in which homology of compact spaces receive a lot of attention, the Steenrod-Sitnikov homology is not even mentioned.[34] Their role in algebraic topology was finally clarified in Borel and Moore (1960), Milnor (1960), Raymond (1961), Chogoshvili (1962) and other works.

The first outline of a homology theory H_* satisfying modern requirements was presented in the publications of Borel's seminar on transformation groups (see Chap. 2 in Borel (1960)). Those ideas were further developed in the well known paper of Borel and Moore (1960). The paper Milnor (1960) also had an essential impact on the development of the theory H_* (in this paper, in particular, free cochains for the cohomology of compacta were introduced for the first time, see the footnote in Sect. 4.2, Chap. 2). In spite of the artificial nature of the proposed definitions (chains for homology were introduced formally by means of the functor Hom applied to specially selected cochains[35], the existence of nontrivial chains in negative dimensions, etc.), after the publication of the papers mentioned above the theory H_* acquired a solid position in the literature. An important achievement of Borel and Moore (1960), crucial from the point of view of applications, was the fact that the chains and cochains described there are the sections of the fine and soft sheaves generated by them, and homology is defined also for nonconstant coefficients (the need for such coefficients arises, for example, in the theory of transformation groups). The last goal, however, is achieved at the price of high formalism and the homology theory, defined formally for the coefficients in sheaves, is no longer completely correct in the case of coefficients in R-modules G that are not finitely generated. This is expressed in the fact that the formulas (2) of Sect. 2.1 above (and consequently, the universal coefficient formulas corresponding to it) are true in Borel-Moore theory for arbitrary G (the fact that this is not natural for infinitely generated G can be illustrated by examples of the simplest zero-dimensional compacta, see Proposition 7 in §5 of

---

[34] Steenrod groups were presented with appreciable care in Sect. 40.6, Chap. 3 of Lefschetz (1942).

[35] The fact that such an approach is natural was discussed in Sect. 2.1 above (see also the end of Sect. 1.3, Chap. 2).

Sklyarenko (1971a)[36]). As a consequence, the homology is not invariant with respect to a change of the ring (see §14, Chap. 5 in Bredon (1967), §1 in Sklyarenko (1969)). This explains why all applications of Borel-Moore homology (see Bredon (1967)) are limited to the case of locally constant coefficients with finitely generated stalks (see also Sect. 3.2, Chap. 7 below).

2.5. Massey Chains. An obvious deficiency of the constructions described at the beginning of this paragraph is the need to prove the independence of homology of the choice of the system of covers, in fact the need to prove their topological invariance (similarly to the situation when one has to prove the independence of the classical homology groups of a polyhedron of its triangulation). As is pointed out in Massey (1978a), this affects primarily the definition of the homomorphism of homology induced by continuous maps of spaces (see the next section). Such difficulties do not arise in the construction of homology proposed in Massey (1978a, 1978b). The chains with coefficients in an R- module G are defined there as $\mathrm{Hom}_R(C^*, G)$, where C^* are free cochains with coefficients in R described earlier in Sect. 4.1, Chap. 2. As with the cochains, they are a functor of X which automatically determines the homology of closed subspaces $Y \subset X$ and pairs (X, Y) (compare with Sect. 4.1, Chap. 2) and the homomorphisms induced by continuous maps. This property of chains is useful also in a description of homology of general spaces (see the following section). This approach (as indicated by its author in Massey (1978a)) is poorly suited for a description of homology with non-constant coefficients[37], and also for the proof of formulas of the type (3) above (the same remark also applies to Borel and Moore (1960) and Bredon (1967)). In addition, it is necessary to verify the *invariance of homology with respect to the change of a ring* (Sect. 4.11 in Massey (1978b)). Such invariance does not exist, for example, in Borel-Moore homology (Sect. 3.2, Chap. 7). With a direct definition of chains (independently of cochains) such problem does not arise. See also Sect. 4.2 of Chap. 2 for a comparison of different approaches.

When we face different approaches it is necessary to make use, in specific situations, of the most suitable ones (see, for example §§3,4 below). Concerning the equivalence of the approaches presented in Sklyarenko (1969) and Massey (1978b) see §6 in Sklyarenko (1979).

---

[36] See also Sect. 2.2, Chap. 7.

[37] Such a description was given recently by L.D. Altshuler and the homology groups which arise in this way are isomorphic with the groups $H_*(X; \mathcal{G})$ of Sect. 2.1 (for locally compact spaces with the groups described in §3 below).

§3. Locally Compact Spaces. The General Case

3.1. Introductory Remarks. The homology considered in §2 is a special case of a more general construction of homology of the "second kind" which is determined by "infinite" chains in the category of locally compact spaces and their proper maps. For compact coefficients such groups were defined for the first time in Kolmogorov (1936c) (who proved in Kolmogorov (1936e) their isomorphism with the simplicial homology of the second kind in the case of infinite manifolds), and shortly after in Alexander (1938). G.S. Chogoshvili used one of the types of chains based on the consideration of all finite partitions of a space (for a similar representation of chains in terms of pairs $(\alpha, \overline{\alpha})$ of covers see Sect. 3.2 below) to establish (see Chogoshvili (1940, 1951)) their isomorphism with a special type of Čech homology (in fact, Čech homology of the one-point compactification of a space, compare Sect. 3.2).

The theory satisfying modern requirements was developed much later (see Borel and Moore (1960), Bredon (1967), Sklyarenko (1969, 1971a, 1971b, 1979, 1984), Massey (1978b)). Finally (primarily due to the arising needs, see Sect. 1.1 above) the reservations towards discrete coefficient groups and discrete topology in the groups of cycles and boundaries (see Sect. 2.4 above) were overcome and correct definitions were discovered (as reported in Algebra, Topology, Geometry: Tbilisi (1985)), also the definition introduced by A.N. Kolmogorov and considered for arbitrary coefficients, leeds to the same homology groups). In addition to the verification of the correctness of the definitions, the indicated developments in principle consisted also of a more thorough study of the theory itself, as well as of the creation of effective methods of its application (suitable constructions in sheaf theory and homological algebra, methods of taking into account properties related to countability and dimension, limit operations, connections with different coefficient groups and many others).

All the constructions and observations of the previous paragraph, as well as the majority of remarks included in it, remain true also in this more general situation (only when we consider the connection with cochains or cohomology it is necessary to consider them with compact supports). Similarly to the way it is done in Sect. 4.2, Chap. 2 we can prove that the homology of a locally compact space X is isomorphic to the reduced homology of the one-point compactification of X (at least for a coefficient system that is constant "at infinity"). This, however, does not mean that homology of the second kind is not worthy of interest in its own right. In addition to the arguments in their favor provided in Sect. 1.4, Chap. 1, there are also further reasons. First of all, the isomorphism $H_*(X, Y; \mathcal{G}) = H_*(X \setminus Y; \mathcal{G}|_{X \setminus Y})$ is very useful (in particular, in connection with the applications of sheaves of chains, see §4 below and §5 in Chap. 8). The role of supports in the description of this homology theory can be clearly visualized only when considering complexes of chains with arbitrary closed supports on locally compact spaces.

3.2. Homology of the Second Kind. Let ω be a fundamental system of canonical covers of a locally compact space X (Sect. 4.2, Chap. 2). The homology $H_*(X; \mathcal{G})$ is determined by the chain complex $C_*^\omega(X; \mathcal{G}) = \varprojlim_{\alpha \in \omega} C_*(\alpha; \mathcal{G})$ (as in §2 it is assumed that the covers of ω are sufficiently fine in comparison to $\mathcal{G}$). We have (see Sect. 2.1) $C_*^\omega(X; R) = \mathrm{Hom}_R(\check{C}_{\omega c}^*(X), G)$, where $\check{C}_{\omega c}^*(X; R)$ is the free complex of cochains with compact supports (Sect. 4.2, Chap. 2). This relation has the same consequences and applications as the formulas presented in §2. In the same way as was done earlier it is possible to construct the exact homology sequences corresponding to short exact sequences of systems of coefficients and to pairs (X, Y) for arbitrary closed $Y \subset X$ (the only difference is that the chain ξ'' which appears in Sect. 2.3 should be represented in the form of a locally finite, in general infinite, sum of chains ξ_i'' with sufficiently small supports).

The proof of the independence of the homology of the choice of the system ω is conducted according to the same strategy as in §2. Since the formulas of the same type as (3) are established not for arbitrary systems of covers (see §1, Chap. 8), it is necessary to use the construction from Sect. 4.2, Chap. 2, according to which $C_*^\omega(X; \mathcal{G})$ coincides with the limit of chain complexes $C_*(\alpha, \tilde{\alpha}; \mathcal{G})$. In fact, this implies that $H_*(X; \mathcal{G}) = H_*(\overline{X}, \infty; \mathcal{G})$, where $\overline{X}$ is the compactification of X with the point ∞. This leads to formulas similar to (3) in which on the left and the right hand sides are the groups $\varprojlim_{(\alpha, \tilde{\alpha})}^1 H_{k+1}(\alpha, \tilde{\alpha}; \mathcal{G})$ and $\varprojlim_{(\alpha, \tilde{\alpha})} H_k(\alpha, \tilde{\alpha}; \mathcal{G})$ (the last groups are called the *reduced Čech homology* of X). The remaining considerations included at the end of Sect. 2.2 remain unchanged (see Theorem 3.1 in Sklyarenko (1984)).

The method applied earlier allows us to describe the *homology of a pair* (X, Y) with coefficients in locally constant systems $\mathcal{G}$ which are determined only on $X \setminus Y$. In order to do this it is necessary to consider canonical covers α of the space X, whose restrictions α' to Y have nerves coinciding with the nerves of subsystems of α consisting of all subsets intersecting Y. For every system ω of covers of such a type, $C_*^\omega(X, Y; \mathcal{G}) = \varprojlim_{(\alpha, \tilde{\alpha})} C_*(\alpha, \tilde{\alpha}; \mathcal{G})$, where $\alpha \in \omega$ and $\tilde{\alpha}$ is any part of α whose complement consists of a finite number of sets contained in $X \setminus Y$. From this, by considerations similar to those in Sect. 4.2, Chap. 2, we obtain the *excision property* $H_*(X, Y; \mathcal{G}) = H_*(X \setminus Y; \mathcal{G})$, and for compact Y also $H_*(X, Y; \mathcal{G}) = H_*(X/Y, y; \mathcal{G})$ (y is the image of Y under the identification map). For compact coefficient groups the first of these equalities was obtained in Kolmogorov (1936f).

With Massey's approach (see Sect. 2.5), the proofs of the indicated properties of the groups $H_*(X; G)$ are more complicated, compare Sect. 4.1, Chap. 2.

For a *proper map* $f : X \to X'$ and for an arbitrary canonical cover α' of the space X' it is easy to construct a canonical cover α of the space X that is inscribed in the inverse image of α' (see Lemma 5 in Sklyarenko (1969)). This guarantees the existence, for any system ω' of such covers of X', of the fundamental system ω of canonical covers of X, for which there exists a well defined map $C_*^\omega(X,;f^*\mathcal{G}) \to C_*^{\omega'}(X',\mathcal{G})$ inducing a map $f_* : H_*(X;f^*\mathcal{G}) \to H_*(X',\mathcal{G})$ (in the case of metrizable X and X' the systems ω and ω' can be replaced by sequences, see Sklyarenko (1969)). As was mentioned in Sect. 2.5, the map f_* is automatically defined in Massey's approach (in which the chains are defined as a functor of X).

To prove that $f_* = g_*$ when there exists a proper homotopy between f and g, it is enough to verify that $i_{1*} = i_{2*}$ for the embedding of X onto the "upper" and the "lower" bases of the space $X \times [0,1]$ (the *homotopy property*). A similar statement for cohomology follows, for example, from the Vietoris-Begle theorem (Sect. 6.3, Chap. 3) and therefore the considerations similar to those presented at the beginning of Sect. 2.2 above can be applied in this case.

3.3. Homology with Supports. Let φ be an arbitrary family of closed subsets of X. The *homology groups $H_p^\varphi(X;\mathcal{G})$ with supports in φ* are defined as $\varinjlim_{F \in \varphi} H_p(F;\mathcal{G})$. This definition does not depend on the selection of the system ω or on whether the homology is described in terms of Massey or Čech chains. Let $C_*^{\omega\varphi}(X;\mathcal{G}) \subset C_*^\omega(X;\mathcal{G})$ be the subcomplex consisting of all chains with supports in φ (see the end of Sect. 2.3). It follows from the exactness of the $\varinjlim$-functor that $H_p^\varphi(X;\mathcal{G}) = H_p(C_*^{\omega\varphi}(X;\mathcal{G}))$. The role of supports will be clarified in great detail in connection with considerations involving sheaves of chains, see §4.

Many usual properties of homology (like the exactness of the homology sequence for pairs of spaces or triples of coefficients, the excision or homotopy properties, the universal coefficient formulas) are established for H_*^φ by the limit operation with respect to $F \in \varphi$.

3.4. Homology with Compact Supports. Of independent interest is the case of compact supports $\varphi = c$, not only for locally compact but also for arbitrary Hausdorff spaces. In fact, the definition of H_*^c was given already in Sect. 1.5, Chap. 2. In the case of Massey chains[38] $H_*^c(X;G) = H_*(\varinjlim C_*(C;G))$, where the C are all the compact subsets of the topological space X. For a compact $Y \subset X$ it is clear that $H_*^c(X,Y;\mathcal{G}) = H_*^c(X/Y,y;\mathcal{G})$.

The argument in favor of H_*^c as homology of general spaces was already presented in Sect. 1.5, Chap. 2 (see also Sects. 5.5 and 5.6, Chap. 8). Another aspect favoring H_*^c is the fact that of the two possible ways $\check{H}_*$ and $\check{H}_*^c$ of defining *Čech homology* of a space (see Sect. 1.1) the second one is more natural (even though the groups $\check{H}_*$ are more frequently used, see Lefschetz (1942), Eilenberg and Steenrod (1952)[39]) since, for example, in the case when

---

[38] Similarly, for *Čech-type chains* indicated in the footnote in Sect. 2.3 above.

[39] In the monograph Lefschetz (1942) they are defined by considering finite covers, compare Sect. 2.1, Chap. 2.

G is a compact group or a field, the theory $\check{H}^c_*$ (by contrast with $\check{H}_*$) is exact (see Sect. 2.4) and coincides with H^c_*. According to similar considerations, in all those cases when for a topological space X it is possible, by taking the inverse limit, to define the *Čech-type chains* $\check{C}_*(X)$, the homology of X is more naturally defined by starting not with $\check{C}_*(X)$, but with the complex $\check{C}^c_*(X) = \varinjlim C_*(C)$, where the C are the compact subspaces of X.

§4. Sheaves of Chains. Cartan's Spectral Sequence

The goal of this paragraph is to show that the chains of a locally compact space are the sections of the flabby sheaves defined by them, to provide a description of the corresponding spectral sequence and to clarify the role of supports in the description of a homology theory. We also sketch a plan for constructing similar sheaves of cochains.

4.1. Sheaves of Cochains Corresponding to a System ω of Covers. Since the complex $\check{C}^*_{\omega c}(X; G)$ is the union $\bigcup_{\alpha \in \omega} C^*_c(\alpha; G)$, for every cochain a we define its *support* $|a|$ as the finite union of all intersections of the elements of a suitable cover $\alpha \in \omega$ (see Sec. 4.2, Chap. 2) on which a is different from zero (as on the simplices of the nerve of α). Here we have taken into account the fact that $C^*_c(\alpha; G) \to C^*_c(\beta; G)$ is a monomorphism for every $\alpha < \beta$. Since the cochains of a closed subset Y are defined by the restrictions α' of the covers $\alpha \in \omega$ to Y, the restriction of a to Y is equal to zero if and only if $|a| \subset X \setminus Y$ (which once more demonstrates that $\check{C}^*_{\omega c}(X, Y; G) = \check{C}^*_{\omega c}(X \setminus Y; G)$, compare with Sect. 4.2, Chap. 2 and Sect. 3.2 above).

Let $\check{C}^*_B(X; G) \subset \check{C}^*_{\omega c}(X; G)$ be the subcomplex consisting of all cochains with supports in $B \subset X$. Consider the differential sheaf $\check{\mathcal{C}}^*_\omega(G)$ generated by the presheaf $U \to \check{C}^*_{\omega c}(X; G)/\check{C}^*_{X \setminus U}(X; G) = C(U)$ (those quotient complexes do not determine $H^*_c(U; G)$ or $H^*(U; G)$, compare Sect. 4.3, Chap. 3). It is true that the sheaves $\check{\mathcal{C}}^*_\omega(G)$ are soft and $\Gamma_c(\check{\mathcal{C}}^*_\omega(G)|_U) = \check{C}^*_{\omega c}(X, X \setminus U; G) = \check{C}^*_{\omega c}(U; G)$.

To verify this, let us notice that the maps $r : C(U) \to \check{\mathcal{C}}^*_\omega(G)(U)$ (compare Sect. 1.2, Chap. 3) are isomorphisms. Indeed, a cochain $a \in \check{C}^*_{\omega c}(X; G)$ defines the zero element in $C(U)$ precisely when $|\alpha| \subset X \setminus U$, and therefore the presheaf $C(U)$ satisfies condition (S1) of Sect. 1.1, Chap. 3. It follows, that to every cochain a with support in U there corresponds a section of $\check{\mathcal{C}}^*_\omega(G)$ with the same support. Conversely, if s is a section of $\check{\mathcal{C}}^*_\omega(G)$ with the compact support $|s| \subset U$, then for every point $x \in |s|$ there exist a neighborhood $V_x \subset U$ and an element $c_x \in C(V_x)$ such that $r(c_x)$ coincides with s on V_x. Let us select a finite cover $V_1, \ldots, V_k$ from the cover $\{V_x\}$. Let $c_i \subset C(V_i)$ be the elements determining s over V_i. Since there are only a finite number of them, there exists $\alpha \in \omega$ such that all the c_i can be represented as cochains $a_i \in C^*_c(\alpha; G)$, and because c_i, c_j coincide on $V_i \cap V_j$, there exists a cochain

$a \in C_c^*(\alpha; G)$ whose restriction to V_i coincides with a_i for all $i = 1, \ldots, k$. We can assume that $a = 0$ on the complement of $\bigcup_i V_i \subset U$. Accordingly, $\alpha \in \check{C}_{\omega c}^*(X; G)$ coincides with s.

Similar considerations prove c-softness of $\check{C}_\omega^*(G)$ (see Sect. 4.4, Chap. 3). Since X is locally compact and paracompact, it is easy to see that c-softness implies softness of sheaves (see Corollary 2.15.6 in Bredon (1967)). The fact that $\check{C}_\omega^*(G)$ is a resolution of G follows from the same argument as in Sect. 1.4, Chap. 3. All these questions are considered in more detail in §7 in Sklyarenko (1969).

4.2. Sheaves of Chains. [40] Let us consider the presheaf $U \to C_*^\omega(X, X \setminus U; \mathcal{G}) = C_*^\omega(U; \mathcal{G})$. The differential sheaf $\mathcal{C}_*^\omega(\mathcal{G})$ generated by it is flabby and its sections over U coincide with $C_*^\omega(U; \mathcal{G})$.

In the case of constant coefficients $C_*^\omega(U; G) = \mathrm{Hom}_R(\check{C}_{\omega c}^*(U; R), G)$, and for $U = \bigcup_\lambda U_\lambda$, in accordance with the previous section $\check{C}_{\omega c}^*(U; R) = \sum_\lambda \check{C}_{\omega c}^*(U_\lambda; R)$, and therefore the presheaf satisfies condition (S1) in Sect. 1.1, Chap. 3. Condition (S2) follows easily from the formula

$$\check{C}_{\omega c}^*(V; R) \cap \check{C}_{\omega c}^*(U; R) = \check{C}_{\omega c}^*(V \cap U; R) = \sum_\lambda (\check{C}_{\omega c}^*(V; R) \cap \check{C}_{\omega c}^*(U_\lambda; R)).$$

Thus for $\mathcal{G} = G$ the chains $C_*^\omega(U; G)$ are sections of $\mathcal{C}_*^\omega(G)$. The analogous statement in the general case follows from the fact that for a sufficiently fine cover $\{V_\lambda\}$, both the complex $C_*^\omega(X; \mathcal{G})$ and the sheaf $\mathcal{C}_*^\omega(\mathcal{G})$ can be obtained from $C_*^\omega(V_\lambda; G)$ (realized in the same way as in Sect. 3.2 with the aid of a single system ω) and from the sheaves of the form $\mathcal{C}_*^\omega(G)$ constructed over V_λ by using the same identifications over the intersections $V_\lambda \cap V_\mu$ of the stalks G_λ of the systems $\mathcal{G}|_{V_\lambda}$.

For arbitrary open $U \subset V$ the module $\check{C}_{\omega c}^*(U; \mathbb{R})$ is the kernel of the map of $\check{C}_{\omega c}^*(V; \mathbb{R})$ onto the free R-module $\check{C}_{\omega c}^*(V \setminus U; \mathbb{R})$ (see Sect. 4.2, Chap. 2) and thus is a direct summand. Accordingly, for sufficiently small V the restrictions of elements of $C_*^\omega(V; \mathcal{G})$ to $U \subset V$ are epimorphisms. This implies the flabbiness of $\mathcal{C}_*^\omega(\mathcal{G})$ (compare Exercise 10 in Chap. 2 in Bredon (1967)). Such sheaves were constructed in Sklyarenko (1969).

4.3. The Role of Supports. Let Y be a subset of a locally compact space X and let θ be a family of closed subsets of X contained in Y. The subcomplex $C_*^{\omega\theta}(X; \mathcal{G}) \subset C_*^\omega(X; \mathcal{G})$ of chains with supports in θ (which are clearly sections of $\mathcal{C}_*^\omega(\mathcal{G})$ with supports in Y) determines the homology $H_*^\theta(Y; \mathcal{G})$ (which coincides with $H_*^\theta(X; \mathcal{G})$) of the subspace Y (with supports in θ). For closed Y we have $H_*^\theta(Y; \mathcal{G}) = H_*(Y; \mathcal{G})$, and $H_*^\theta(Y; \mathcal{G})$ coincides with the "ordinary" homology $H_*^c(Y; \mathcal{G})$ when Y has a compact closure. The quotient complex determines the *homology $H_*(X, Y; \mathcal{G})$ of the pair* (X, Y). Since the sheaves $\mathcal{C}_*^\omega(\mathcal{G})$ are flabby, in typical situations (for example for closed or open Y) it

---

[40] For standard applications, see Sects. 1.5, 2.1, 3.3 and §5, Chap. 8.

coincides with the complex of sections of the restriction of the sheaf $\mathcal{C}_*^\omega(\mathcal{G})$ to $X \setminus Y$. Let us point out (this relates also to the remaining part of this section) that the picture which appears in the description of the homology of subspaces and pairs is, in a sense, opposite to the one for cohomology (Sects. 4.1-4.3, Chap. 3).

Since *homology with supports* is used no less frequently than the original H_* (for example, when $\varphi = c$), we shall indicate how it is possible to determine for it the homology of subspaces or pairs.[41] The general point of view is to define the homology of $Y \subset X$ as $H_*^{\theta(\varphi)}(Y; \mathcal{G})$, where $\theta(\varphi) = \varphi|_Y$ (all $\Phi \in \varphi$ contained in Y), while homology of (X, Y) is determined by the corresponding quotient complex of chains $C_*^{\omega\theta}(X; \mathcal{G})$ with respect to $C_*^{\omega\theta(\varphi)}(X; \mathcal{G})$. Clearly, the last complex coincides with the restrictions of sections in $\Gamma_\varphi(\mathcal{C}_*^\omega(\mathcal{C}))$ to $X \setminus Y$, and in the case of an open or closed Y (due to the fact that the sheaves $\mathcal{C}_s^\omega(\mathcal{G})$ are flabby) with the sections of the restriction of $\mathcal{C}_*^\omega(\mathcal{G})$ to $X \setminus Y$ that have supports in $\varphi \cap (X \setminus Y)$ (in the case of an open Y the family φ should be paracompactifying). This leads to the following *exact sequence of the homology of a pair with paracompactifying family φ of supports* (see also Sect. 5.4, Chap. 8)

$$\ldots \to H_k^{\theta(\varphi)}(Y; \mathcal{G}) \to H_k^{\varphi}(X; \mathcal{G}) \to H_k^{\varphi}(X, Y; \mathcal{G}) \to H_{k-1}^{\theta(\varphi)}(Y; \mathcal{G}) \to \ldots$$

For $\varphi = c$ we also have $\theta(\varphi) = c$. If φ is the family of all closed subsets, then in the case of a closed Y it is possible to omit the symbols φ and $\theta(\varphi)$ in this sequence. At the same time, for an arbitrary Y it is possible to omit only the symbol φ (in this case $\theta(\varphi) = \theta$, see above). In the latter case the homology of Y obviously depends both on the topological structure of Y itself and on the way in which Y is embedded in X (for example, it coincides with $H_*^c(Y; \mathcal{G})$ generally speaking only under the condition that the closure of Y in X is compact). For a closed Y the restriction of $\mathcal{C}_*^\omega(\mathcal{G})$ to $X \setminus Y$ is the sheaf of chains on $X \setminus Y$, and $H_*^{\varphi}(X, Y; \mathcal{G}) = H_*^{\varphi \cap (X \setminus Y)}(X \setminus Y; \mathcal{G})$ (in particular, $H_*^c(X, Y; \mathcal{G}) = H_*(X \setminus Y; \mathcal{G})$ if the closure of $X \setminus Y$ is compact). In the general case this is not true (for example, for open Y and $\varphi = c$).

4.4. Sheaves of Local Homology. These sheaves are denoted by $\mathcal{H}_k(\mathcal{G})$ and are determined by the presheaves $U \to H_k^c(X, X \setminus U; \mathcal{G})$. The stalk of $\mathcal{H}_k(\mathcal{G})$ at a point $x \in X$ is the limit $\varinjlim H_k^c(X, X \setminus U; \mathcal{G})$ with respect to the neighborhoods U of this point and this limit clearly coincides with the *local homology module* $H_k^x(G) = H_k^c(X, X \setminus x; \mathcal{G})$. For a locally compact X and for sufficiently small neighborhoods we have $H_k^c(X, X \setminus U; \mathcal{G}) = H_k(U; \mathcal{G})$, and the sheaves $\mathcal{H}_k(\mathcal{G})$ are the *derived sheaves of the differential sheaf* $\mathcal{C}_*^\omega(\mathcal{G})$. In this case it is also possible to verify in a different way that the stalks of $\mathcal{H}_k(\mathcal{G})$ coincide with the local homology. Namely, the stalks of $\mathcal{C}_*^\omega(\mathcal{G})$ at the point $x \in X$ are the

---

[41] See also Sect. 5.8, Chap. 8.

sections of $\mathcal{C}^\omega_*(\mathcal{G})$ over $x \in X$ which define (according to the previous section) homology of the pair $(X, X \setminus x)$.

4.5. Spectral Sequence of a Differential Sheaf of Chains. As we have seen, by contrast with sheaves of cochains, sheaves of chains do not form an exact sequence (a resolution). In accordance with Sect. 6.1 of Chap. 3, to every flabby differential sheaf $\mathcal{C}^*$ defined as $\mathcal{C}^k = \mathcal{C}^\omega_{-k}(\mathcal{G})$ there corresponds a spectral sequence with the second term $E_2^{pq} = H^p_\varphi(X; \mathcal{H}_{-q}(\mathcal{G}))$. If the *homological dimension* n over the group $G = \mathcal{G}_x$ of a space X is finite, then $\mathcal{H}_k(\mathcal{G}) = 0$ for $k > n$ (see §3, Chap. 8), and therefore the spectral sequence is convergent to the homology $H^\varphi_{-p-q}(X; \mathcal{G})$ and there exists a natural map $\nu : H^p_\varphi(X; \mathcal{H}_n(\mathcal{G})) \to H^\varphi_{n-p}(X; \mathcal{G})$ (see Sect. 6.1, Chap. 3). The necessity of the condition $n < \infty$ is clearly visible if we consider the example of the spaces X defined as a countable product of circles (for this space all sheaves $\mathcal{H}_k(\mathcal{G})$ are trivial).

The spectral sequence described above was first obtained by Cartan in the language of the singular theory (see Cartan (1950-51)). For a more complete description see §8, Chap. 5 in Bredon (1967). One of its implications, mentioned in practically every exposition dealing with this question, is the *Poincaré duality*. In the case when X is a (generalized) manifold, $\mathcal{H}_p(\mathcal{G}) = 0$ for $p \neq n$, and the spectral sequence degenerates to the isomorphism $H^p_\varphi(X; \mathcal{H}_n(\mathcal{G})) = H^\varphi_{n-p}(X; \mathcal{G})$. However, more complete results can be obtained in a much simpler way (in particular, without using spectral sequences) as follows (see Chap. 8): the sheaves $\mathcal{C}^\omega_p(\mathcal{G})$ numbered in reverse order can be interpreted in this particular situation as a resolution of the *"orientation"* sheaf $\mathcal{H}_n(\mathcal{G})$.

Chapter 6
Products

§1. Alexander-Kolmogorov Product ($\smile$-Product)

The product operation in cohomology was first defined (in the framework of a specific cohomology theory) by A.N. Kolmogorov (see Kolmogorov (1936b)) and Alexander (see Alexander (1935)). Below we describe a general construction of a product.

1.1. Definitions. Let $\mathcal{R}$ be a sheaf of rings (with an identity). A sheaf $\mathcal{C}$ is called a *left (right) $\mathcal{R}$-module* if for every open $U \subset X$ the groups $\mathcal{G}(U)$ are left (right) $\mathcal{R}(U)$-modules (and the restrictions $\mathcal{G}(U) \to \mathcal{G}(V)$, for $V \subset U$, are homomorphisms of modules compatible with the ring homomorphisms

$\mathcal{R}(U) \to \mathcal{R}(V))$. We shall denote by $\mathcal{G}_1 \otimes_{\mathcal{R}} \mathcal{G}_2$ the sheaf generated by the presheaf $U \to \mathcal{G}_1(U) \otimes_{\mathcal{R}(U)} \mathcal{G}_2(U)$, and we shall call it the *tensor product* of right and left $\mathcal{R}$-modules. Let $0 \to \mathcal{G}_1 \to \mathcal{L}^*$, $0 \to \mathcal{G}_2 \to \mathcal{M}^*$ be resolutions consisting of right and left $\mathcal{R}$-modules. Similarly to Sect. 2.5, Chap. 3 their tensor product $\mathcal{L}^* \otimes_{\mathcal{R}} \mathcal{M}^*$ is a *bigraded differential sheaf* which is naturally considered in full gradation and with total differential.

If we take as $\mathcal{L}^*$ and $\mathcal{M}^*$ the Godement resolutions $\mathcal{C}^*(\mathcal{G}_1)$ and $\mathcal{C}^*(\mathcal{G})$ (Sect. 2.4, Chap. 3), then $\mathcal{L}^* \otimes_{\mathcal{R}} \mathcal{M}^*$ is a resolution of $\mathcal{G}_1 \otimes_{\mathcal{R}} \mathcal{G}_2$. Indeed, the exact $\underrightarrow{\lim}$-functor commutes with the operation $\otimes$ and therefore for the stalks we have $(\mathcal{L}^* \otimes_{\mathcal{R}} \mathcal{M}^*)_x = \mathcal{L}^*_x \oplus_{\mathcal{R}_x} \mathcal{M}^*_x$. Moreover, for $x \in U$ there exists a projection $\mathcal{C}^0(\mathcal{G}_1)(U) \to (\mathcal{G}_1)_x$ (the sections of $\mathcal{C}^0(\mathcal{G}_1)$ over U are assigned their values at the point x) and accordingly the embeddings $(\mathcal{G}_1)_x \subset \mathcal{C}^0(\mathcal{G}_1)_x$ of stalks are split. By a similar argument all other homomorphisms in the exact sequence of stalks $\mathcal{C}^*(\mathcal{G}_1)_x$ are split and this property is called the *stalkwise homotopic triviality of the Godement resolution*. Since the same property is satisfied also for $\mathcal{M}^*$, a simple diagram chasing (compare Sect. 2.5, Chap. 3) proves that the sequence of stalks $0 \to (\mathcal{G}_1 \otimes_{\mathcal{R}} \mathcal{G}_2)_x \to (\mathcal{L}^* \otimes_{\mathcal{R}} \mathcal{M}^*)_x$, considered with its complete gradation, is exact.

The homomorphisms $r : \mathcal{G}_1(U) \otimes_{\mathcal{R}(U)} \mathcal{G}_2(U) \to (\mathcal{G}_1 \otimes_{\mathcal{R}} \mathcal{G}_2)(U)$ (compare Sect. 1.2, Chap. 3) applied to $\mathcal{L}^* \otimes_{\mathcal{R}} \mathcal{M}^*$ define a map $\Gamma_\varphi(\mathcal{L}^*) \otimes_{\mathcal{R}(X)} \Gamma_\psi(\mathcal{M}^*) \to \Gamma_{\varphi \cap \psi}(\mathcal{L}^* \otimes_{\mathcal{R}} \mathcal{M}^*)$ which together with the comparison homomorphism γ corresponding to the resolution $\mathcal{L}^* \otimes_{\mathcal{R}} \mathcal{M}^*$ (see Sect. 2.1, Chap. 3) defines a homomorphism (the *cup product*) $\smile : H^p_\varphi(X; \mathcal{G}_1) \otimes_{\mathcal{R}(X)} H^q_\psi(X; \mathcal{G}_2) \to H^{p+q}_{\varphi \cap \psi}(X; \mathcal{G}_1 \otimes_{\mathcal{R}} \mathcal{G}_2)$. In the case of a constant sheaf $\mathcal{R} = R$ the symbol $\mathcal{R}(X)$ can clearly be replaced by R (see also Sect. 5.3., Chap. 8).

When $\mathcal{G}_1 = \mathcal{G}_2 = \mathcal{R}$ we have $\mathcal{R} \otimes_{\mathcal{R}} \mathcal{R} = \mathcal{R}$ and the cup product converts $H^*_\varphi(X; \mathcal{R})$ into *cohomology ring* (with an identity if φ is the family of all closed subsets). Moreover, for $\varphi \subset \psi$ the isomorphism $\mathcal{R} \otimes_{\mathcal{R}} \mathcal{G} \to \mathcal{G}$ induces in $H^*_\varphi(X; \mathcal{G})$ the structure of a $H^*_\psi(X; \mathcal{R})$-module.[42].

Frequently products are considered for cohomology of subspaces or pairs (for example in Massey (1978b), $H^p_c(A) \otimes H^q_c(B) \to H^{p+q}_c(A \cap B)$, $H^p(X, A) \otimes H^q(X, B) \to H^{p+q}(X, A \cup B)$ for $A, B \subset X$, etc.). However, let us recall (see §4, Chap. 3) that such cohomology can be reduced to cohomology of X by a suitable choice of the coefficient sheaf or supports (to a great extent this is true also for the homology considered in the following section).

1.2. Fundamental Properties. a) It is clear that for $p = q = 0$ the operation $\smile$ coincides with the map $r : \Gamma_\varphi(\mathcal{G}_1) \otimes_{\mathcal{R}(X)} \Gamma_\psi(\mathcal{G}_2) \to \Gamma_{\varphi \cap \psi}(\mathcal{G}_1 \otimes_{\mathcal{R}} \mathcal{G}_2)$. b) If $\mathcal{E}' : 0 \to \mathcal{G}_1' \to \mathcal{G} \to \mathcal{G}_1'' \to 0$ is a short exact sequence that remains exact after the operation $\otimes_{\mathcal{R}} \mathcal{G}_2$ then, since after taking the tensor product with $\mathcal{M}^* = \mathcal{C}^*(\mathcal{G}_2)$ the triple of Godement resolutions corresponding to $\mathcal{E}'$ does not loose the property of being exact, for $a \in H^p_\varphi(X; \mathcal{G}_1'')$ and $b \in H^q_\psi(X; \mathcal{G}_2)$ we have $(\delta' a) \smile b = \delta(a \smile b)$ (δ' and δ are the connecting homomorphisms

---

[42] When $\mathcal{R}$ is a sheaf of commutative rings.

corresponding to $\mathcal{E}'$ and $\mathcal{E}'\otimes_\mathcal{R}\mathcal{G}_2$). c) If $\mathcal{E}'' : 0 \to \mathcal{G}_2' \to \mathcal{G}_2 \to \mathcal{G}_2'' \to 0$ is a short exact sequence for which the sequence $\mathcal{G}_1 \otimes \mathcal{E}''$ is also exact, then (for suitable a and b) $a \smile (\delta''b) = (-1)^p\delta(a \smile b)$. The appearance of the sign $(-1)^p$ can be explained by the relationship $\delta'\delta'' = -\delta''\delta'$ for the composition of connecting homomorphisms corresponding to the sides of the "square" $\mathcal{E}'\otimes_\mathcal{R}\mathcal{E}''$ (see Sect. 4.1, Chap. 3 in Cartan and Eilenberg (1956)) and by the fact that on passing to δ both sides should determine the composition $\delta\delta$.

In the case when $\mathcal{R}$ is a sheaf of commutative rings, there exists a natural isomorphism between the sheaves $\mathcal{G}_1 \otimes_\mathcal{R} \mathcal{G}_2$ and $\mathcal{G}_2 \otimes_\mathcal{R} \mathcal{G}_1$. By a standard argument, properties a)–c) imply the relationship d) $a \smile b = (-1)^{pq}b \smile a$. In particular such a relationship (for the indicated $\mathcal{R}$) is also true for the cohomology ring $H^*_\varphi(X;\mathcal{R})$ (*skew commutativity*).

The cup product can be defined also by using other resolutions $\mathcal{L}^*$ and $\mathcal{M}^*$ (acyclic with respect to corresponding families of supports) for which $\mathcal{L}^* \otimes_\mathcal{R} \mathcal{M}^*$ is a resolution of $\mathcal{G}_1 \otimes_\mathcal{R} \mathcal{G}_2$, or in some other way. It is true that it is uniquely determined by the properties a)-c) described above. A simple proof of this (see §7, Chap. 2 in Bredon (1967)) is based essentially on the fact that cohomology can be treated as a derived functor. In the case of a sheaf $\mathcal{R}$ of commutative rings any of the conditions b) or c) can be replaced by d). It is also easy to establish some other properties of the $\smile$-product (for example its distributivity with respect to addition, and for commutative $\mathcal{R}$ its associativity).

In the language of Alexander-Spanier cochains (Sect. 3.1, Chap. 2) the ring structure in $H^*(X;R)$ is defined by the relationship

$$(\xi_1 \otimes \xi_2)(x_0,\dots,x_{p+q}) = \xi_1(x_0,\dots,x_p) \cdot \xi_2(x_p,\dots,x_{p+q}).$$

In a similar way it is described in terms of continuous Alexander-Spanier cochains (Korobov (1969), see Sect. 3.2, Chap. 2), and in terms of singular and simplicial cochains. In the category of smooth manifolds it is defined by the exterior product of differential forms.

1.3. Cartesian Product of Cohomology Classes.

In the case when $\mathcal{G}_1$ is a sheaf of right R-modules on X and $\mathcal{G}_2$ is a sheaf of left R-modules on Y there exists a so called *total tensor product* $\mathcal{G}_1\hat{\otimes}_R\mathcal{G}_2 = \pi_X^*\mathcal{G}_1 \otimes_R \pi_Y^*\mathcal{G}_2$, namely, the product of the inverse images of $\mathcal{G}_1$ and $\mathcal{G}_2$ under the natural projections π_X and π_Y of the product $X \times Y$ onto X and Y. It is clear that for $X = Y$ the sheaf $\mathcal{G}_1 \otimes_R \mathcal{G}_2$ coincides with the restriction of $\mathcal{G}_1\hat{\otimes}_R\mathcal{G}_2$ to the diagonal of $X \times X$. The constructions described above can be applied (see §6, Chap. 2 in Godement (1958)) to define the Cartesian product of cohomology

$$H^p_\varphi(X;\mathcal{G}_1) \otimes_R H^q_\psi(Y;\mathcal{G}_2) \to H^{p+q}_{\varphi\times\psi}(X \times Y; \mathcal{G}_1\hat{\otimes}_R\mathcal{G}_2).$$

This operation is used in computing the cohomology of $X \times Y$. In the case of locally compact spaces with $\varphi = c$ and in some other cases we can obtain in this way the so-called *Künneth formulas* (see for example, §18, Chap. 2 and §8, Chap. 4 in Bredon (1967)).

§2. The $\frown$-Product

2.1. Basic Definitions. There exists also a dual construction in which homology groups are used. For an arbitrary locally compact space X it leads to the following operation, called the $\frown$-*product*.

$$\frown: H^{\varphi}_{p+q}(X;G_1) \otimes_R H^p_{\psi}(X;G_2) \to H^{\varphi\cap\psi}_q(X;G_1 \otimes_R G_2).$$

This operation is defined by the $\frown$-product of chains and cochains

$$\frown: C_{p+q}(X;G_1) \otimes_R C^p(X;G_2) \to C_q(X;G_1 \otimes_r G_2),$$

described by the formula $< \xi \frown a, v >=< \xi \otimes G_2, a \smile v >$ for $\xi \in C_{p+q}(X;G_1) = \mathrm{Hom}_R(C^{p+q}_c(X;R),G_1)$, and $a \in C^p(X;G_2)$, where $\xi \frown a \in C_q(X;G_1 \otimes_R G_2) = \mathrm{Hom}_R(C^q_c(X;R),G_1 \otimes_R G_2)$ and v is an arbitrary element of $C^q_c(X;R)$. Here, by cochains $C^p(X;G_2)$ we mean the sections of a flabby resolution of the constant sheaf G_2 on X, and as $C^k_c(X;R)$ we use the free cochains introduced in Sect. 4.1 or 4.2 in Chap. 2. The symbol $\xi \otimes G_2$ denotes the homomorphism $\xi \otimes_R \mathbb{1}$, where $\mathbb{1}$ is the identity isomorphism of G_2. It is easy to verify that this operation satisfies the following *boundary formula*

$$\delta(\xi \frown a) = (-1)^p\{(\delta\xi) \frown a - \xi \frown (da)\}.$$

Since the support of $\xi \frown a$ is clearly the intersection of the supports of ξ and a, it is possible to define the required product.[43]

The $\frown$-product operation is applied basically in the case when φ is the family of all closed subsets. In the case when ψ is the family of all closed subsets and $\varphi = c$ it is defined for an arbitrary (paracompact) space X. Indeed, for every $h \in H^c_{p+q}(X;G_1)$ there exists a compact subset C of X and an element $h' \in H_{p+q}(C;G_1)$ whose image is equal to h. The restriction α' to C of an element $\alpha \in H^p(X;G_2)$ defines $h' \frown \alpha' = k' \in H_q(C;G_1 \otimes_R G_2)$. Let $h \frown \alpha = k$ be the image of k' under the embedding $C \subset X$. It is easy to show that k does not depend on the choice of C and h' (see §9.4 in Massey (1978b)).

In a similar way, in addition to supports, we can take into account also some other factors. In this way it is possible to define a $\frown$-product of the form

$$H^{\varphi}_{p+q}(X,Y_1 \cup Y_2;G_1) \otimes_R H^p_{\psi}(X,Y_1;G_2) \to H^{\varphi\frown\psi}_q(X,Y_2;G_1 \otimes_R G_2)$$

(see §9.5 in Massey (1978b)). All these constructions can be extended also to the case of locally constant systems of coefficients $\mathcal{G}_1, \mathcal{G}_2$.

2.2. The Case of Manifolds. Let X be an orientable n-dimensional (generalized) manifold over R (the sheaf $\mathcal{H}_n(R)$ is equal to R), and let $\xi = \mathbb{1} \in H_n(X;R) = R$ be the fundamental homology class (determined by the

---

[43] See also Sect. 5.3, Chap. 8.

"orientation" cycle). In this case for $p + q = n$ the map $\frown: a \to \xi \frown a$, $\frown: H^p_\varphi(X; R) \to H^\varphi_q(X; R)$ is an isomorphism which is the *Poincaré duality*, see Chap. 11 in Massey (1978b). A similar description of this duality is possible also in the case when $\mathcal{H}_n(R)$ is a locally constant sheaf with stalk R (see §10, Chap. 5 in Bredon (1967)). For a general approach see Chap. 8 below.

Chapter 7
Axiomatic Approach

In this chapter all homology and cohomology are considered only with constant coefficients.

§1. Axioms in the Category of Polyhedra

1.1. Uniqueness Theorem of Eilenberg-Steenrod. The theorem, published for the first time in 1945 in Eilenberg and Steenrod (1945), states that in the category $\mathcal{A}_0$ of compact polyhedra homology theory is uniquely determined by the following properties, which since then have acquired the status of axioms:

1) To each pair $(X, Y) \in \mathcal{A}_0$ (where Y is a subcomplex in some triangulation of X) there corresponds an exact homology sequence

$$\ldots \to H_{p+1}(X, Y) \xrightarrow{\delta} H_p(Y) \xrightarrow{i} H_p(X) \xrightarrow{j} H_p(X, Y) \to \ldots$$

which is a covariant functor of the argument (X, Y) (a single polyhedron X is identified with the pair $(X, \emptyset)$; the maps i and j are induced by the corresponding inclusions).

2) Homotopic maps $(X, Y) \to (X', Y')$ induce the same homomorphisms of homology sequences (the *homotopy axiom*).

3) The inclusion $(X \setminus \operatorname{Int} Y, Y \setminus \operatorname{Int} Y) \subset (X, Y)$ induces an isomorphism $H_p(X \setminus \operatorname{Int} Y, Y \setminus \operatorname{Int} Y) \to H_p(X, Y)$ (the *excision axiom*).

4) For a one point space P we have $H_q(P) = 0$ for all $q \neq 0$ (the *dimensionality axiom*).

A similar result is also true for cohomology (all arrows in Axiom 1 are reversed, and the cohomology sequence arising in this way is a contravariant functor of (X, Y)).

In the axiomatic approach the group $G = H_0(P)$ (or $H^0(P)$) is called the *coefficient group* of the particular theory. The meaning of the theorem is that for theories with identical coefficients and satisfying Axioms 1-4 every isomorphism between the coefficients can be extended (in the category $\mathcal{A}_0$) in a unique way to an isomorphism between these theories.

Properties 1-4 are clearly the most elementary ones. In addition, there are many other properties and the importance of the Eilenberg-Steenrod theorem also consists of the fact that all the properties of homology (and cohomology) can be obtained (in the category $\mathcal{A}_0$) from these four. The axiomatic approach was developed in Eilenberg and Steenrod (1952).

1.2. Additivity Axioms. The axiomatic approach attained its further development in the works of Milnor (see Milnor (1960, 1962)). In Milnor (1962) it is proved that in the category $\mathcal{K}$ of (cell) polyhedra the homology H_*^c (cohomology H^*) is uniquely determined by the Eilenberg-Steenrod axioms and the property of $\sum$-*additivity* (respectively $\prod$- *additivity*) (see Sect. 1.6, Chap. 1). A similar result was obtained by C.V. Petkova (see Petkova (1972, 1973a)) for the theories H_* and H_c^* of the second kind in the category $\mathcal{A}$ of locally compact polyhedra and their proper maps (for homology a proof of this fact is actually included in Milnor (1960) even though the result is not explicitly stated). In this case the Eilenberg-Steenrod axioms are complemented by $\prod$-additivity for the H_* theory and by $\sum$-additivity for the H_c^* theory (also the excision axiom can be considered in a stronger form as an isomorphism corresponding to the inclusions $(X \setminus Y, \emptyset) \subset (X, Y)$, see Sects. 4.1, 4.2, Chap. 2 and 3.2, Chap. 5).

The fundamental scheme in the proof of these results, as well as in the proof of 1.1 is presented at the end of Sect. 2.4, Chap. 1. By contrast with the category $\mathcal{A}_0$, in the proofs of the uniqueness theorems in the categories $\mathcal{K}$ and $\mathcal{A}$ it is necessary to deal with the difficulties (see Milnor (1962), Petkova (1973)) associated with the presence of infinite-dimensional polyhedra.

1.3. Discussion of the Axiomatic Approach. The fact that the list of the Eilenberg-Steenrod axioms is not complete outside of the limits of the category $\mathcal{A}_0$ is highlighted not only by examples of functors that in fact have nothing in common with homology and cohomology and yet satisfy the axioms (see James and Whitehead (1958)), but also by the presence of the $\sum$-additivity of the theory H_*^c that is quite different from the crucial $\prod$-additivity of the theory H_*. Nevertheless, from the time they were introduced, the Eilenberg-Steenrod axioms began to play the role of a standard with which every new theory was compared. It is clear that the conditions contained in them constitute the minimum requirements expected of an arbitrary homology or cohomology. Conversely, the properties (axioms) of additivity did not receive the attention they deserve (possibly because they were not reflected in any monograph of the type similar to (Eilenberg and Steenrod (1952)), and accordingly their ro le was somewhat less visible (see Sect. 1.6, Chap. 1). In this situation the importance of the Eilenberg-Steenrod axioms was to a certain degree exaggerated and it became customary to call a homology (cohomology) theory any functors satisfying these axioms (see James and Whitehead (1958)). Sometimes it happens that such homology (less frequently cohomology) theories, created for arbitrary categories of topological spaces, are different from the simplicial theory on the subcategory of polyhedra.

By considering such "theories" as homology (or cohomology), we remove automatically the original idea consisting primarily in the fact that homology and cohomology are the result of a consideration of the boundary operation (1) or the coboundary operation (2) introduced in Chap. 1.

In the following sections we shall see that the additivity properties play a decisive role in the axiomatic approach to homology and cohomology theories in general categories of topological spaces.

1.4. Another Approach to the Axioms. For homology and cohomology of the second kind a different approach was given in Sklyarenko (1971a). Let K_λ be closed subpolyhedra of a locally finite polyhedron K, whose complements have compact closures. It can be proved that the groups of the second kind become *trivial at infinity*: $\varprojlim_\lambda H_p(K_\lambda; G) = 0, \varinjlim_\lambda H^p_c(K_\lambda; G) = 0$.[44] As it happens, the above property in the category $\mathcal{A}$, together with the Eilenberg-Steenrod axioms and the additivity properties assumed only for discrete spaces D and for zero-dimensional groups determines the homology H_* and cohomology H^*_c uniquely. The condition that the homology H_* vanishes at infinity is not equivalent to the axiom of $\prod$-additivity, for example, for non-finitely generated modules of coefficients G (for example, if we define $H_*(K; G)$ by using chain complexes $C_*(K; R) \otimes_R G$, compare Sect. 1.7, Chap. 1). The $\sum$-additivity of the cohomology H^*_c is a consequence of the fact that these groups are trivial at infinity (see Sect. 2.3 below).

§2. Compact Spaces

2.1. Additivity Properties. The Eilenberg-Steenrod axioms form the basis for any collection of axioms for homology and cohomology theories in general categories of topological spaces. It turns out, however, that outside the category of polyhedra it is always necessary to make the *excision axiom* more precise. In the category $\mathcal{B}_0$ of compact Hausdorff spaces we shall use the following form of this axiom:

3^c) The natural maps $(X, Y) \to (X/Y, y)$ induce an isomorphism of homology (and cohomology) of a pair onto a *reduced homology* (cohomology) of the quotient space (that is, the groups of the pair $(X/Y, y)$, compare with Sect. 3.2, Chap. 5).

For Čech homology and cohomology this property was established in §5, Chap. 10 in Eilenberg and Steenrod (1952) (see also §4, Chap. 2 above), and for Steenrod-Sitnikov homology in Milnor (1960) (see also Sect. 3.2, Chap. 5).

---

[44] In accordance with §4, Chap. 2 and Sect. 3.2, Chap. 5 the groups appearing in this formulation are the same as the homology and cohomology of closed neighborhoods of the point added in the one-point compactification of K. Therefore these relations are a special case of the local conditions considered in Sect. 2.3 below.

We shall say that a compact space $X = \bigcup_\lambda X_\lambda$ is a *compact bouquet* of its closed (possibly disconnected) subspaces X_λ, if $X_\lambda \cap X_\mu = x_0 \in X$ for $\lambda \neq \mu$ and each neighborhood of x_0 contains all but a finite number of the spaces X_λ (such X clearly coincides with the subspace of the product $\prod_\lambda X_\lambda$ consisting of all the points with at most one coordinate different from x_0). A functor H is called $\prod_c$-*additive* (respectively $\sum_c$-*additive*) if its values on the compact bouquets $X = \bigcup_\lambda X_\lambda$ are isomorphic with the products $\prod_\lambda H(X_\lambda)$ (respectively, direct sums $\sum_\lambda H(X_\lambda)$).

It turns out that in the category $\mathcal{B}_0$ cohomology is $\sum_c$-additive while homology is $\prod_c$-additive. The simplest way to verify this is to use the complexes $C_*^\omega(X; G)$ and $C_\omega^*(X; G)$ in defining homology and cohomology (§4, Chap. 2 and §2, Chap. 5). Indeed, we can consider a system ω of covers α containing as its elements all but a finite number of $X_\lambda \subset X$. It is then not difficult to observe that for a compact bouquet we have $C_\omega^*(X, x_0; G) = \oplus_\lambda C_\omega^*(X_\lambda, x_0; G)$ and $C_*^\omega(X, x_0; G) = \prod_\lambda C_*^\omega(X_\lambda, x_0; G)$ and that these equalities imply the indicated properties. Using Massey's approach (see Massey (1978b)) the result is obtained in a different way. Namely, the reduced homology and cohomology of the bouquet X coincide with $H_c^*(X \setminus x_0; G)$ and $H_*(X \setminus x_0; G)$ (see Theorem 2.8 in Massey (1978b)), while the chain and cochain complexes of $X \setminus x_0$ can be represented respectively as the direct sum and the direct product of the chain and cochain complexes of the open subsets $X_\lambda \setminus x_0$ of $X \setminus x_0$ (see the end of §4.7 in Massey (1978b)).[45]

2.2. Metrizable Case. In the category $\mathcal{B}_0' \subset \mathcal{B}_0$ of compact metric spaces the theories H_* and H^* are the unique theories with the properties listed above. Accordingly, in order to guarantee uniqueness as in the categories $\mathcal{A}$ and $\mathcal{K}$ it is enough to add to the Eilenberg-Steenrod axioms (strengthened by 3^c) just one requirement of $\prod_c$-additivity for homology or $\sum_c$-additivity for cohomology. This result was obtained in Milnor (1960). Another proof was given in Petkova (1972, 1973a).

The idea of the proof is the following. By passing to the one-point compactifications (or by adding a single point to finite polyhedra, compare with Sect. 1.4, Chap. 2) it is possible to represent the category $\mathcal{A}$ of locally finite polyhedra and their proper maps as a (not full) subcategory of $\mathcal{B}_0'$. Accordingly, homology (cohomology) in the category $\mathcal{B}_0'$ is an extension to $\mathcal{B}_0'$ of the unique theory in the category $\mathcal{A}$ and is the same as homology (cohomology) of the second kind (the requirements of $\prod_c$- and $\sum_c$-additivity in the category

---

[45] It is also known (see Edwards and Hastings (1976), Kahn, Kaminker and Schochet (1977)) that an arbitrary *extraordinary homology theory* (that is a theory satisfying only the axioms 1-3 of Sect. 1.1) on a category of finite polyhedra admits an extension to the category of metrizable compacta (with the excision property in the form presented above) and that this extension has the $\prod$-additivity property. Similar results were obtained in Sanebelidze (1986) without the assumption of metrizability. Such extensions of extraordinary theories have certain additional properties of Steenrod homology (see, for example, the footnote to §1 in Chap. 8).

$\mathcal{B}_0$ clearly induce the requirements of $\prod$- and $\sum$-additivity in the category $\mathcal{A}$). However, such an extension (with the excision property in the form 3^c) is unique (see Sect. 1.3, Chap. 2).

As an example of a non $\prod_c$-additive homology we can take Borel-Moore homology with coefficients in an infinitely generated group G. According to the universal coefficient formulas considered in Sect. 2.4, Chap. 5, the reduced zero- dimensional homology of the one-point compactification $\overline{D}$ of a countable discrete set D is $H_0(\overline{D}, x_o; \mathbb{Z}) \otimes G = (\prod \mathbb{Z}) \otimes G \neq \prod G$, see Sect. 1.7, Chap. 1 (for the same reasons the condition 3^c is true for this homology).

2.3. Local Properties.

Such properties can also be used in the axiomatic approach (see Sklyarenko (1971a)). Let us recall (Sect. 1.4, Chap. 3), that the (reduced) cohomology is locally trivial. For every element $h \in H^n(X; \mathcal{G})$ and every $x \in X$ there exists a neighborhood U of x such that the image of h in $H^n(U; \mathcal{G})$ is equal to zero, and therefore $\varinjlim_U H^n(U, x; \mathcal{G}) = 0$, where the U are neighborhoods of the point x. For an element $h \in H_n^c(X; \mathbb{Z})$ (even in the category $\mathcal{B}_0'$) it may be possible to find, in general, points $x \in X$, such that h belongs to the image of $H_n^c(U; \mathbb{Z}) \to H_n^c(X; \mathbb{Z})$ for arbitrarily small neighborhoods of x (see the remark following Lemma 8 in Sklyarenko (1969)). However, it is not difficult to show that the homology H_*^c of spaces with the first axiom of countability is locally trivial in the following sense: $\varprojlim_{x \in U} H_n^c(U, x; G) = 0$ (Lemma 8 in Sklyarenko (1969)).[46]

Note that the *local triviality of homology (cohomology)* can be viewed as a strengthening the dimensionality axiom from Sect. 1.1. Together with the Eilenberg-Steenord axioms strengthened by condition 3^c), this property, when combined with the additional requirement that $H_0(\overline{D}, x_0; G) = \prod G$ (and $H^0(\overline{D}, x_0; G) = \sum G$), uniquely determines homology and cohomology in the category $\mathcal{B}_0'$, see Sect. 2.2. In the case of cohomology, the last requirement is a consequence of their local triviality property (it is enough to take the direct limit with respect to n in exact sequences of zero-dimensional cohomology of the pairs $(\overline{D}, \overline{D} \setminus D_n)$, where D_n is a finite subset of D), and therefore for H^* the requirement of local triviality is equivalent to the axiom of $\sum_c$ additivity. In §4 of Petkova (1973a) it is proved that a similar result is true for homology with coefficients in finitely generated groups. For groups G that are not finitely generated this is not the case: Borel-Moore homology is locally trivial (for the same reasons as H_*), but the zero-dimensional homologies of $(\overline{D}, x_0)$ are different from $\prod G$ (see Sect. 2.1) and they form a subgroup in $\prod G$ that contains $\sum G$ (see §5 in Sklyarenko (1971a)).

By applying the inverse limit operation to the exact sequences of the reduced zero-dimensional Borel-Moore homology of the pairs $(\overline{D}, \overline{D} \setminus D_n)$ and taking into account the left exactness of the $\varprojlim$-functor we obtain the ex-

---

[46] As a corollary a similar result can be easily established at the point x_0 of the compact bouquet $X = \bigcup_\lambda X_\lambda$ when $X_\lambda \in \mathcal{B}_o'$.

act sequence $0 \to \varprojlim_{n} H_0(\overline{D} \setminus D_n, x_0; G) = 0 \to H_0(\overline{D}, x_0; G) \to \prod G \to$ $\varprojlim_{n}^1 H_0(\overline{D} \setminus D_n, x_0; G) \to 0$. It follows that for Borel-Moore homology the term on the right is different from zero (G is not finitely generated). For the Steenrod-Sitnikov homology H_*^c of first countable spaces it is always true that $\varprojlim_{x \in U}^1 H_n^c(U, x; G) = 0$ (see Sect. 2.1, Chap. 8).

2.4. The General Case. The attempts to describe in an axiomatic way the theory in the category $\mathcal{B}_0$ of compact spaces without the assumption of metrizability were made primarily for homology. V.I. Kuz'minov (see Kuz'minov (1980)) characterized H_* as the theory whose chains can be obtained as duals to arbitrary cochains which yield the integral Čech cohomology. The proof uses the construction obtained by him and B.I. Botvinnik (see Botvinnik and Kuz'minov (1979)) of a continuous map $\omega : N(X) \to X$ that produces an isomorphism of Čech cohomology, where $N(X)$ is the inverse limit of nerves of finite covers of X with unique projections. By using the filtration of $N(X)$ with the limits $N_p(X)$ of skeletons of dimension p of the nerves of covers, N.A. Berikashvili (see Berikashvili (1980)) proved that the Eilenberg-Steenrod axioms together with condition 3^c) determine H_* uniquely under the following conditions: B) if $S = \varprojlim S_\lambda$ is the inverse limit of a system of finite bouquets of spheres of the same dimension in which the restrictions of the projections $S_\lambda \to S_\mu$ to individual spheres are either homeomorphisms or constant maps into the base point, then $\varprojlim H_k(S_\lambda) = H_k(S)$; C) ω induces the isomorphisms $H_k(N_p(X)) \to H_k(X)$ for $k \leq p - 1$. The dual axioms are clearly true also for cohomology. In place of conditions B) and C) one can require (see Berikashvili (1980)) the presence of a (functorial with respect to X) universal coefficient formula of the type (8) from Sect. 1.7, Chap. 1. The same two conditions can be replaced also (see Berikashvili (1986)) by the requirement of $\prod_c$-additivity and the following condition C'): the fact that the integral Čech cohomology $H^*(X, Y; \mathbb{Z})$ is trivial implies that $H_*(X, Y) = 0$. The fact that we use cohomology in this formulation increases interest in the question relating to the existence of an axiomatic system also for cohomology.

In general, in the case of cohomology the requirement of additivity in the category $\mathcal{B}_0$ can be replaced by the following *continuity axiom* (in addition to the axioms 1-4): $H^n(X) = \varinjlim_{\lambda} H^n(X_\lambda)$ for each n, provided that $X = \varprojlim_{\lambda} X_\lambda$ (see Eilenberg and Steenrod (1952)). Such an approach leads to the Čech theory, but it cannot be applied to homology (since $\check{H}_* \neq H_*$, see Sect. 1.1, Chap. 5 or §1, Chap. 8; for $\check{H}_*$ the condition of exactness contained in Axiom 1 is not satisfied, see Sect. 2.4, Chap. 5). However, the continuity axiom can be successfully replaced (see II.N. Inasaridze and L.D. Mdzinarishvili (1980)) by the requirement of the existence (for constant coefficients) of the exact sequences, natural with respect to X, of the type (3) from Sect. 2.2, Chap. 5. The proof also makes use of the connection between X and $N(X)$. An equivalent formulation of this requirement (but under the condition that the groups $H_*(X;G)$ are considered as a bifunctor) is the following: for H_* there exist exact homology sequences (Sect. 2.3, Chap. 5) corresponding to the exact triples of coefficient groups (and depending on them in a natural way) and the theory H_* is continuous for every divisible group G (see H.I. Inasaridze (1986)). Let us recall that there always exists a transformation $H_* \to \check{H}_*$, see Sect. 2.2, Chap. 5 and §1, Chap. 8, which is an isomorphism only for algebraically compact groups G (see also Sect. 2.4, Chap. 5)

§3. Axiomatic Approach to the Theories of the Second Kind

3.1. The Role of Additivity Property. Let $\mathcal{B}$ be the category of locally compact spaces and their proper maps. In accordance with §4, Chap. 2 and Sect. 3.2, Chap. 5, the excision property for homology H_* and cohomology H_c^* is true in the form:

(3^{abs}) For a closed subset Y of X the groups $H_*(X,Y;G)$ and $H_c^*(X,Y;G)$ are naturally isomorphic to $H_*(X \setminus Y;G)$ and $H_c^*(X \setminus Y;G)$.

The cochains determining H_c^* (§4, Chap. 2) are clearly $\sum$-additive and in accordance with Sect. 3.2, Chap. 5 the chains determining H_* are $\prod$-additive (for Massey's chains and cochains this is mentioned at the end of §4.7 in Massey (1978b)). Therefore, the cohomology H_c^* is $\sum$-additive and the homology H_* is $\prod$- additive (see also Theorem 2.13 and Section 6, §4.3 of Massey (1978b)).

In accordance with Sect. 1.4, Chap. 2, every homology (or cohomology) theory defined on the subcategory $\mathcal{B}_0 \subset \mathcal{B}$ admits a natural extension to $\mathcal{B}$. We note that the property of $\prod_c$-additivity ($\sum_c$-additivity) in the subcategory $\mathcal{B}_0 \subset \mathcal{B}$ is a simple consequence of the $\prod$-additivity ($\sum$-additivity) and the property (3^{abs}) stated above. Similarly the excision property 3^c is a consequence of (3^{abs}). Therefore in the subcategory $\mathcal{B}'$ of spaces satisfying the second axiom of countability the homology H_* and cohomology H_c^* are the unique theories satisfying the Eilenberg-Steenrod axioms (together with (3^{abs})) that are respectively $\prod$-and $\sum$-additive. This result was obtained in Petkova (1972, 1973a). The possibility of such approach to an axiomatic description of H_* in the category $\mathcal{B}'$ was briefly indicated in Milnor (1960).

3.2. Borel-Moore Homology. This theory is defined directly in the category $\mathcal{B}$ and it is an example of a theory for which the natural connection between homology in $\mathcal{B}_0$ and $\mathcal{B}$, described in Sect. 1.4, Chap. 2, does not exist. Let us recall (see Sect. 2.4, Chap. 5 and Sect. 2.2 above) that Borel-Moore homology H_*' is different from H_* in particular for coefficients in modules that are not finitely generated. Precisely for this reason the groups H_*' are not invariant under a change of ring (homology H_*' with coefficients in the Abelian group

$\mathbb{Q}$ of rational numbers is different from homology with coefficients in the ring $\mathbb{Q}$, see the end of Sect. 2.2).

In accordance with its definition (§3, Chap. 5 in Bredon (1967)) Borel-Moore homology is $\prod$-additive. Since the addition of this property to the usual Eilenberg-Steenrod axioms in the category of locally compact polyhedra and their proper maps determines H_* in a unique way (see Sect. 1.2 above and Petkova (1973a)), in this category H'_* coincides with the simplicial homology of the second kind (this result does not follow from (Bredon (1967)).[47] Since H'_* is not $\prod_c$-additive (see Sect. 2.2 above), the homology H'_* of the one-point compactification of X does not have to coincide with the homology of X. In particular, for H'_* the excision property in the same form as (3^{abs}) does not hold. However, for compact Y the homology of (X, Y) is naturally isomorphic to the reduced homology of the quotient space X/Y (compare with Sect. 2.1). In the subcategory $\mathcal{B}_0 \subset \mathcal{B}$ this follows from the universal coefficient formulas (Sect. 3.10, Chap. 5 in Bredon (1967)) which represent homology with coefficients in G in terms of homology with coefficients in a basic ring which is equivalent to H_* (the validity of these formulas for homology of pairs is a consequence of the description of homology of subspaces and pairs from §5 in Chap. 5 of Bredon (1967)). For $X \in \mathcal{B}$ the ana;ogous result can be obtained by considering the map of the exact *homology sequence of a triple* (X, X_0, Y), in which $X_0 \in \mathcal{B}_0$ and $Y \subset \mathrm{Int}\, X_0$, into the exact sequence of the pair $(X/Y, X_0/Y)$, and by using the isomorphism $H'_*(X, X_0; G) = H'_*(X/Y, X_0/Y; G)$ (the existence of this isomorphism is guaranteed by Theorem 5.1, Chap. 5 in Bredon (1967)). See also the end of Sect. 2.3 above and Sects. 5.1, 5.4, Chap. 8.

3.3. Other Approaches. The requirement of additivity in an axiomatic description of H_* and H_c^* in the subcategory $\mathcal{B}' \subset \mathcal{B}$ in Sect. 3.1 can be replaced by conditions of local triviality of homology and cohomology of the space X (see Sect. 2.3). However, this condition is imposed not only at the points $x \in X$ but also (as in Sect. 1.4) at infinity. In doing so it is necessary (in general) to impose the requirement of additivity for the zero-dimensional groups of a countable discrete set D (this requirement is a consequence of local conditions also in the cases considered in Sect. 2.3). Such an approach to the axiomatic description in the category $\mathcal{B}'$ was used in Sklyarenko (1971a). It can be verified by considerations similar to those in Sect. 3.1.

It was demonstrated in the editor's remarks to Chap. 6 in the Russian translation of Massey (1978b) that cohomology in the category $\mathcal{B}$ is uniquely determined by Axioms 1-4 together with the so called *"tautness" property* with respect to all closed subsets $Y \subset X$. It states that $H_c^p(Y; G) = \varinjlim_\lambda H_c^p(Y_\lambda; G)$, where the Y_λ are the closed neighborhoods of Y in X. The analogous property is not satisfied by H_* and therefore such axiomatic approach is useful only for H_c^* (compare with the remarks concerning the continuity axiom in Sect. 2.4).

---

[47] For a more general result see Sect. 5.3, Chap. 8.

§4. More General Categories

4.1. Cohomology. In the category of paracompact Hausdorff spaces the theory $H^*(X; G)$ is the unique $\prod$-additive cohomology theory satisfying, together with the usual Eilenberg–Steenrod axioms, the following requirements: a) $H^q = 0$ for $q < 0$; b) the groups H^q vanish locally at the points $x \in X$ (see Sect. 2.3). This result was obtained in Bacon (1974). If we restrict our attention to locally compact spaces satisfying the second axiom of countability then condition a) can be removed and condition b) can be weakened to the assumption of $\sum_c$- additivity (equivalent to it in the subcategory of compact spaces) under the requirement, however, that the condition 3^c) in Sect. 2.1 is satisfied (see §3 in Petkova (1973a)). A result analogous to the one obtained by Bacon is also true in the category of pairs of *weakly locally contractible* paracompact Hausdorff spaces. Namely, any $\prod$-additive cohomology satisfying the Eilenberg–Steenrod axioms and such that $H^q = 0$ for $q < 0$ is equivalent to the singular cohomology (see Petkova (1973b)).

4.2. Homology with Compact Supports. It is clear that every axiomatic system of homology of metric compact spaces can be considered as an axiomatic system of homology H_*^c with compact supports in the category of all topological spaces every compact subspace of which is metrizable. It is clear that this category contains metric spaces, cell complexes and typical spaces of functions (in the compact-open topology). Another example is the fairly natural category of topological spaces described in Hyman (1968).

If we restrict our attention to locally compact spaces that are countable at infinity (in particular, spaces which satisfy the second axiom of countability), then the condition of compact supports follows from the weaker requirement of $\sum$-additivity for homology (see Petkova (1973b)). In the category of pairs of weakly locally contractible paracompact Hausdorff spaces every theory with compact supports (satisfying the Eilenberg–Steenrod axioms), for which $H_q = 0$ when $q < 0$, coincides with the theory H_*^c considered earlier (Theorem 8 in Sklyarenko (1971a)). In particular, in this case the groups H_*^c coincide with the singular groups, which confirms once more (compare with Sect. 5.2, Chap. 3 and Sect. 4.1 above) the naturality of the requirement relating to local contractibility in considerations of the singular theory.

With the restrictions relating to metrizability and local compactness for homologically locally connected spaces (see Chap. 8) some other theories coincide with H_*^c (see, for example, Petkova (1972, 1982), Borel-Moore homology with compact supports, theories related to $\check{H}_*$ and $\check{H}_*^c$, see §1 in Chap. 5). Accordingly, unlike cohomology, differences in the initial approach to a the definition of homology (compare with §1, Chap. 5) are not reflected in the final result only if we impose certain essential restrictions (for a survey of such results see §3 of Sklyarenko (1979)).

Chapter 8
Special Properties and Results

In addition to the properties reflected in the axioms, in applications of homology and cohomology it is also necessary to use many other, more specific properties. It is clear that in the corresponding categories of topological spaces all of them can be obtained from the axioms without appealing to concrete definitions of groups in terms of chains and cochains (as is done, for example, in the opening chapters of Eilenberg and Steenrod (1952)). Such an approach to the development of a theory is not completely justified. First of all, the facts under consideration very often are true also outside of the category in which the axiomatic system exists. Secondly, the direct proofs of such facts are typically simpler than in the axiomatic approach. Several such facts will be considered in this chapter.

§1. Properties Relating to Tautness. Limit Operations

1.1. A General Construction. One of the reasons obstructing the development of homology theory by comparison with cohomology theory (see Sect. 1.1 Chap. 5 and the end of Sect. 4.2. Chap. 7) was the necessity to overcome the difficulties caused by the non-exactness of the inverse limit functor. Since (by contrast with the exact $\varinjlim$-functor) the $\varprojlim$-functor does not commute with the operation H_p of homology, this factor is reflected in the majority of typical formulas involving limits. In particular, the properties of tautness and continuity in the same form as for cohomology (Sect. 1.5, Chap. 3 and Sects. 2.4, 3.3, Chap. 7) are not true for homology. There are many situations in which it is necessary to apply limit operations and therefore it is natural to attempt to find a unified approach to such situations. In the most typical instances, in order to obtain a full picture it necessary, first of all, to explain the connection between the groups $H_n(\varprojlim_\lambda C_*^\lambda)$ and $\varprojlim_\lambda H_n(C_*^\lambda)$ corresponding to the inverse system C_*^λ of chain complexes.

In the category of inverse systems the functor $F = \varprojlim$ is clearly left exact (Sect. 1.5, Chap. 3). As always, the F^p denote the right derived functors of F, and Z_n and B_n are the subgroups of cycles and boundaries of the complex C_*. The inverse system $\{C_*^\lambda\}$ will be considered as a complex $\overline{C}_*$ of inverse systems $\{C_n^\lambda\}$. We shall assume that $F^p(\overline{C}_n) = 0$, for $p > 0$ (for inverse systems this is established in each case separately), from which it follows that $F^p(\overline{B}_{k-1}) = F^{p+1}(\overline{Z}_k)$, $p > 0$. Corresponding to the relation $H_n(\overline{C}_*) = \overline{Z}_n/\overline{B}_n$ is the following exact sequence

$$0 \to F(\overline{B}_n) \to F(\overline{Z}_n) \to F(H_n(\overline{C}_*)) \xrightarrow{\delta} F^1(\overline{B}_n) \to F^1(\overline{Z}_n) \to \ldots$$

It is quite typical (see the following section) that its terms, for $p \geq 1$, can be grouped together into triples $0 \to F^p(\overline{Z}_n) \to F^p(H_n(C_*)) \to F^{p+1}(\overline{B}_n) \to 0$ which are split. In this situation the map δ is an epimorphism. Independently of this, in view of the left exactness of F, the $F(\overline{Z}_n)$ are the cycles of the complex $F(\overline{C}_*)$, and the map $\overline{C}_{n+1} \to \overline{B}_n$ determines an embedding $B_n(F(\overline{C}_*)) \subset F(\overline{B}_n)$. This in turn determines the map $H_n(F(\overline{C}_*)) \to F(H_n(\overline{C}_*))$ with kernel $F(\overline{B}_n)/B_n(F(\overline{C}_*))$ that is isomorphic to $F^1(\overline{Z}_{n+1})$ (this follows easily from the exact sequence $F(\overline{C}_{n+1}) \to F(\overline{B}_n) \to F^1(\overline{Z}_{n+1}) \to F^1(\overline{C}_{n+1}) = 0$). Therefore, (compare with Lemma 1.2 of Sklyarenko (1984)) in the situation under consideration we obtain the following *exact Künneth sequence*

$$0 \to F^1(\overline{Z}_{n+1}) \to H_n(F(\overline{C}_*)) \to F(H_n(\overline{C}_*)) \to F^2(\overline{Z}_{n+1}) \to 0 \qquad (1)$$

(which guarantees, for $F = \varprojlim$, the connections mentioned above), and also the isomorphisms $F^p(H_n(\overline{C}_*)) \simeq F^p(\overline{Z}_n) \oplus F^{p+2}(\overline{Z}_{n+1})$. (By using the projection of $F^p(H_n(\overline{C}_*))$ onto $F^{p+2}(\overline{Z}_{n+1}) \subset F^{p+2}(H_{n+1}(\overline{C}_*))$ and $F^p(\overline{Z}_n) \subset F^{p-2}(H_{n-1}(\overline{C}_*))$ with the kernels $F^p(\overline{Z}_n)$ and $F^{p+2}(\overline{Z}_{n+1})$ respectively, the sequence (1) can be extended to a sequence stretching infinitely in both directions by the groups of the form $F^{2k}(H_{n+k}(\overline{C}_*))$ on the right and of the form $F^{2k-1}(H_{n+k}(\overline{C}_*))$ on the left. In this version the result under consideration is presented in Mdzinarishvili (1982, 1984)).

In the "classical" case when F^2 (and consequently all F^p, for $p \geq 2$) is equal to zero, we obtain the following exact sequence

$$0 \to F^1(H_{n+1})(\overline{C}_*)) \to H_n(F(\overline{C}_*)) \to F(H_n(\overline{C}_*)) \to 0 \qquad (2)$$

1.2. A Construction for the Inverse Limit Functor. For $F = \varprojlim$ the conditions of Sect. 1.1 are satisfied for the inverse systems $\overline{C}_* = \{C_*^\lambda\}$ of the form $C_*^\lambda = \mathrm{Hom}(C_\lambda^*, G)$, where $\{C_\lambda^*\}$ is the direct system of free chain complexes with the free limit $C^* = \varinjlim C_\lambda^*$. This situation arises, for example, for homology of the second kind of locally compact spaces, see Sects. 2.1, 3.2, Chap. 5.[48]. Indeed, since the coboundaries are free, the exact sequences $0 \to H^n(C_\lambda^*) \to C_\lambda^n/B_\lambda^n \to B_\lambda^{n+1} \to 0$ are split and after application of the functor Hom we obtain the following exact sequences $0 \to \mathrm{Hom}(B_\lambda^{n+1}, G) \to Z_n^\lambda \xrightarrow{j} \mathrm{Hom}(H^n(C_\lambda^*), G) \to 0$. Since the groups $\varinjlim_\lambda B_\lambda^{n+1}$ are free, the groups $\varprojlim_\lambda^p \mathrm{Hom}(B_\lambda^{n+1}, G) = P\,\mathrm{ext}^p(\varinjlim_\lambda B_\lambda^{n+1}, G)$ (compare with Theorem 2 in Kuz'minov (1967), or with the Proposition 1.3 in Sklyarenko (1984)) are equal to zero for $p \geq 1$ (for the same reasons, $\varprojlim_\lambda^p C_n^\lambda = 0$,

---

[48] It is not necessary, therefore, to consider complexes C_λ^*, C^* that are not free, see Kuz'minov and Shvedov (1975). In this case, in place of C_λ^* and $C_* = \varprojlim C_*^\lambda$ one considers the complexes $\mathrm{Hom}(P_\lambda^*, G)$ and $\mathrm{Hom}(P^*, G)$, where P_λ^*, P^* are free covers of C_λ^* and C^* (with the same cohomology), and $P^* = \varinjlim P_\lambda^*$.

for $p \geq 1$), from which it follows that $\varprojlim_{\lambda}^{p} Z_n^{\lambda} = \varprojlim_{\lambda}^{p} \mathrm{Hom}(H^n(C_\lambda^*), G)$. Since, according to the universal coefficient formulas, the map j can be represented as the composition $Z_n^{\lambda} \to H_n(C_*^{\lambda}) \to \mathrm{Hom}(H^n(C_\lambda^*), G)$, for $p \geq 1$ the maps $\varprojlim_{\lambda}^{p} Z_n^{\lambda} \to \varprojlim_{\lambda}^{p} H_n(C_*^{\lambda})$ are embeddings onto direct summands. Thus (1) becomes the following exact sequence:

$$0 \to \varprojlim_{\lambda}^{1} \mathrm{Hom}(H^{n+1}(C_\lambda^*), G) \to H_n(\varprojlim_{\lambda} C_*^{\lambda}) \overset{\sigma}{\longrightarrow} \varprojlim_{\lambda}(H_n(C_*^{\lambda}) \to$$

$$\to \varprojlim_{\lambda}^{2} \mathrm{Hom}(H^{n+1}(C_\lambda^*), G) \to 0. \tag{3}$$

In addition

$$\varprojlim_{\lambda}^{p} H_n(C_*^{\lambda}) = \varprojlim_{\lambda}^{p} \mathrm{Hom}(H^n(C_\lambda^*), G) \oplus \varprojlim_{\lambda}^{p+2} \mathrm{Hom}(H^{n+1}(C_\lambda^*), G).$$

These formulas (in the form of a doubly infinite exact sequence, see Sect. 1.1) were obtained by L.D. Mdzinarishvili (see Mdzinarishvili (1984)) by applying certain consequences of the spectral sequence of Roos to the inverse systems formed by the universal coefficient formulas. However, with the aid of analogous methods, formula (3) (even in a slightly more general case) was established earlier by V.I. Kuz'minov and I.A. Shvedov (see Lemma 1 and the beginning of §2, Proposition 1 of Theorem 5 in Kuz'minov and Shvedov (1975)). They studied in depth the map σ (in particular, they demonstrated in the general case the nontriviality of the fourth term in (3), see Example 2 in Kuz'minov and Shvedov (1975)).

It is known that $\varprojlim_{\lambda}^{p} \mathrm{Hom}(A_\lambda, G) = 0$, for $p \geq 2$ for finitely generated groups A_λ. Therefore, in the case of finitely generated C_λ^n (for example in the case of cochains of compact polyhedra, in particular, in cohomology theory of compacta considered in Sect. 2.1, Chap. 5) the Künneth sequence (5) is transformed into the following exact sequence:

$$0 \to \varprojlim_{\lambda}^{1} H_{n+1}(C_*^{\lambda}) \to H_n(\varprojlim_{\lambda} C_*^{\lambda}) \to \varprojlim_{\lambda} H_n(C_*^{\lambda}) \to 0. \tag{4}$$

Under these general assumptions this formula was obtained by V.I. Kuz'minov and I.A. Shvedov (see part 3 of Theorem 2, and Proposition 1 of Theorem 3 in Kuz'minov and Shvedov (1975); the term $P\,\mathrm{ext}^1(H^{n+1}(C^*, G))$ which appears in those statements coincides with $\varprojlim^1\{H_{n+1}(C_*^{\lambda})\}$ by Theorem 6 in Kuz'minov (1967)) and later by L.D. Mdzinarishvili (see Mdzinarishvili (1980)). The case of countable inverse systems is of "classical" nature (see Sect. 1.1): $\varprojlim^{p} = 0$ for $p \geq 2$ and $\varprojlim^1 = 0$ for the systems with epimorphic projections (see, for example, Kharlap (1975)). Formula (4) for such a case was known for a long time (see, for example, Milnor (1960)).

1.3. Connection with Čech Homology. In the case of non-constant coefficients the following exact sequence (used already in Sect. 2.2, Chap. 5, see also the editor's comments to the Russian translation of Massey (1978b) and §3 in Sklyarenko (1984))

$$0 \to \varprojlim_{\alpha}{}^1 H_{n+1}(\alpha;\mathcal{G}) \to H_n(X;\mathcal{G}) \xrightarrow{\sigma} \check{H}_n(X;\mathcal{G}) \to 0 \qquad (5)$$

does not fall into the realm of the construction in Sect. 1.2. In this sequence $\{\alpha\}$ is a fundamental system of canonical covers of a compactum X and $\mathcal{G}$ is a system of locally constant coefficients (in a modified form, see Sect. 3.2 Chap. 5, the result is also true for locally finite covers of a locally compact X). The proof is complicated, even though it makes use of the same ideas (that is, the $\varprojlim$-acyclicity of inverse systems of chain complexes of covers and the triviality for the subgroups of cycles $\varprojlim_{\alpha}{}^p Z_n(\alpha;\mathcal{G}), p \geq 2$, which can be established directly without appealing to cochains or cohomology).

Formula (5) allows us to unscover the conditions (on X and $\mathcal{G}$) under which $H_n = \check{H}_n$. The isomorphism holds for $n = \dim X$ for homologically locally connected X (see Sklyarenko (1969)) (in the case of a locally compact space - for the groups H_n^c and $\check{H}_n^c$, see Kharlap (1975)), in the case when $H^{n+1}(X;\mathbb{Z})$ is the direct sum of a free and a torsion group (see Kuz'minov and Shvedov (1975)), when the groups G and $H_n(X;G)$ are countable (see the following subsection). For a fixed group G the isomorphism holds for all compact X precisely when G is algebraically compact (see Kuz'minov and Shvedov 1975)) (in particular, $H_* = \check{H}_*$ for finite and compact G, for divisible groups and for a group G that is the additive group of a field, etc.). The reason is that algebraically compact groups are precisely the purely injective groups and therefore the kernel of σ coinciding with $\varprojlim_{\alpha}{}^1 \mathrm{Hom}(H^{n+1}(\alpha), G) = P\,\mathrm{ext}^1(H^{n+1}(X), G)$ (see Sect. 1.2) is trivial for such groups G.

Accordingly, for algebraically compact groups G, by taking into account the identity of the theories $H_* = \check{H}_*$, it is possible to use, in the category of compact spaces, the following *continuity property*: $H_n(X;G) = \varprojlim_{\lambda} H_n(X_\lambda;G)$, if $X = \varprojlim_{\lambda} X_\lambda$. For arbitrary G there exists only a map $H_n(X;G) \to \varprojlim_{\lambda} H_n(X_\lambda;G)$ that can be included in an exact sequence of the type (3) which becomes a sequence of the type (4) in the case when the cohomology groups $H^{n+1}(X_\lambda;\mathbb{Z})$ are finitely generated, or for arbitrary countable systems $\{X_i\}$[49] (see §4 in Kuz'minov and Shvedov 1975)).

1.4. Typical Limit Formulas. The Countable Case. Let $Y = \bigcup_i Y_i$ be a union of open or closed subspaces, and let $Y_i \subset \mathrm{Int}\, Y_{i+1}$. In this case, for the cochains realized as sections of a flabby resolution, we have $C^*(Y;\mathcal{G}) = \varprojlim_i C^*(Y_i;\mathcal{G})$

---

[49] In the last case the result is true in the category of metrizable compacta and for *extraordinary theories of the Steenrod type*, see §8.5 in Edwards and Hastings 1976) (compare with the footnote to Sect. 2.1 in Chap. 7 above).

with epimorphic restriction homomorphisms, and this leads to the following exact sequence (see Kharlap 1975)):

$$0 \to \varprojlim_i{}^1 H^{p-1}(Y_i; \mathcal{G}) \to H^p(Y; \mathcal{G}) \xrightarrow{\sigma} \varprojlim_i H^p(Y_i; \mathcal{G}) \to 0. \tag{6}$$

In the category of CW-complexes this formula for constant coefficients was obtained in Milnor 1962). The sequence is of interest in the case, for example, when Y_i is a compact subspace of a locally compact Y. The fact that σ is an epimorphism in this case was established in Grothendieck (1957).

For a descending sequence of closed subsets $B_i \subset X$ and $B = \bigcap_i B_i$ the inverse limit applied to the exact sequences of cochains $0 \to C^*(X, Y_i; \mathcal{G}) \to C^*(X; \mathcal{G}) \to C^*(Y_i; \mathcal{G}) \to 0$, where $Y_i = X \setminus B_i, Y = X \setminus B$, preserves exactness. Therefore the fact that $\varprojlim{}^1 C^n(X, Y_i; \mathcal{G}) = 0$ together with (4) yield the following exact sequence (see Kharlap (1975))

$$0 \to \varprojlim_i{}^1 H^{p1}(X, Y_i; \mathcal{G}) \to H^p(X, Y; \mathcal{G}) \to \varprojlim_i H^p(X, Y_i; \mathcal{G}) \to 0. \tag{7}$$

The analogue of (6) and (7) for the homology H_*^c is the obvious isomorphism of the middle terms with the direct limits of homology of the spaces Y_λ and the pairs (X, Y_λ) respectively.

Exactly in the same way, when we realize the chains of a locally compact space as sections of flabby differential sheaves (Sect. 4.2, Chap. 5), we obtain the following two exact sequences (obtained in Kharlap (1975) in the case of constant coefficients). For $Y = \bigcup_i U_i$, where $U_i \subset U_{i+1}$ and the sets U_i are open:

$$0 \to \varprojlim_i{}^1 H_{p+1}(U_i; \mathcal{G}) \to H_p(Y; \mathcal{G}) \to \varprojlim_i H_p(U_i; \mathcal{G}) \to 0 \tag{8}$$

and for closed subsets B_i of a (locally compact) space X, such that $B_{i+1} \subset B_i$ and $B = \bigcap_i B_i$:

$$0 \to \varprojlim_i{}^1 H_{p+1}(B_i; \mathcal{G}) \to H_p(B; \mathcal{G}) \to \varprojlim H_p(B_i; \mathcal{G}) \to 0. \tag{9}$$

If we replace H_* by cohomology H_c^* with compact supports it is easy to see that both formulas reduce to the equality between the cohomology of the middle term and the direct limit of the cohomology of the space on the right.

The term containing $\varprojlim{}^1$ of all of the formulas presented above vanishes under the following *Mittag-Leffler stability condition of images* which must be satisfied for the corresponding inverse system of groups: for every i the images in the group with number i of the groups following it stabilize (beginning with some $i_0 > i$). In certain cases the triviality of $\varprojlim{}^1$ (for a countable G) is equivalent to this condition (see Kharlap (1975) and Sklyarenko (1980a)). The following result (which hasmany consequences) is also true: for countable groups A_i either $\varprojlim{}^1 A_i = 0$ or the cardinality of $\varprojlim{}^1 A_i$ is not less than that of the continuum (Lemma 3 in Kharlap (1975) and a more general result in Lemma 1.6 in Sklyarenko (1980a)). Since for a countable group G the homology of covers of a metrizable compactum is countable, it follows, for example, that the map σ in Sect. 1.3 is an isomorphism for all indices n for which the groups $H_n(X; G)$ are countable. Another consequence is the fact that for a locally compact X satisfying the second axiom of countability (in particular, for a polyhedron or a compactum) and a finitely generated module G over a countable ring R the modules $H_p(X; G)$ are either finitely generated of have the cardinality of the continuum. A similar statement is also true for cohomology $H^p(X; G)$ under the additional condition of the local homological connectedness of X (see the end of §1 in Sklyarenko (1984)).

1.5. The General Case. Let $Y = \bigcup_\lambda Y_\lambda, \lambda \in \Lambda$ be the union of a system of subspaces Y_λ, ordered by inclusion, and let $C^*(Y; \mathcal{G})$ be the cochains of Y realized as sections of a flabby resolution $\mathcal{C}^*$ of the coefficient sheaf $\mathcal{G}$. Clearly, there exists an inclusion $C^*(Y; \mathcal{G}) \subset \varinjlim_\lambda \Gamma(\mathcal{C}^*|_{Y_\lambda})$ that is an isomorphism at least in the following cases: a) the Y_λ are open in Y; b) if $\lambda < \mu$ then $Y_\lambda \subset \operatorname{Int} Y_\mu$; c) the Y_λ are finite unions of sets from a locally finite cover of a hereditarily paracompact space Y.

Under the indicated assumptions $\varprojlim_{\lambda \in M} \Gamma(C^*|_{Y_\lambda}) = \Gamma(C^*|Y')$ (where $Y' = \bigcup_{\lambda \in M} Y_\lambda$, and M is a directed subsystem in Λ) is an epimorphic image of $C^*(Y; \mathcal{G})$ which in turns implies the $\varprojlim$-acyclicity of the inverse systems $\{\Gamma(C^*|_{Y_\lambda})\}$ (Theorem 1.8 in Jensen (1972)). The application of any of the standard $\varprojlim$-acyclic resolutions to these systems (see Jensen (1972)) and the application of the $\varprojlim$-functor to the resulting bigraded differential system guarantees the existence of a spectral sequence (the second of the two situations described in Sect. 6.1, Chap. 3) with the term $E_2^{pq} = \varprojlim_\lambda{}^p H^{pq}(Y_\lambda; \mathcal{G})$ converging to $H^{p+q}(Y; \mathcal{G})$. In particular, $H^0(Y; \mathcal{G}) = \varprojlim_\lambda H^0(Y_\lambda; \mathcal{G})$ and we obtain the following exact sequence:

$$0 \to \varprojlim_\lambda{}^1 H^0(Y_\lambda; \mathcal{G}) \to H^1(Y; \mathcal{G}) \to \varprojlim_\lambda H^1(Y_\lambda; \mathcal{G}) \to$$

$$\to \varprojlim_\lambda{}^2 H^0(Y_\lambda; \mathcal{G}) \to H^2(Y; \mathcal{G}).$$

It also follows from the relations $H^q(Y_\lambda; \mathcal{G}) = 0$ for $q \le n$ (at least for a cofinal part of Λ) that $H^q(Y; \mathcal{G}) = 0$ for $q \le n$ (the cohomology in the dimension zero is reduced). In the situations, when for one reason or another $\varprojlim_\lambda{}^p = 0$, for $p \ge 2$ (for example for a countable set of indices or when $H^q(Y_\lambda; \mathcal{G})$ are finitely generated modules over a countable principal ideal domain, see Proposition 1.6 in Sklyarenko (1984)), we obtain exact sequences similar to (6) above (for $\lambda \in \Lambda$). Such is the case when the coefficients $\mathcal{G}$ are locally constant and when the Y_λ are finite subcomplexes of a (cell) complex Y (this can be established by using exactly the same construction as in the proof of formulas (5) in Sect. 1.3). In particular, for $H^n(Y;\mathcal{G}) \neq 0$ every finite subcomplex of Y is contained in a larger subcomplex Y_λ for which $H^q(Y_\lambda;\mathcal{G}) \neq 0$, for some $q \leq n$ (a manifestation of the *compactness factor* in the case of cohomology, compare with Sect. 1.5, Chap. 2 and Sect. 1.1, Chap. 5).

In the case when Y is a subspace of a paracompact space X, exactly the same argument (first noticed by A.E. Kharlap) with the same restrictions on Y and Y_λ is true for the cohomologies of the pair $H^n(X, Y;\mathcal{G})$ and $H^q(X, Y_\lambda;\mathcal{G})$ ($\varprojlim$-acyclicity of the inverse systems of cochains of (X, Y_λ) is a consequence of the existence of the exact sequences $0 \to C^*(X, Y_\lambda;\mathcal{G}) \to C^*(X;\mathcal{G}) \to \Gamma(C^*|Y_\lambda) \to 0$ and the lim-acyclicity of the systems for Y_λ).

Similar results are also true for homology. Let Y be an arbitrary paracompact subspace of a Hausdorff compactum X, and let $B = X \setminus Y, B_\lambda = X \setminus Y_\lambda$, where the Y_λ are of the type indicated above. The most typical case occurs when $\{B_\lambda\}$ is a system, directed with respect to intersection, of closed or open subsets with compact $B = \bigcap_\lambda B_\lambda$. By applying to a description of homology with locally constant coefficients $\mathcal{G}$ on X the sheaf of chains described in Sect. 4.2, Chap. 5, we obtain, under the assumption that $\dim X = n < \infty$, two spectral sequences converging to $H^c_{n-p-q}(B;\mathcal{G})$ and $H^c_{n-p-q}(X, B;\mathcal{G})$ in which $E_2^{p,q} = \varprojlim_\lambda{}^p H^c_{n-q}(B_\lambda;\mathcal{G})$ and $E_2^{p,q} = \varprojlim_\lambda{}^p H^c_{n-q}(X, B_\lambda;\mathcal{G})$, respectively. In particular, $\tilde{H}^c_n(B;\mathcal{G}) = \varprojlim_\lambda H^c_n(B_\lambda;\mathcal{G})$ and we obtain the following exact sequence

$$0 \to \varprojlim_\lambda{}^1 H^c_n(B_\lambda;\mathcal{G}) \to H^c_{n-1}(B;\mathcal{G}) \to \varprojlim_\lambda H^c_{n-1}(B_\lambda;\mathcal{G}) \to$$

$$\to \varprojlim_\lambda{}^2 H^c_n(B_\lambda;\mathcal{G}) \to H^c_{n-2}(B;\mathcal{G}).$$

The acyclicity of B_λ (for a cofinal part of $\lambda \in \Lambda$) implies the acyclicity of B. Similar considerations are valid for the homologies of the pair (X, B) and (X, B_λ).

§2. Local Conditions. Local Homology and Cohomology

2.1. Homological Local Connectedness. As we know (Sect. 2.3, Chap. 7), locally cohomology is trivial; $\varinjlim_U H^p(U, x;\mathcal{G}) = 0$ (the U form the family of neighborhoods of x). In fact this is the tautness property (Sect. 1.5, Chap. 3) applied to the points $x \in X$ (due to which the sheaves of cochains form resolutions, see Sect. 1.4 Chap. 3). A topological space is called *cohomologically locally connected at a point $x \subset X$* over a coefficient ring R if, in addition, for every natural number n and every neighborhood U of x there exists a smaller neighborhood V of x such that the homomorphism $H^p(U, x; R) \to H^p(V, x; R)$ is trivial for $p \leq n$. If this condition is satisfied at all points of X, then the space is called *cohomologically locally connected over R*, or briefly a clc_R-space (clc-space if $R = \mathbb{Z}$). Examples are polyhedra, manifolds, neighborhood retracts of Euclidean or Hilbert spaces, standard function spaces and arbitrary locally contractible spaces.

If the above requirement is satisfied for a fixed n, then X is called *locally cohomologically connected up to the dimension n* (briefly clc_R^n-space). Such spaces appear less frequently. For $n = 0$ the requirement is equivalent to topological local connectedness (Borel and Moore (1960)). Similar definitions can be considered in the case of an arbitrary coefficient group (module) leading to the concepts of clc_G- and clc_G^n-spaces.

Locally compact clc_R-spaces are characterized by each of the following properties (see Borel and Moore (1960), or Theorem 16.4, Chap. 2 in Bredon (1967)): a) for every $N \subset X$ and every M whose closure is compact in $\text{Int}\, N$, the images $H^p(N; R) \to H^p(M; R)$ are finitely generated (over R) for every p; b) for the same M, N that are in addition open, the images $H_c^p(M; R) \to H_c^p(N; R)$ are finitely generated (for each p). In particular, for a compact X (since we can put $M = N = X$) the groups $H^p(X; R)$ are finitely generated.

By comparison with cohomology, the local behavior of homology is slightly more complicated (even in the category of metrizable compacta, see Sect. 2.3, Chap. 7). Nevertheless for first countable spaces $\varprojlim_U{}^p H_n^c(U, x; G) = 0$ (the U are the neighborhoods of x) not only for $p = 0$ (see Lemma 8 in Sklyarenko (1969)), but also for $p > 0$ (see Lemma 8 in Kharlap (1975)) (for $p \geq 2$ the result is implied by countability of inverse systems). In the category of locally compact spaces the class of clc_R-spaces coincides, (see Borel and Moore (1960)), with the class of *homologically locally connected spaces* over R (briefly hlc_R-spaces) the definition of which (given in terms of H_*^c) is dual to that of cohomologically locally connected spaces. Dual characterizations (in terms of H_p^c and H_p in place of H^p and H_c^p) are also true. In particular, for a compact hlc_R-space X the modules $H_p(X; R)$ are finitely generated. In connection with the above, see Borel and Moore (1960), Bredon (1967), Sklyarenko (1980a).

In the case of a metrizable locally compact space X of finite cohomological dimension over R the hlc_R requirement is equivalent to any of the following conditions: 1) the sections of sheaves $\mathcal{H}_p$ of local homology (section 4.4, Chap. 5) over a compact $C \subset X$ are countably generated modules (when R is countable or Artin, see Sklyarenko (1980a); 2) the stalks of the sheaves $\mathcal{H}_p$ (that is, the $H_c^p(X, X \setminus x; R)$) are countable (for the case of a countable principal domain R, see Mitchell (1979). There exist a variety of characterizations of homological local connectedness in terms of the local Betti numbers (at the points $x \in X$), in terms of behavior of homology and cohomology of the sets $U \setminus x$ for arbitrarily small neighborhoods U of x, and others (for more information on this subject see Sklyarenko (1980a)) For conditions guaranteeing the hlc property for the one-point compactification of X see Dydak (1986).

2.2. Local Groups. The local homological structure of a space can be characterized not only in terms of the behavior of homology and cohomology of neighborhoods U of points $x \in X$, but also by groups of subspaces of the form $U \setminus x$ and of the pairs $(X, X \setminus U)$ or $(X, X \setminus x)$. For a long time different types of *local Betti numbers* were considered instead of such groups (in fact they were determined by these groups).

In Sect. 4.4 of Chap. 5 we already introduced the *local homology* $H_k^x(G) = H_k^c(X, X \setminus x; G) = \varinjlim H_k^c(X, X \setminus U; G)$. The definition of these groups goes back to P.S. Aleksandrov (see §6 in Sklyarenko (1969)). By taking the inverse limit with respect to the neighborhoods U of a point x, the connecting homomorphisms $H_k^x(G) \to H_{k-1}^c(U \setminus x; G)$ of the pairs $(U, U \setminus x)$ define the maps

$$\delta_x : H_k^x(G) \to I_{k-1}^x(G) = \varprojlim_U H_{k-1}^c(U \setminus x; G)$$

(for $k-1 = 0$ reduced homology is considered), which were studied for the first time in Sklyarenko (1969). For the first countable spaces we have the following exact sequence (see Kuz'minov and Shvedov (1975), Kharlap (1975)):

$$0 \to \varprojlim_U{}^1 H_k^c(U \setminus x; G) \to H_k^x(G) \to I_{k-1}^x(G) \to 0$$

(the proof does not depend on the relationships introduced in Sect. 1.4 above). It is clear that in the case of polyhedra or manifolds the map δ_x is an isomorphism. This is also the case for metric hlc_G-spaces (see Kharlap (1975)) since the inverse systems of groups $H_k^c(U \setminus x; G)$ satisfy the stability condition of Sect. 1.4 and the first term of the exact sequence is equal to zero. In the general case the maps δ_x are not isomorphisms (see Kharlap (1975)).[50]

In a similar way we define the *local cohomologies* $I_x^k(G) = \varinjlim_U H^k(U \setminus x; G)$ and $H_x^k(G) = \varprojlim_U H^k(X, X \setminus U; G)$. These local groups were applied by Conner, Raymond and Borel (see Borel (1960) and §6 in Sklyarenko (1969)). By applying the operation of the direct limit with respect to the neighborhoods U of a point x to the cohomological sequences of pairs $(U, U \setminus x)$, we obtain an isomorphism $I_x^k(G) = H^{k+1}(X, X \setminus x; G)$ (see Benyaminov and Sklyarenko (1967)) and therefore there always exists a transformation $\beta_x : I_x^k(G) \to H_x^{k+1}(G)$. In the case when X is a first countable paracompact space, the exact sequence (7) of Sect. 1.4 yields the following exact sequence (see Kharlap (1975)):

$$0 \to \varprojlim_U{}^1 H^p(X, X \setminus U; G) \to I_x^p(G) \to H_x^{p+1}(G) \to 0.$$

The above relationship is true also for cohomology with coefficients in a sheaf.

The transformation β_x in general is not an isomorphism, and this is the case even in sufficiently "good" spaces. There exists a contractible, locally

---

[50] For relationship between the groups H_k^x and I_{k-1}^x see also Sect. 6.2 below.

contractible compactum X, for which $I_x^2(G) = H_x^3(X, X \setminus x; G) \neq 0$ at a point x whose complement is a locally finite polyhedron and such that $\dim X = 2$ (Example 4.6 in Sklyarenko (1980a)). (Another example can be constructed by using Exercise 23 to Chap. 2 in Bredon (1967)). Conditions implying that β_x is an isomorphism are considered in the following section.

2.3. Peripheral Homological Local Connectedness. The example presented in Sect. 2.2 shows that membership of X in the class of hlc-spaces (even to the class of locally contractible spaces) does not imply its simple local character. It is likely that there should exist some additional natural local conditions that guarantee this. Such conditions do exist, but they are much less known.

In Benyaminov and Sklyarenko (1967), where the homomorphisms β_x were considered for the first time, it was observed that for an arbitrary point x of a first countable space X it is true that $\varprojlim_U H^p(X \setminus x, X \setminus U; G) = 0$. It can be shown also that $\varinjlim_U H_p^c(X - x, X \setminus U; G) = 0$) (let us point out that for polyhedra and manifolds such groups are equal to zero even before the application of the operations of the inverse or direct limit). This justifies the definitions introduced in §5 of Sklyarenko (1979). Namely, a space X is called *peripherally homologically locally connected over* G (briefly phlc$_G$-space), if for every n and every neighborhood U of an arbitrary point $x \in X$ there exists a smaller neighborhood V of x, such that for $p \leq n$ the maps $H_p^c(X \setminus x, X \setminus U; G) \to H_p^c(X \setminus x, X \setminus V; G)$ are trivial. We can obtain, in terms of the homomorphisms $H^p(X \setminus x; X \setminus V; G) \to H^p(X \setminus x, X \setminus U; G)$, a similar definition of the peripheral cohomological local connectedness (pclc$_G$-spaces). These conditions are satisfied, for example, by spaces in which each point has arbitrarily small neighborhoods U such that $X \setminus U$ is a deformation retract of $X \setminus x$.

Locally compact phlc- and pclc-spaces were considered in Rothe (1983b). He showed, for example, that these conditions are independent of hlc$_R$. For first countable pclc$_G$-spaces the maps β_x in Sect. 2.2 above are isomorphisms, and at the same time for countable G the pclc$_G$ requirement is equivalent to the countability of all the groups $I_x^k(G)$. In the class of hlc$_R$-spaces the phlc$_R$ and pclc$_R$ requirements are equivalent, and for a countable ring R the hlc$_R$ and phlc$_R$ condittions together are equivalent to the requirement that the local modules $H_p^x(R)$ (or $H_p^x(R)$ and $I_x^p(R)$) are finitely generated for all points x, provided that the homological dimension of X over R is finite. There are also many other useful observations in Rothe (1983b).

2.4. Connection Between Theories of the Second Kind and the Ordinary Theories. In the case when X is a locally compact Hausdorff space (for example, a locally finite polyhedron), the chains (and also cochains) with compact supports form a subcomplex of the complex of all chains (cochains) and this leads to the following exact sequences[51]:

---

[51] For more general exact sequences see Sect. 6.2.

$$\ldots \to H_n^c(X;G) \to H_n(X;G) \to H_n^\infty(G) \to H_{n-1}^c(X;G) \to \ldots$$

$$\ldots \to H_c^n(X;G) \to H^n(X;G) \to I_\infty^n(G) \to H_c^{n+1}(X;G) \to \ldots,$$

in which $H_n^\infty(G) = H_n(\overline{X}, X; G)$, and $I_\infty^n = H^{n+1}(\overline{X}, X; G)$ are local groups at the point added to X in the one-point compactification $\overline{X}$ (see the arguments given in Sect. 1.4, Chap. 1). These sequences were considered in §6 in Sklyarenko (1969), the second one earlier in Godbillon (1967) (see also §10.2 in Massey (1978b)). By taking into account the isomorphisms $H_*(X;G) = H_*(\overline{X}, \infty; G)$, $H_c^*(X;G) = H^*(\overline{X}, \infty; G)$ both sequences can be treated as exact sequences of the pair $(\overline{X}, X)$.

The connection between theories of the second kind and the ordinary theories is also reflected in the existence of the isomorphisms $H_*^c(X, X \setminus U; \mathcal{G}) = H_*(U, \mathcal{G})$, $H^*(X, X \setminus U; \mathcal{G}) = H_c^*(U; \mathcal{G})$, which exist for open U with a compact closure (see the end of Sect. 4.3, Chap. 5 or Sect. 3.2, Chap. 5, §4 Chap. 2 or Sects. 4.2, 4.3, Chap. 3), and also by the universal coefficient formulas considered in §4 below. To a great extent the importance of theories of the second kind is described in Sect. 1.3, Chap. 2.

§3. Homological Dimension

3.1. Basic Definitions. One of the ways of proving that Euclidean spaces of different dimensions are not homeomorphic is to notice that they have different homology H_* or cohomology H_c^*. This observation serves as a foundation of the definition of dimension in terms of homology (or cohomology) proposed by P.S. Aleksandrov already at the beginning of the thirties.

The *homological dimension* $h \dim_G X$ of a space X with respect to a coefficient groups G is defined as the largest integer n such that $H_n^c(X, A; G) \neq 0$ for some closed subset A of X. If such n does not exist, then we put $h \dim_G X = \infty$. In a similar way we define the *cohomological dimension* $\dim_G X$ in terms of H^* (with respect to an arbitrary coefficient sheaf $\mathcal{G}$). Since forming the the one-point compactification does not change dimension, in defining the dimension of locally compact spaces we can use the theories H_* and H_c^* in place of H_*^c and H^*.

It follows from a local characterization of $h \dim_G X$ (see Sect. 3.3 below) that the homological dimension of locally compact spaces can be determined by the pairs (X, A) with open A. This was indicated in §5 of Sklyarenko (1980a). By contrast, the cohomological dimension $\dim_G X$ can be determined only by closed pairs (X, A). In §§4, 5 in Sklyarenko (1980a) there are examples of metric compacta X of dimension n for which $H^{n+1}(X, X \setminus x; G) \neq 0$ at some point x (see the end of Sect. 2.2 above for $n = 2$, Miryuk (1973) for $n = 0$ and Kharlap (1975) for $n = 1$).

The condition $\dim_G X \leq n$ for a paracompact X is equivalent to the requirement that every soft resolution $0 \to \mathcal{G} \to \mathcal{L}^*$ remains soft after replacing $\mathcal{L}^n$ by $\mathcal{Z}^n = \operatorname{Ker} d$ and removing all the $\mathcal{L}^p$, for $p > n$ (Proposition 2.15.2.in

Bredon (1967)). This characterization reveals the related nature of the concept of the cohomological dimension of a topological space and the dimension of objects in homological algebra, in particular the *global dimension of a ring* (see Cartan and Eilenberg (1956)). In exactly the same way, the condition $H^{n+1}(X, U; \mathcal{G}) = 0$, for all open $U \subset X$ can be characterized in terms of flabby resolutions (compare with Exercise 22 in Chap. 2 of Bredon (1967); see also Kuz'minov and Shvedov (1976)). It is clear that any flabby resolution of $\mathcal{G}$ cannot be shorter than any of its soft resolutions. In the case of a hereditarily paracompact X the difference in their length does not exceed 1 (that is, $H^p(X, U; \mathcal{G}) = 0$ for open U and $p > \dim_{\mathcal{G}} X + 1$), and it is easy to construct sheaves $\mathcal{G}$ such that $H^p(X, U; \mathcal{G}) \neq 0$ for $p = \dim_{\mathcal{G}} X + 1$ (§5 in Sklyarenko (1980a)). This is impossible, however, for any locally constant sheaf $\mathcal{G}$ in the case when X is a cell complex, for a locally constant sheaf with finitely generated stalks in the case when X is a generalized manifold and for a locally constant sheaf $\mathcal{G}$ whose stalks are any field in the case when X is a hlc-space over this field (§5 in Sklyarenko (1979b)).

3.2. Fundamental Relations. As a corollary to the universal coefficients formula, for a locally compact space X we always have $\dim_G X \leq \dim_{\mathbb{Z}} X$. In the case of a paracompact space X of finite Lebesgue dimension $\dim X$, it always coincides with the cohomological dimension $\dim_{\mathbb{Z}} X$ with respect to the integers (P.S.Aleksandrov). Whether the same statement is true for infinitely dimensional X (even for metric compacta) is still an open question (*problem of Aleksandrov*)[52] A partial negative answer was obtained recently in Dranishnikov (1987) who constructed infinite-dimensional compacta X for which $\dim_G X = 2$ with respect to every finite cyclic group $G = \mathbb{Z}_m$. The problem of Aleksandrov is equivalent to the question whether cell-like maps can raise dimension (see Dranishnikov and Shchepin 1986)).

When $\dim X < \infty$ the equality $\dim_G X = \dim X$ for a given compactum X and every G is equivalent to the *dimensional fullness* of X (in the sense of P.S. Aleksandrov), that is to the condition that $\dim X \times Y = \dim X + \dim Y$, for all compacta Y (theorem of V.G. Boltyanskij). Examples of compacta, for which $\dim_G X < \dim X$ were constructed by L.S. Pontryagin, and later by V.G. Boltyanskij, Williams, Raymond and others. Possible relations between the dimensions of a compactum with respect to different coefficient groups, different methods of computing $\dim_G X$ for G based on other known cases and associated with the formulas expressing the dimension of a product in terms of dimensions of factors, were studied thoroughly by M.F. Bockstein. A detailed account of these and related questions associated with the theory of cohomological dimension, as well as typical problems of this theory can be found in the very complete survey Kuz'minov (1968). In Dranishnikov (1987) it is proved that the difference $\dim X - \dim_{\mathbb{Z}_m} X$ (for finite dimensional X) can be arbitrarily large. Using this result Dranishnikov constructed dimen-

---

[52] This problem was solved recently by A.N. Dranishnikov.

sionally deficient compact ANR's (a negative solution to *Borsuk's problem*). The problems of dimensional fullness and the question of explaining possible relations between cohomological dimensions of compacta with respect to different coefficients are closely related to some typical problems in the theory of transformation groups, in particular to the famous *Hilbert-Smith problem* (for more information on these subject see §9, Chap. 1 of Kuz'minov's survey).

The invariant $h \dim_G X$ was studied in great detail in Kharlap (1975). It follows easily from the universal coefficient formulas relating H_* with H_c^* that for every module G over a hereditary ring R it is always true that $h \dim_G \le h \dim_R X$ and that $h \dim_R X = \dim_R X$ for locally compact X when R is a field. The same considerations lead to $\dim_{\mathbb{Z}} X - 1 \le h \dim_{\mathbb{Z}} X \le \dim_{\mathbb{Z}} X$. If $h \dim_{\mathbb{Z}} X = \dim_{\mathbb{Z}} X$, then X is dimensionally full, and if X is hlc over $\mathbb{Z}$, then the converse is also true (this property is not true in general). For details see Kharlap (1975).

3.3. Local Characterizations. A fairly easy consideration of the sheaves of chains $\mathcal{C}_*$ shows (see Sklyarenko (1971b), or Sect. 5.1 below), that for $p > n = h \dim_G X < \infty$ they form an exact sequence, $H_p(X, U; G) = 0$ for every open $U \subset X$, and the presheaf $U \to H_n(U; G)$ is a sheaf and n is the largest number for which $\mathcal{H}_n = \mathcal{Z}_n / \mathcal{B}_n \ne 0$ (here $\mathcal{Z}_n = \operatorname{Ker} d, \mathcal{B}_n = \operatorname{Im} d$ in $\mathcal{C}_*$). Let us recall (Sect. 4.4, Chap. 5), that the stalks of $\mathcal{H}_n$ are the local groups $H_n^x(G) = H_n(X, X \setminus x; G)$. The homological dimension of the set of all $x \in X$ for which $H_n^x(G) \ne 0$ is also equal to n (Theorem 4 in Kharlap (1975)), while the dimension of its complement does not exceed $n-1$ (Corollary 8 in Kharlap (1975)). The condition $h \dim_G X < \infty$ is essential (Sect. 4.5, Chap. 5).

As was shown in Nguyen Le Anh (1981), the condition $h \dim_G X \le n$ can be characterized by the fact that the presheaf $U \to H_n(U; G)$ is a sheaf. Similarly, in the case of a locally compact space $\dim_G X \le n$ if and only if the assignment $U \to H_c^n(U; G)$ is a *cosheaf* (V.G. Miryuk and B.L. Okun').

For cohomology the situation is more complicated, even in the category of metric compacta. As was demonstrated by V.G. Miryuk and A.E. Kharlap (§3 in Kharlap (1975)), the dimension $\dim_G X$ is not characterized by local cohomology of the type $I_x^{p-1}(G)$ or $H_x^p(G)$ (the unsuitability of $I_x^{p-1}(G)$ for this purpose is illustrated, in particular, by examples at the end of Sect. 2.2; see also Sect. 3.1). However for locally compact spaces satisfying the second axiom of countability it coincides with the largest number n, for which at some point x the sum $H_x^n(G) \oplus I_x^n(G)$ is non-trivial under the assumption that G is countable and $\dim_G X < \infty$, and at the same time the set of such x has cohomological dimension equal to n (Theorem 6 in Kharlap (1975)). In the case when X is a metrizable, locally compact pclc-space of finite cohomological dimension over the ring $\mathbb{Z}$ of integers, then $\dim_G X$ coincides with the largest n, for which $I_x^{n-1}(G) = H_x^n(G) \ne 0$ at some points $x \in X$ (see Rothe (1983c)).

§4. Universal-Coefficient Formulas. Functorial Dependence

In this paragraph R denotes a commutative hereditary ring (for example a principal ideal domain), and G is a module over R.

4.1. Relations Following from Standard Relationships. Let Y be a closed subspace of a space X and let φ be a paracompactifying family of supports. For every finitely presented module G we have the following exact sequences

$$0 \to H_\varphi^n(X, Y; R) \otimes_{\mathcal{R}} G \to H_\varphi^n(X, Y; G) \to$$

$$\to \mathrm{Tor}_R(H_\varphi^{n+1}(X, Y; R), G) \to 0. \tag{10}$$

In the case when φ is the family of all closed subsets of X, in view of the fact that the functors $\otimes$ and Tor commute with direct limits and $\varinjlim$ preserves exactness, the above formula follows from formula (10) of Sect. 1.7, Chap. 1 by applying the operation of the direct limit with respect to open covers α of the space X (we assume that $H^* = \check{H}^*$). From the very description of the subcomplexes $C_\varphi^*(\alpha; G) \subset C^*(\alpha; G)$ of cochains with supports in φ it is clear that $C_\varphi^*(\alpha; G) = C_\varphi^*(\alpha; R) \otimes_{\mathcal{R}} G$, and therefore similar arguments apply also to $H_\varphi^* = \check{H}_\varphi^*$ (see Sect. 2.2, Chap. 4). These conclusions can also be obtained using the Alexander-Spanier cochains. The necessity of the restrictions imposed on G is explained in Sect. 1.7, Chap. 1.

For the same reason, it is only for finitely presented G (in contrast to (5) in Sect. 1.7, Chap. 1), that we have the following exact sequences (for a Hausdorff space X):

$$0 \to H_n^c(X, Y; R) \otimes_{\mathcal{R}} G \to H_n^c(X, Y; G) \to$$

$$\to \mathrm{Tor}_R(H_{n-1}^c(X, Y; R); G) \to 0, \tag{11}$$

while if X is locally compact, then for arbitrary φ also

$$0 \to H_n^\varphi(X, Y; R) \otimes_{\mathcal{R}} G \to H_n^\varphi(X, Y; G) \to$$

$$\to \mathrm{Tor}_R(H_{n-1}^\varphi(X, Y; R), G) \to 0 \tag{12}$$

(compare with formula (9) in Sect. 1.7, Chap. 1). For compact X they are a consequence of the description of the chains given in §2, Chap. 5, and for homology H_* of locally compact spaces, they follow from the results presented in §3, Chap. 5. The validity of formulas in the general case is a result of the description of H_*^c and H_*^φ and the fact that the functors Tor and $\otimes$ commute with the direct limit functor, see Sects. 3.3, 3.4, Chap. 5.

Formulas (11) and (12) were obtained in Sklyarenko (1969) (see also Sklyarenko (1979)). They can be obtained also by taking into account Massey's chains (Sect. 2.5, Chap. 5, see also editor's note to Chapter 4 in the Russian translation of Massey (1978b)). The fact that these formulas are not natural for non-finitely presented G is illustrated (appart from the remarks with respect to formula (9) in Sect. 1.7, Chap. 1) by the example of the one-point compactification of a countable discrete space. For Borel Moore homology only formula (11) is true, but for arbitrary modules G (see the end of Sect. 2.4, Chap. 5 above or Sect. 3.10, Chap. 5 in Bredon (1967)).

If X is a locally compact space then for every R-module G we have the following exact split sequences (compare with (7) in Sect. 1.7, Chap. 1):

$$0 \to H_c^n(X, Y; R) \otimes_{\mathcal{R}} G \to H_c^n(X, Y; G) \to$$

$$\to \mathrm{Tor}_R(H_c^{n+1}(X, Y; R), G) \to 0. \tag{13}$$

There exists a splitting of this sequence that is natural with respect to G (but not with respect to (X, Y)). The validity of (13) is a simple consequence of the description of the cochains with compact supports which was given in Sect. 4.2, Chap. 2, and according to which the complex $\check{C}_{\omega c}^*(X, Y; R)$ is free and $\check{C}_{\omega c}^*(X, Y; G)$, as a direct limit of finite cochains of covers and in view of the fact that the operation of the direct limit commutes with $\otimes$, is isomorphic to $\check{C}_{\omega c}^*(X, Y; R) \otimes_{\mathcal{R}} G$. The above formula was obtained in Sklyarenko (1969). Its derivation by using Massey's chains (Sect. 4.1, Chap. 2) is more complicated (see §4.11 in Massey (1978b) for $R = \mathbb{Z}$ and the editor's note to Chapter 4 in the Russian translation of Massey (1978b) for the general case).

We obtain from the description of the chains and of homology of pairs of locally compact spaces (Sect. 3.2, Chap. 5; see also the description of cochains in §4, Chap. 2) the following exact sequences for an arbitrary R-module G:

$$0 \to \mathrm{Ext}_R(H_c^{n+1}(X, Y; R), G) \to H_n(X, Y; G) \to$$

$$\to \mathrm{Hom}_R(H_c^n(X, Y; R), G) \to 0 \tag{14}$$

(compare with (8) in Sect. 1.7, Chap. 1). These sequences are split and the splitting is natural with respect to G. These formulas were obtained in Sklyarenko (1969, 1979). For compact spaces and for $R = \mathbb{Z}$ they were obtained in Milnor (1960) (and in terms of the original definition of Steenrod homology by MacLane and Eilenberg already in 1942, see the survey Sklyarenko 1979)). For Abelian groups the formulas were considered also in §4.8 of Massey (1978b). The construction of Borel-Moore homology is such, that (taking into account (11) and (12)) the formulas (14) need not be true, at least in the case when G is not finitely presented. In Borel and Moore (1960) as well as in Bredon (1967) they are obtained only for $G = R$. For formula (14) expressed in terms of Massey's chains see the editor's remarks to Chapter 4 in the Russian translation of Massey (1978b).

4.2. Nonstandard Relationships. All the formulas in the previous section were obtained as a consequence of standard connections between chains and cochains corresponding to different coefficients. As an example of a different nature we can consider the following (split) exact sequences:

$$0 \to \mathrm{Ext}_R(H_{n-1}^c(X, Y; R), G) \to H^n(X, Y; G) \to$$

$$\to \operatorname{Hom}_R(H_n^c(X,Y;R),G) \to 0, \tag{15}$$

which exist for a locally compact space X and arbitrary G (Theorem 6.5 in Borel and Moore (1960) or Sect. 5.12.7 in Bredon (1967)), but under the condition that X and Y are hlc_R-spaces (Y can be both open or closed). While in appearance they do not differ from (6) in Sect. 1.7, Chap. 1, nevertheless they are obtained by using the more delicate tools of homological algebra and sheaf theory. This is because the chain complex C_*^c defining homology is not free and the cochain complex defining cohomology does not coincide with $\operatorname{Hom}_R(C_*^c, G)$. Without the hlc_R assumption the formulas are not true (for example for the one-point compactification of a countable discrete space).

We shall provide some more examples in which the universal coefficient formulas (with suitable restrictions) arise in spite of the lack of a connection between chains or cochains expressed in terms of the functors $\otimes$ and Hom. Let X be a locally compact space satisfying the second axiom of countability, and let Y be a closed subspace of X (for example a pair of locally finite polyhedra or manifolds), and finally let R be a countable ring. If the groups $H_p(X,Y;R)$ are countable (for Noetherian R they are then finitely generated according to Theorem 1.5 of Sklyarenko (1984)) for $p = n, n-1, n-2$, then for every R-module G there are the following exact sequences:

$$0 \to H_n(X,Y;R) \otimes_R G \to H_n(X,Y;G) \to \operatorname{Tor}_R(H_{n-1}(X,Y;R),G) \to 0,$$

$$0 \to \operatorname{Ext}_R(H_{n-1}(X,Y;R),G) \to H_c^n(X,Y;G) \to$$

$$\to \operatorname{Hom}_R(H_n(X,Y;R),G) \to 0.$$

This result was obtained in §2 of Sklyarenko (1984) by applying methods of homological algebra and the functors $\varprojlim$ and $\varprojlim^1$. Since the chains defining H_* can be identified with the complex of the homomorphisms of free cochains with compact coefficients into R, this result is close in nature to Theorems 10 and 12 from §5, Chap. 5 in Spanier (1966), which are proved under similar assumptions for abstract chain and cochain complexes.

The conditions mentioned above are also satisfied by pairs of hlc_R-spaces. C.V. Petkova (see Petkova (1977)) proved that the homology H_*^c of pairs of metrizable, locally compact hlc_R-spaces coincides with Borel-Moore homology with compact supports and therefore for such pairs (11) is true for an arbitrary G.

Similarly, (§2 in Sklyarenko (1984)), if for a locally compact X satisfying the second axiom of countability and a closed $Y \subset X$, the groups $H^p(X,Y;R)$ are countable for $p = n, n+1, n+2$, then for an arbitrary R-module G we have the following exact sequences:

$$0 \to H^n(X,Y:R) \otimes_R G \to H^n(X,Y:G) \to$$

$$\to \operatorname{Tor}_R(H^{n+1}(X,Y;R),G) \to 0,$$

$$0 \to \operatorname{Ext}_R(H^{n+1}(X,Y;R),G) \to H_n^c(X,Y;G) \to$$

$$\to \mathrm{Hom}_R(H^n(X, Y : R), G) \to 0.$$

The countability of $H^p(X, Y; R)$ in this case does not imply, in general, the existence of a finite system of generators.

All typical universal coefficient formulas are also true for *local homology and cohomology* of locally compact spaces, but with restrictions of the type hlc_R and pclc_R, see Rothe (1983b). Under similar restrictions, complete information about local groups of products (*Künneth formulas*) is also obtained in Rothe (1983b).

4.3. Functorial Dependence. By allowing the "computation" of homology and cohomology with coefficients in a module G in terms of homology and cohomology with coefficients in a base ring R by means of purely algebraic operations and by ignoring the properties of spaces, the universal coefficient formulas thus reflect the connections not depending on the specific structure of the underlying spaces. Bredon (see Bredon (1968)) indicated the existence of connections in which such dependence is even less essential.

Let F, F_1 be two functors defined on the same category. We shall say that F_1 *depends functorially* on F (we shall denote this briefly by $F \Rightarrow F_1$) if every time a map induces an isomorphism $F(f)$, the map $F_1(f)$ is also an isomorphism. Clearly, prime examples of such dependence are provided by arbitrary universal coefficient formulas. Under the assumption that R is a countable principal ideal domain, Bredon established the following functorial dependence of homology and cohomology of topological spaces (the singular theory): $H^*(X; R) \Rightarrow H^c_*(X; G), H^*(X; R) \Rightarrow H^*(X; G)$ (G is an arbitrary R-module). The arguments in Bredon (1968) are automatically applicable to the category of locally compact spaces and their proper maps and to the theories H^*_c and H_* considered by us, and they yield (in addition to the universal coefficients formulas obtained above) the following dependences: $H_*(X; R) \Rightarrow H^*_c(X; G), H_*(X; R) \Rightarrow H_*(X; G)$. The question of dependence for the functors H^c_* and H^* was solved recently by L.D. Altusher (see Altshuler (1985)) in the category of locally compact spaces satisfying the second axiom of countability and their arbitrary (that is, not only proper) continuous maps. For a countable principal ideal domain and for an arbitrary G it was established that $H^*(X; R) \Rightarrow H^*(X; G), H^*(X; R) \Rightarrow H^c_*(X; G), H^c_*(X; R) \Rightarrow H^c_*(X; G), H^c_*(X; R) \Rightarrow H^*(X; R)$. There exist examples proving that these dependencies cannot be the result of any relationships expressing one functor in terms of another.

In all the above cases the argument X can be replaced by a pair (X, Y) in which Y is a closed subspace of X.

§5. Duality

The existence of a natural isomorphism between the homology and cohomology of a manifold in complementary dimensions which is expressed in the *Poincaré duality* and which has numerous applications in analysis, geometry and topology was considered for a long time to be a consequence of the particular combinatorial structure of the manifold. The development of sheaf theory resulted in a radical change in this perception by shedding light on the nature of this phenomenon that in essence turns out to be only a consequence of formal algebraic constructions depending not on the global combinatorial structure of a manifold, but exclusively on its local topological (even homological) nature.

The first such proofs of duality were obtained as consequences of Cartan's spectral sequence whose meaning is, in general, that the homology of a locally compact space is not independent of the cohomology, but is in a way expressed by it. In terms of singular homology such a proof was given in Cartan (1950-51) and Swan (1958), and in terms of Borel-Moore homology in Borel and Moore (1960) and Bredon (1967). The most complete and definite results can be obtained directly in a much simpler way, without using the apparatus of spectral sequences, by considering suitable sheaves of chains (see Sklyarenko (1971b)). In this approach, the homology of a (generalized) manifold turns out to be in fact the same as the cohomology numbered in reverse order. At the same time the Poincaré duality itself can be interpreted as one of the manifestations of a more general construction in homological algebra, namely the duality between derived functors. Another special case of this construction is the well known duality between the functors Ext^k and Tor_m (see Sklyarenko (1980b)). This reveals the central place occupied specifically by the Poincaré duality: all the other dualities are either simple consequences or different formulations of it.

Clearly, the possibility of identityfying concrete homology and cohomology groups which appear in practical applications is not always simple. Most frequently such identification is realized through the operation of the $\frown$-product.

5.1. Poincaré Duality. In accordance with Sect. 4.2, Chap. 5, the chains with coefficients in a locally constant system $\mathcal{G}$ determining the homology of a locally compact space X form flabby differential sheaves $\mathcal{C}_*$ and are sections of them. The sections of $\mathcal{C}_*$ over an open set $U \subset X$ coincide with the chains $C_*(X, X \setminus U; \mathcal{G}) = C_*(U, \mathcal{G})$ (Sect. 3.2, Chap. 5). This is impossible for singular chains, see Exercises 12 to Chap. 1 and 32(b) to Chap. 2 in Bredon (1967). The Borel-Moore construction satisfies these conditions,[53] but its deficiency is of different nature. Its chains are constructed only for $\mathcal{G} = R$, and homology with other coefficients is defined by complexes of sections of certain differential sheaves constructed especially with this purpose in mind (in particular, the

---

[53] However, the sheaves of chains when $\mathcal{G} \neq R$ are fine.

construction for homology with coefficients in an R-module $G \neq R$ is not independent of the original construction for coefficients in R), see §3 in Chap. 5 in Bredon (1967) and also Borel and Moore (1960).

Let us assume that $n = h\dim_G X < \infty$ (the essential role of this assumption is discussed in Sect. 4.5, Chap. 5). Here G denotes the stalk of the system $\mathcal{G}$. For $p > 0$ the sheaves $\mathcal{Z}_{n+p}$ (see Sect. 3.3 above) are flabby: being a cycle of dimension $n + p$ of the pair $(X, X \setminus U)$ an arbitrary section of $\mathcal{Z}_{n+p}|_U$ is covered by a section of a flabby sheaf $\mathcal{C}_{n+p+1}$ and therefore can be extended to all of X. For the same reason $\mathcal{Z}_{n+p} = \mathcal{B}_{n+p}$ and n is the largest number for which the *sheaf of local homology* $\mathcal{H}_n(\mathcal{G})$ (Sect. 4.4, Chap. 5) is different from zero (it is easy to show, see Lemma 9 in Kharlap (1975), that $\Gamma(\mathcal{H}_n(\mathcal{G})|_U) = H_n(X, X \setminus U; \mathcal{G})$). Let $\widetilde{\mathcal{C}}_*$ be the flabby differential sheaf obtained from $\mathcal{C}_*$ by deleting all $\mathcal{C}_p$ for $p > n$ and by factoring $\mathcal{C}_n$ by the flabby sheaf $\mathcal{B}_n = \mathcal{C}_{n+1}/\mathcal{Z}_{n+1}$. Since $\widetilde{\mathcal{C}}_*$ is a quotient sheaf of $\mathcal{C}_*$ with respect to a flabby differential sheaf generated by the exact sequence $\mathcal{K}_* : 0 \leftarrow \mathcal{B}_n \leftarrow \mathcal{C}_{n+1} \leftarrow \ldots$ with flabby $\mathcal{Z}_{n+p} = \mathcal{B}_{n+p}$ and since flabby sheaves are φ-acyclic for arbitrary families φ of supports (Sect. 2.3, Chapter 3) and consequently the complexes $\Gamma(\mathcal{K}_*|_U)$ are acyclic for arbitrary U, all the homologies associated with X can be defined in terms of sections of the differential sheaf $\widetilde{\mathcal{C}}_*$. We note that $\widetilde{\mathcal{B}}_n = 0$ and $\mathcal{Z}_n = \mathcal{H}_n(\mathcal{G})$.

Let us assume now that $\mathcal{H}_p(G) = 0$ also for $p < n = h\dim_G X$. In this case clearly $\mathcal{H}_p(\mathcal{G}) = 0$ for $p \neq n$. Such a space X, by analogy with Bredon (1967), will be called a $(G - n)$-*space*. As examples we can consider not just the ordinary manifolds; the same condition is satisfied by the union of the sphere S^n and n-dimensional diametrical disc. For $(G - n)$-spaces $\widetilde{\mathcal{C}}_*(\mathcal{G})$ is clearly the same as the flabby resolution (in reverse order) of its *orientation sheaf* $\mathcal{H}_n(\mathcal{G})$, and therefore for an arbitrary family φ of supports we have $H_k^\varphi(X; \mathcal{G}) = H_\varphi^{n-k}(X; \mathcal{H}_n(\mathcal{G}))$ (in this case the transformation ν in Sect. 4.5, Chap. 5 is an isomorphism). The last statement is precisely the *Poincaré duality*.

Poincaré duality preserves supports. In particular, if X is not compact, then the ordinary homology H_*^c is dual to cohomology H_c^* of the second kind and conversely, the ordinary cohomology H^* is dual to homology H_* of the second kind.

5.2. Generalized Manifolds. In the relationships obtained above the coefficients for cohomology $\mathcal{H}_n(\mathcal{G})$ are determined by the coefficients $\mathcal{G}$ for homology. However, in the case of manifolds, in place of $\mathcal{H}_n(\mathcal{G})$ we can use arbitrary locally constant coefficients.

A $(G - n)$-space is called a *generalized* (or *homology) manifold* over G if the stalks of the sheaf $\mathcal{H}_n(\mathcal{G})$ are isomorphic to G. Typically we consider generalized manifolds over a ring R. In Kharlap (1975) it is proved that a generalized manifold X (satisfying the second axiom of countability) over a principal ideal domain R is locally connected, $\dim_R X = h\dim_R X$, and if R

226   E.G. Sklyarenko

is countable (or is a field), then X is a hlc$_R$-space.[54] In Bredon (1969) it is proved that generalized manifolds are always *locally orientable* (that is the sheaves $\mathcal{H}_n(R)$ are locally constant). Originally, dating back to R.L. Wilder, conditions of such nature were included in the definition of a generalized manifold. It follows from the universal coefficient formulas for local homology (Theorem 4 in Rothe (1983b)) that $\mathcal{H}_k(\mathcal{G}) = \mathcal{H}_k(R) \otimes_{\mathcal{R}} \mathcal{G}$ for hlc$_R$-spaces (if the stalks of $\mathcal{G}$ are finitely generated this is true without the assumption of hlc$_R$). Let $\mathcal{H}_n(R)^{-1}$ be the unique, up to an isomorphism, (locally constant) sheaf, for which $\mathcal{H}_n(R)^{-1} \otimes_{\mathcal{R}} \mathcal{H}_n(R) = R$, see Borel and Moore (1960) or §9, Chapter 5 in Bredon (1967), and let $\mathcal{G}'$ be an arbitrary locally constant sheaf of R-modules. Then, in the situation under consideration it is clearly true that $\mathcal{G}' = \mathcal{H}_n(\mathcal{G})$ for $\mathcal{G} = \mathcal{H}_n(R)^{-1} \otimes_{\mathcal{R}} \mathcal{G}'$.

In describing Poincaré duality for (generalized) manifolds X with boundary K the restriction of $\mathcal{H}_n(\mathcal{G})$ to K is trivial, and therefore one considers either the formulas of Sect. 5.1, or for a given system $\mathcal{G}'$ (that should be equal to zero on K) the isomorphisms $H_k^\varphi(X, K; \mathcal{G}) = H_k^\varphi(X \setminus K; \mathcal{G}) = H_\varphi^{n-k}(X \setminus K; \mathcal{G}')$, since the sheaf $\mathcal{G} = \mathcal{H}_n(R)^{-1} \otimes_{\mathcal{R}} \mathcal{G}'$ is defined only over $X \setminus K$ (see Sect. 3.2, Chap. 5; the above reflects the fact that the homology of X is defined only in the case when locally constant coefficients are defined over all of X). By a *generalized manifold with a boundary* we shall understand an $(R-n)$-space for which only nontrivial stalks of $\mathcal{H}_n(R)$ are isomorphic to R. For such X, in addition to the results listed above, the following is true (see Kharlap (1975)): a) the set K is closed and nowhere dense in X[55] (in particular, $X \setminus K$ is a generalized manifold); b) in the case of a countable ring, $\dim_R K \leq n - 1$ (for $R = \mathbb{Z}$ this becomes equality) and we have isomorphisms of local cohomology $I_x^{p-1}(R) = H_x^p(R)$; c) for $R = \mathbb{Z}$ the boundary K is a generalized $(n-1)$-dimensional manifold over $\mathbb{Z}_2$. (Kharlap (1975) also contains certains equivalent definitions of generalized manifolds in terms of local homology and cohomology of all four types.)

5.3. Duality and the $\frown$-Product. The constructions of duality that do not make use of sheaf theory are typically based on its description in terms of the $\frown$-product (Sect. 2.2, Chap. 6). One of them, for (orientable) topological manifolds was described in Chap. 11 of Massey (1978b). Known proofs of the equivalence of such descriptions using sheaves are quite cumbersome (the proof in Bredon (1967), in addition, depends heavily on the construction of Borel-Moore homology, also with coefficients in sheaves). The argument given below is much simpler (and allows additional insight into the nature of the $\frown$ product).

Let $\mathcal{C}_*(\mathcal{R})$ be a differential sheaf of chains with locally constant coefficients $\mathcal{R}$ of a locally compact space X (with stalk R). From the description of the chains in §§2 and 3, Chap. 5, we obtain $\mathcal{C}_*(\mathcal{R} \otimes_R \mathcal{G}) = \mathcal{C}_*(R) \otimes_R (\mathcal{R} \otimes_R \mathcal{G}) = \mathcal{C}_*(\mathcal{R}) \otimes_R \mathcal{G}$ whenever the stalk G of the system $\mathcal{G}$ is a finitely presented R-

---

[54] See also Mitchell (1977).
[55] K is the set of all $x \in X$ for which $\mathcal{H}_n(R)_x = 0$.

module. In the general case such a relationship does not exist, but the sheaf $\mathcal{C}_*(\mathcal{R}) \otimes_\mathcal{R} \mathcal{G}$ nevertheless defines homology of X with coefficients in $\mathcal{R} \otimes_R \mathcal{G}$, provided that X is a metrizable hlc_R-space. Indeed, the sections of $\mathcal{C}_*(\mathcal{R}) \otimes_R \mathcal{G}$ define the Borel-Moore homology which, under the above restrictions, is the same as the homology considered by us. This follows from the transformations in the Borel-Moore and H_* theories of the auxiliary homology considered in §5 of Sklyarenko (1971a) and defined by the chain complexes of the type $C_*(X; R) \otimes_R G$ by means of formulas (8) in Sect. 1.4 (in these formulas the groups $H_p(U_i; \mathcal{G})$ should be replaced with $H_p^c(X, X \setminus \overline{U}_i; \mathcal{G})$, where $\overline{U}_i$ is the closure of U_i), and in addition we have to take into account the isomorphisms for the leftmost and the rightmost terms that follows from the result of S.V. Petkova mentioned in Sect. 4.2.

Let $0 \to \mathcal{G} \to \mathcal{L}^*$ be an arbitrary resolution of $\mathcal{G}$ defining the cohomology $H_\psi^*(X; \mathcal{G})$. The sections of the bigraded differential sheaf $\mathcal{C}_*(\mathcal{R}) \otimes_R \mathcal{L}^* = \{\mathcal{C}_p(\mathcal{R}) \otimes_R \mathcal{L}^q\}$, considered as a differential sheaf with respect to the total gradation $n = p - q$ (see Sect. 2.5, Chap. 3) define a homology of X with coefficients in $\mathcal{R} \otimes_R \mathcal{G}$. Indeed, since the stalks of $\mathcal{C}_p(\mathcal{R})_x$ are torsion free and since $(\mathcal{C}_p(\mathcal{R}) \otimes_R \mathcal{L}^*)_x = \mathcal{C}_p(\mathcal{R})_x \otimes_R \mathcal{L}_x^*$ (see Sect. 1.1, Chap. 6), the $\mathcal{C}_p(\mathcal{R}) \otimes_\mathcal{R} \mathcal{L}^*$ are the resolutions of the sheaves $\mathcal{C}_p(\mathcal{R}) \otimes_\mathcal{R} \mathcal{G}$ and since, furthermore the sheaves $\mathcal{C}_*(\mathcal{R})$ are flabby, all the indicated sheaves are c-soft (Exercise 18, Chap. 2 in Bredon (1967)). Consequently, since X is locally compact, they are also soft (compare with Theorem 3.4.1, Chap. 2 in Godement (1958)), and the statement can be proved by a standard diagram chasing (see Sect. 2.5, Chap. 3). For paracompactifying families φ and ψ we obtain the following chain map:

$$C_*^\varphi(X; \mathcal{R}) \otimes_R \Gamma_\psi(\mathcal{L}^*) \to \Gamma_{\varphi \cap \psi}(\mathcal{C}_*(\mathcal{R}) \otimes_R \mathcal{L}^*),$$

that in turn defines the $\frown$-product (see Sect. 2.1, Chap. 6)

$$\frown: H_{p+q}^\varphi(X; \mathcal{R}) \otimes_R H_\psi^p(X; \mathcal{G}) \to H_q^{\varphi \cap \psi}(X; \mathcal{R} \otimes_R \mathcal{G}).$$

The construction of the $\frown$-product through this approach is in fact no different from the construction of the $\smile$-product, see Sect. 1.1, Chap. 6 (the resolution $\mathcal{L}^* \otimes_\mathcal{R} \mathcal{M}^*$ appearing in this construction is acyclic in the case when X is locally compact or (since injective sheaves are fine, see Exercise 17, Chap. 2 in Bredon (1967)) for injective sheaves $\mathcal{L}^*$ and $\mathcal{M}^*$, compare with Theorem 3.7.3, Chap. 2 in Godement (1958) or Corollary 2.9.13 in Bredon $(1967)^{56}$), and in the case of generalized manifolds both constructions can be identified via duality. Indeed, the property of being an hlc_R-space for such spaces is either satisfied automatically (for typical situations see the beginning of Sect. 5.2), or is included in the original definition (see Bredon (1969)). For $a \in H_\psi^k(X; \mathcal{G})$ and $\xi \in H_m^\varphi(X; \mathcal{R})$, by taking into account the fact that $\widetilde{\mathcal{C}}_*(\mathcal{R})$ with the terms numbered in reverse order is the resolution of $\mathcal{H}_n(\mathcal{R})$, we have clearly $\nu^{-1}(\xi \frown a) = \nu^{-1}(\xi) \smile a \in H_{\varphi \cap \psi}^{n-m+k}(X; \mathcal{H}_n(\mathcal{R}) \otimes_R \mathcal{G})$, where ν is the duality isomorphism from Sect. 5.1.

---

[56] See also Sect. 5.4, Chap. 3.

Let X be connected and let $\mathcal{R} = \mathcal{H}_n(R)^{-1}$. In accordance with Sect. 5.2 this sheaf is locally constant. Let $\xi \in H_n(X; \mathcal{H}_n(R)^{-1})$ be the image of $1 \in R = H^0(X; R)$ under the isomorphism ν. It is called the *fundamental homology class of the manifold* X. In this case it is clear that $\xi \frown a = \nu(a)$, compare with Sect. 2.2, Chap. 6.

5.4. Lefschetz Duality. Since the homology of a subspace $A \subset X$ with supports in a family φ coincides with the homology of X with supports in $\varphi|_A$ (Sect. 4.3, Chap. 5), the sheaves $\widetilde{C}_*$ from Sect. 5.1 define homology of arbitrary subspaces of X. In addition, by the five lemma corresponding to the map $C_* \to \widetilde{C}_*$ and applied to the homology sequence of a pair, the same is true for the homology of arbitrary pairs (X, A). Therefore, without a loss of generality, in what follows we shall assume that for the sheaves of chains defining homology, $C_p = 0$ for $p > n = h \dim_G X$.

Let A be an open or a closed subset of a paracompact $(G-n)$-space X, and if X is hereditarily paracompact, then A can be arbitrary, and let $B = X \setminus A$. In accordance with Sect. 4.3, Chap. 5 the homology of A is determined by all sections of C_* with supports in $\theta(\varphi) = \varphi|_A$, and the groups $H_*^\varphi(X, A; \mathcal{G})$ by the quotient complex $\Gamma_\varphi(C_*)/\Gamma_{\theta(\varphi)}(C_*)$, which in the case of a paracompactifying family φ, coincides with the sections $\Gamma_{\varphi \cap B}(C_*|_B)$ with supports in $\varphi \cap B$ of the restriction of C_* to B. However, under a reversal of order of numbering, C_* becomes a flabby resolution of the "orientation" sheaf $\mathcal{H}_n(\mathcal{G})$ and the indicated operations correspond exactly to the definition of cohomology $H_\varphi^*(X, B; \mathcal{H}_n(\mathcal{G}))$ and $H_{\varphi \cap B}^*(B; \mathcal{H}_n(\mathcal{G}))$ (see Sect. 4.3, Chap. 3). Accordingly, for these A and B and for an arbitrary paracompactifying family φ the homology sequence of the pair (X, A)

$$\ldots \to H_p^{\theta(\varphi)}(A; \mathcal{G}) \to H_p^\varphi(X; \mathcal{G}) \to H_p^\varphi(X, A; \mathcal{G}) \to \ldots \qquad (16)$$

coincides with the exact cohomology sequence of the pair (X, B)

$$\ldots \to H_\varphi^{n-p}(X, B; \mathcal{H}_n(\mathcal{G})) \to H_\varphi^{n-p}(X; \mathcal{H}_n(\mathcal{G})) \to$$

$$\to H_{\varphi \cap B}^{n-p}(B; \mathcal{H}_n(\mathcal{G})) \to \ldots \qquad (17)$$

The result is true also for arbitrary φ in the case when A is closed since the restriction $j : \Gamma_\varphi(C_*) \to \Gamma_{\varphi \cap B}(C_*|_B)$ remains an epimorphism (every section of $C_*|_B$ with support in $\Phi \cap B$, $\Phi \in \varphi$ can be extended by the zero section to the complement of the closure of $\Phi \cap B$, and therefore, since C_* is flabby, to all of X). In the general case the requirement that φ be paracompactifying is essential. For example, if $A = X \setminus x$ and $\varphi = x$, then the map j defined above is not an epimorphism (the complex $\Gamma_x(C_*)$ defines the homology of a point while C_* determines the local homology $H_*^x(X; X \setminus x; \mathcal{G})$ at this point) and the definitions of $H_*^\varphi(X, A; \mathcal{G})$ and $H_\varphi^*(B; \mathcal{H}_b(\mathcal{G}))$ in this case are different.

Since $H_\varphi^*(X, B; \mathcal{H}_n(\mathcal{G})) = H_{\theta(\varphi)}^*(X; \mathcal{H}_n(\mathcal{G}))$ (Sect. 4.3, Chap. 3), in fact the only new isomorphism, among those recently obtained, is $H_p^\varphi(X, A; \mathcal{G}) =$

$H^{n-p}_{\varphi \cap B}(B; \mathcal{H}_n(\mathcal{G}))$. For a subcomplex A of a compact combinatorial (orientable) manifold X this isomorphism was obtained by Lefschetz. By taking into account character theory it is contained, for example, in Proposition (33.2) (c) Chap. 5, and also in (19.2) Chap. 8 of Lefschetz (1942)). For a compact B in a topological manifold X this isomorphism (in terms of Čech cohomology and singular homology) was established in §2, Chap. 6 in Spanier (1966), in §5, Chap. 17 in Husemöller (1966), Chap. 8 in Dold (1971) and other sources (in terms of sheaf theory in Swan (1958). The isomorphism of the sequences (16) and (17) in terms of Borel-Moore homology was considered in Raymond (1961), and also in Bredon (1967). However, the sheaves of chains in this theory are flabby only for $\mathcal{G} = R$, in connection with which the isomorphism of sequences in full generality is obtained only for open A, and for closed subset with essential restrictions (including, in particular, the requirement that the stalks of $\mathcal{G}$ be finitely generated, see Theorem 5.9.3 in Bredon (1967)). The proof presented here was given in Sklyarenko (1971b).

An obvious consequence of the equality of the sequences (16) and (17) is the duality theorems for topological manifolds (including manifolds with boundary), from Chap. 11 in Massey (1978b) (in connection with the above, see also the editor's remarks to Chap. 11 of the Russian translation of Massey (1978b)).

5.5. Duality for Arbitrary Subsets. Let A be an arbitrary subset of a hereditarily paracompact $(G - n)$-space X, $B = X \setminus A$ and let φ be the family of all closed subsets of X. In accordance with Sect. 3.3, Chap. 5, $H^\theta_p(A; \mathcal{G}) = \varinjlim_{\Phi \in \theta} H_p(\Phi, \mathcal{G})$ (θ denotes the family of all closed subsets Φ of X such that $\Phi \subset A$). According to the five lemma, the whole sequence (16) coincides with the direct limit of homology sequences of the pairs (X, Φ), in particular $H_p(X, A; \mathcal{G}) = \varinjlim_{\Phi \in \theta} H_p(X, \Phi; \mathcal{G})$. By a similar argument, the sequence (17) is the direct limit of cohomology sequences of the pairs $(X, X \setminus \Phi)$. Indeed, for $\Phi \in \theta$ the sets $X \setminus \Phi$ form all possible neighborhoods of B and the identity at the third term of (17) is the tautness property for cohomology adjusted by us (Sect. 1.5, Chap. 3), while the isomorphisms $H^{n-p}(X, B; \mathcal{H}_n(\mathcal{G})) = \varinjlim_{\Phi \in \theta} H^{n-p}(X, X \setminus \Phi; \mathcal{H}_n(\mathcal{G}))$ are a consequence of the five lemma.

Accordingly, the duality relationships from Sect. 5.4 for arbitrary subsets A and $B = X \setminus A$ are a consequence of the analogous relationships for the special case of a closed A. This illustrates, in addition, the duality between the property of tautness for cohomology and the property of having closed supports (in particular the property of having compact supports when X is compact) for homology of a "non-closed" subspace $A \subset X$ (the homology of A and the pair (X, A) in the sequence (16) in this case is simply the same as $H^c_p(A; \mathcal{G})$ and $H^c_p(X, A; \mathcal{G})$, while the symbol φ in (17) should be omitted). This provides further evidence (compare with Sect. 1.5, Chap. 2, Sect. 3.4,

Chap. 5, Sect. 4.2, Chap. 7) that our perception of homology of noncompact spaces as homology with compact supports is natural.

We note that duality formulas serve also as a peculiar "test" for Steenrod homology and Čech cohomology (equivalently, sheaf theories). If a metrizable compactum Φ is finite-dimensional, then for every embedding of Φ into the Euclidean space $\mathbb{R}^n$, the set $\mathbb{R}^n \setminus \Phi$ is a polyhedron, and therefore the homology groups $H_k(\mathbb{R}^n, \mathbb{R}^n \setminus \Phi)$ and the cohomology groups $H^k(\mathbb{R}^n, \mathbb{R}^n \setminus \Phi)$ are determined, for all practical purposes, uniquely (as the singular groups), namely they should be identical with the dual groups $H^{n-k}(\Phi)$ and $H_{n-k}(\Phi)$ (compare with Sect. 1.3, Chap. 2). This once more supports the point of view, expressed in Sect. 1.2, Chap. 2, concerning the uniqueness of the theory.

5.6. Alexander-Pontryagin-Steenrod-Sitnikov Duality.

Let A be an arbitrary subset of the sphere S^n and let $B = S^n \setminus A$. Let us agree to consider only the reduced groups in dimension 0. Then $H_p^c(A; G) = H_{p+1}^c(S^n, A; G)$ (for $p = n - 1$ the last group should be factored by the image of $H_n(S^n; G) = G$). Therefore (according to Sect. 5.4), for $p + q = n - 1$ there always exists an isomorphism $H_p^c(A; G) = H^q(B; G)$. The result is also true for an arbitrary compact (orientable) manifold M^n for which $H_p(M^n; G) = H_{p+1}(M^n; G) = 0$ (for $p = n - 1$, it is enough merely to assume $H_p(M; G) = 0$ only).

In the case when A and B are polyhedra, the result was obtained by Alexander, see Lefschetz (1942) Sect. 36, Chap. 3 and (by taking into account character theory) Proposition (33.5), Chap. 5 and Theorem (19.4), Chap. 8. For compact B it was accomplished (by using character theory) in Pontryagin (1934) (see also Theorem (19.4), Chap. 8 of Lefschetz (1942)). For compact A the result was obtained in Steenrod (1940),[57] and in full generality in Sitnikov (1951) (see also Sitnikov (1954)).[58]

Since for a compact A the set $B = S^n \setminus A$ is a polyhedron whose cohomology is determined uniquely, Steenrod homology is the unique theory satisfying the above duality (see also Sect. 5.5). In essence, Steenrod while solving a specific problem in Steenrod (1940), was able to select from among a large number of possibilities (see Sect. 2.1, Chap. 2 and Sects. 1.1, 2.4, Chap. 5) the only correct one (compare with Sect. 1.3, Chap. 2), In the general case B is the intersection of its polyhedral neighborhoods of the form $S^n \setminus \Phi$, where Φ is an arbitrary compact subset of A. In addition, the homology H_*^c with compact supports (considered for the first time by K.A. Sitnikov), according to the arguments in Sect. 5.5 is the unique admissible theory for subspaces of Euclidean spaces and spheres and therefore it is the most natural for arbitrary noncompact spaces (compare with Sect. 1.5, Chap. 2 and Sect. 3.4, Chap. 5).

---

[57] This result was in fact obtained in Kolmogorov also (1936f) (who considered only compact coefficients, compare Sect. 3.1, Chap. 5).

[58] In the special case when A is a neighborhood retract (for example, a homeomorphic image of a non-compact polyhedron), the result was obtained in Chogoshvili (1946).

5.7. Remarks. The existence of a $(G - n)$-structure in X is, in general, also a necessary condition for the existence of duality in X. In order to verify this it is enough to compare sequences (16) and (17) in Sect. 5.4 in the case when $A = X \setminus x$ is the complement of a single point and φ is the family of all closed subsets of X. It is clear that a necessary condition for these sequences to coincide, is that $H_p(X, X \setminus x; \mathcal{G}) = H^{n-p}(x; \mathcal{H}_n(\mathcal{G})) = 0$ for $p \neq n$, or equivalently, that $\mathcal{H}_p(G) = 0$ for $p \neq n$.

The first version of the Poincaré duality consisted in the assertion concerning the equality (for a compact orientable triangulated manifold) of the Betti numbers in the complementary dimensions p and $n - p$ and the torsion numbers in dimensions p and $n-p-1$. The most natural modern formulation in the form of an isomorphism in complementary dimensions of homology and cohomology groups was given by A.N. Kolmogorov (see Kolmogorov (1936e)), and practically at the same time he introduced a definition of cohomology. This was done for arbitrary orientable triangulated manifolds and both for groups with compact and closed supports. Kolmogorov also gave the modern formulation of the Alexander-Pontryagin duality (see Kolmogorov (1936f)). The need to obtain clear connections between homological groups (and not merely between their Betti and torsion numbers) led to the application in duality problems of the Pontryagin's *character theory*, which was already receiving a considerable attention at this time. Here, in the case of Poincaré duality, for example (and similarly in all other remaining situations), it was in fact necessary to prove a little more, namely, the existence of an isomorphism $H_p(X; G) = H_{n-p}(X; G^*)^*$ which entails (implicitly) the isomorphism $H_{n-p}(X; G^*) = H^{n-p}(X; G)^*$. The last equality (see Sect. 2.4, Chap. 5) is not true, generally speaking (in contrast to duality itself), for arbitrary X and G (G^* is the *character group* of the group G). Nevertheless, by inertia, the tendency to combine duality with constructions of character theory which played an important role initially, persisted for a long time and became (apparently) common enough, so that the relationships (laws) of duality in manifolds for some time were not even formulated without using the language of character theory. In the formulations given by A.N. Kolmogorov the fundamental theorems are presented in Chap. 14 of Aleksandrov (1947) and §2, Chap. 4, and Chap. 5 of Aleksandrov (1955) (Sitnikov's theorem is included in Chap. 3 of Aleksandrov (1959)). Duality theory from the point of view of the character theory is considered in Lefschetz (1942), in Chap. 6 of Aleksandrov (1955), in Chaps. 4, 5 of Aleksandrov (1959) (and in some later works).

Not contributing anything new to the Poincaré duality itself, the search for its formulation in the language of character theory, which became a goal in itself, hampered the clarification of the true picture by obscuring in various degrees (and sometimes replacing) the fundamental connections, and by introducing into duality theory difficulties that were not related to it. It also created an illusion of the existence of a large number of logically possible versions of Poincaré duality (because of the necessity to separate the coefficient groups into discrete and compact, because of the aspiration to include not only homology, but also cohomology, etc.). Without a doubt the formula $H_k(X; G^*) = H^k(X; G)^*$ in itself (in the terminology of Kolmogorov (1936a)– the reciprocity relation), when true, is useful in topology (not only in homological dimension theory) However conditions under which it is true must be clarified separately, without considering an embedding of X into a manifold. For theories of the second kind, in the case of a discrete group G, it is a consequence of the exactness of the functor $G \to G^*$ and the description of the chains and cochains in §4, Chap. 2 and Sect. 3.2, Chap. 5, and it coincides with the "reciprocity " theorem of A.N. Kolmogorov (see Kolmogorov (1936d)). In the general case the existence of such a relationship is prevented by the very nature of the chain and cochain complexes (as is demonstrated by examples of zero-dimensional spaces), and for H_*^c and H^* also by the limit formulas of the type (6) in Sect. 1.4 above. Constructions by several authors, with the intention of replacing the relationship under consideration (adaptable to duality theory as well as independent of it) and based typically on certain modifications of homology and cohomology groups are presented in Aleksandrov (1959) and Chogoshvili (1962, 1966).

For a long time (in some works until very recently) the term "duality" was mistakenly used by several authors also in connection with the isomorphisms $H_p(X \setminus A) = H_{p-1}(A)$ or $H^{p-1} = H_c^p(X \setminus A)$ in the case of a closed subset A of a space X that is acyclic in dimensions p and $p-1$. Expressed in the language of character theory in the monograph Lefschetz (1942) (Sects. 38, 39, 41, Chap. 3, Sect. 15, Chap. 6, Sect. 9, Chap. 7) this formula is referred to as the *duality theorem of Alexander type*. The reason for this is that isomorphisms of such a type, being the result of exact sequences of pairs (and excision property, §4, Chap. 2 and Sect. 3.2, Chap. 5), were noticed before these exact sequences were introduced and were considered initially as a manifestation of the duality theory. Moreover, this was reinforced by the fact that they were used as a step in the proof of Alexander's duality. Such isomorphisms are obtained also in Kolmogorov (1936f) (where they were, however, clearly distinguished from the Poincaré duality) which explains the term "*Kolmogorov duality*" which appeared during this period (and whose use is not justified after the publication of Eilenberg and Steenrod (1952)). In generalizations of these formulas to the case of sets A that are not closed in X, in accordance with the exact sequences of Sects. 4.2, 4.3 of Chap. 3 and Sect. 4.3, Chap. 5, the groups "dual" to A can only be the corresponding groups of the pair (X, A).

5.8. Duality for Sequences of Triples. Let $A_1 \subset A_2$ be a pair of subsets of a $(G - n)$-space X (with the same restrictions as in Sect. 5.4) and let φ be a paracompactifying family of supports. The homology of the pair (A_2, A_1) is defined by the chains that are the sections of $\mathcal{C}_*(\mathcal{G})$ with supports in $\varphi|_{A_2}$ restricted to $X \setminus A_1$ (compare with Sect. 4.3, Chap. 5), that is, in fact, the sections of $\mathcal{C}_*(\mathcal{G})$ over $X \setminus A_1$ with supports in the family $(\varphi \cap (X \setminus A_1))|_{A_2} = (\varphi \cap B_1)|_{(X \setminus B_2)}$, where $B_i = X \setminus A_i$, for $i = 1, 2$ (the independence of this definition of thea method of construction of chains, compare with Sects. 2.2

and 3.2, Chap. 5, is a standard corollary of this statement for A_i and the five lemma). When numbered in reverse order, $C_*(\mathcal{G})$ is a flabby resolution of $\mathcal{H}_n(\mathcal{G})$ and therefore, according to Sect. 4.3, Chap. 3, we have the identity $H_p^{\varphi|_{A_2}}(A_2, A_1; \mathcal{G}) = H_{\varphi \cap B_1}^{n-p}(B_1, B_2; \mathcal{H}_n(\mathcal{G}))$. For compact subsets B_1, B_2 of a topological manifold, the existence of this isomorphism is the content of Proposition 7.2 of Chap. 8 in Dold (1971) (obtained for the ring of coefficients in terms of Čech cohomology and for the singular homology with the aid of the $\frown$-product; a connection with this operation can be established by the methods of Sect. 5.3).

Let $A_2 \subset A_3$ and let $B_3 = X \setminus A_3$. As a standard consequence of our observations we obtain the coincidence of the *exact homology sequence of a triple* (A_3, A_2, A_1)

$$\ldots \to H_p^{\varphi|_{A_2}}(A_2, A_1; \mathcal{G}) \to H_p^{\varphi|_{A_3}}(A_3, A_1; \mathcal{G}) \to$$

$$\to H_p^{\varphi|_{A_3}}(A_3, A_2; \mathcal{G}) \to H_{p-1}^{\varphi|_{A_2}}(A_2, A_1; \mathcal{G}) \to \ldots$$

with the *cohomological sequence of a triple* (B_1, B_2, B_3)

$$\to H_{\varphi \cap B_1}^{n-p}(B_1, B_2, \mathcal{H}_n(\mathcal{G})) \to H_{\varphi \cap B_1}^{n-p}(B_1, B_3; \mathcal{H}_n(\mathcal{G})) \to$$

$$\to H_{\varphi \cap B_2}^{n-p}(B_2, B_3; \mathcal{H}_n(\mathcal{G})) \to H_{\varphi \cap B_1}^{n-p+1}(B_1, B_2; \mathcal{H}_n(\mathcal{G})) \to \ldots$$

5.9. Duality for Sequences of Triads. Let $A_i \subset X_i$ be pairs of subspaces of a $(G-n)$-space X (with the same restrictions as in Sect. 5.4), let φ be a paracompactifying family of supports and let $B_i = X \setminus X_i$ and $Y_i = X \setminus A_i$. By using the chains the sections described in the previous section of the differential sheaf $C_*(\mathcal{G})$ over sets corresponding to specific situations, we obtain the following exact sequence of chain complexes

$$0 \to C_*^{\varphi|_{X_1 \cap X_2}}(X_1 \cap X_2, A_1 \cap A_2; \mathcal{G}) \xrightarrow{\mu} C_*^{\varphi|_{X_1}}(X_1, A_1; \mathcal{G}) \oplus C_*^{\varphi|_{X_2}}(X_2, A_2; \mathcal{G}) \xrightarrow{j}$$

$$\xrightarrow{j} C_*^{\varphi|_{X_1 \cup X_2}}(X_1 \cup X_2, A_1 \cup A_2; \mathcal{G}) \to 0,$$

in which $\mu = \mu_1 - \mu_2$, and the maps μ_i are induced by the inclusions $(X_1 \cap X_2, A_1 \cap A_2) \subset (X_i, A_i)$ and j by the inclusions $(X_i, A_i) \subset (X_1 \cup X_2, A_1 \cup A_2)$. It is easy to verify its exactness in the first two terms. Since the sheaves $C_*(\mathcal{G})$ are flabby, the fact that j is an epimorphism follows from the conditions considered in §13, Chap. 2 in Bredon (1967), in particular in the case of open or closed sets (compare also with the proof of Theorem 5.5 and with the Exercises 8 and 9 of Chapter 5 in Bredon (1967). In such situations we obtain the exact *Mayer-Vietoris type homology sequence*

$$\ldots \to H_p^{\varphi|_{(X_1 \cap X_2)}}(X_1 \cap X_2, A_1 \cap A_2; \mathcal{G}) \to$$

$$\to H_p^{\varphi|_{X_1}}(X_1, A_1; \mathcal{G}) \oplus H_p^{\varphi|_{X_2}}(X_2, A_2; \mathcal{G}) \to$$

$$\to H_p^{\varphi|_{X_1 \cup X_2}}(X_1 \cup X_2, A_1 \cup A_2; \mathcal{G}) \to H_{p-1}^{\varphi|_{X_1 \cap X_2}}(X_1 \cap X_2, A_1 \cap A_2; \mathcal{G}) \to \dots$$

According to the previous section, it coincides with the *exact cohomology sequence*

$$\dots \to H_{\varphi \cap (Y_1 \cup Y_2)}^{n-p}(Y_1 \cup Y_2, B_1 \cup B_2; \mathcal{H}_n(\mathcal{G})) \to$$

$$\to H_{\varphi \cap Y_1}^{n-p}(Y_1, B_1; \mathcal{H}_n(\mathcal{G})) \oplus H_{\varphi \cap Y_2}^{n-p}(Y_2, B_2; \mathcal{H}_n(\mathcal{G})) \to$$

$$\to H_{\varphi \cap (Y_1 \cap Y_2)}^{n-p}(Y_1 \cap Y_2, B_1 \cap B_2; \mathcal{H}_n(\mathcal{G})) \to$$

$$\to H_{\varphi|_{(Y_1 \cup Y_2)}}^{n-p+1}(Y_1 \cup Y_2, B_1 \cup B_2; \mathcal{H}_n(\mathcal{G})) \to \dots$$

We shall consider separately two standard situations. First, let $X_1 = X_2 = X$. Then the *additive homology sequence of the triad* $(X; A_1, A_2)$

$$\dots \to H_p^{\varphi}(X, A_1 \cap A_2; \mathcal{G}) \to H_p^{\varphi}(X, A_1; \mathcal{G}) \oplus H_p^{\varphi}(X, A_2; \mathcal{G}) \to$$

$$\to H_p^{\varphi}(X, A_1 \cup A_2; \mathcal{G}) \to H_{p-1}^{\varphi}(X, A_1 \cap A_2; \mathcal{G}) \to \dots$$

coincides with the usual *cohomology sequence of the triad* $(Y; Y_1, Y_2)$ (here $Y = Y_1 \cup Y_2$)

$$\dots \to H_{\varphi \cap Y}^{n-p}(Y; \mathcal{H}_n(\mathcal{G})) \to H_{\varphi \cap Y_1}^{n-p}(Y_1; \mathcal{H}_n(\mathcal{G})) \oplus H_{\varphi \cap Y_2}^{n-p}(Y_2; \mathcal{H}_n(\mathcal{G})) \to$$

$$\to H_{\varphi \cap (Y_1 \cap Y_2)}^{n-p}(Y_1 \cap Y_2; \mathcal{H}_n(\mathcal{G})) \to H_{\varphi \cap Y}^{n-p+1}(Y; \mathcal{H}_n(\mathcal{G})) \to \dots$$

In the second case, let $A_1 = A_2 = \emptyset$ with $\widetilde{X} = X_1 \cup X_2$. Then the *exact homology sequence of the triad* $(\widetilde{X}; X_1, X_2)$

$$\dots \to H_p^{\varphi|_{(X_1 \cap X_2)}}(X_1 \cap X_2; \mathcal{G}) \to H_p^{\varphi|_{X_1}}(X_1; \mathcal{G}) \oplus H_p^{\varphi|_{X_2}}(X_2; \mathcal{G}) \to$$

$$\to H_p^{\varphi|_{\widetilde{X}}}(\widetilde{X}; \mathcal{G}) \to H_{p-1}^{\varphi|_{(X_1 \cap X_2)}}(X_1 \cap X_2; \mathcal{G}) \to \dots$$

coincides with the *additive cohomology sequence of the triad* $(X; B_1, B_2)$

$$\dots \to H_{\varphi}^{n-p}(X, B_1 \cup B_2; \mathcal{H}_n(\mathcal{G})) \to H_{\varphi}^{n-p}(X, B_1; \mathcal{H}_n(\mathcal{G})) \oplus$$

$$\oplus H_{\varphi}^{n-p}(X, B_2; \mathcal{H}_n(\mathcal{G})) \to H_{\varphi}^{n-p}(X, B_1 \cap B_2; \mathcal{H}_n(\mathcal{G})) \to$$

$$\to H_{\varphi}^{n-p+1}(X, B_1 \cup B_2; \mathcal{H}_n(\mathcal{G})) \to \dots$$

§6. Some Further Results

We shall mention briefly certain other results associated either with the development of the theory or with its most typical applications and also with some generalizations which have not been reflected in the main part of this survey.

6.1. Applications. The theorem on the invariance of dimension, the fixed point theorem of Lefschetz, the Borsuk-Ulam theorem, etc., are the most typical examples illustrating the effectiveness of classical homology (and cohomology). Several typical applications of homology and cohomology theories are directly associated with certain far reaching generalizations of such examples.

In particular, the homological dimension theory described in §3, belongs to the category of such generalizations. Characterization of the homological dimension $n = h \dim_G X$ of compacta in terms of the triviality of the groups $H_k(X, A; G)$, for $k > n$, $H_{n+1}(X, A; G)$, in terms of the injectivity of the map $H_n(A; G) \to H_n(X; G)$, for $A \subset X$ and in terms of the presheaf $U \to H_n(X, X \setminus U; G)$ being a sheaf, among others, were obtained in Nguyen Le Anh (1981) (without the assumption of metrizability, some of them were obtained in Mdzinarishvili (1987)). New characterizations of $h \dim_G X$ and $\dim_G X$, in particular, in terms of the structure of the exact homology and cohomology sequences in the highest dimension of a triad were found recently by B.L. Okun'. Some properties of $h \dim_G X$ and $\dim_G X$, as well as the connections between local groups and the local connectedness of a space X, were studied in Kharlap (1975). The behavior of $\dim_G X$ under limit operations is considered in Deo and Muttepawar (1983).

A large number of papers have been devoted to generalizations of the *Lefschetz fixed point theorem*. Topological spaces and their maps are selected in these papers in such way that the induced maps of homology (or cohomology) have a properly defined *Lefschetz number* (in particular, the homology groups or their images under the induced maps should be finitely generated, in connection with which, in the case of non-compact spaces, it is customary to consider maps with relatively compact images, etc.). Such maps should also have a properly defined index of a fixed point (this happens in the categories of ANR spaces and AANR-spaces, where AANR is an approximate ANR, for special intersections of ANR-spaces or special inverse limits of polyhedra, for open subsets of Hausdorff locally convex spaces and some other more general topological vector spaces, etc.). The Lefschetz theorem is typically obtained as a corollary to the fact that the Lefschetz number coincides with the sum of indices (instead of the coincidence with the Lefschetz number, described in terms of simplicial chains in the classical version). Frequently the theorem was established for multi-valued maps, typically with acyclic images of points. Among the large number of authors contributing to this area are Kinoshita, Bourgin, Dyer, Leray, Conner, Floyd, Browder, Dold, Granas, Fadell, Brown, Bredon, G.C. Skordev, Gorniewicz, Dugundji, Jiang, Yu. G. Borisovich. The most general results, together with detailed discussion of earlier research and a complete bibliography, are given in Skordev (1973, 1976a, 1976b, 1981a), Dugundji (1981), Eisensack and Fenske (1978). A variety of selected questions of this theory is discussed in Dyer (1981) (homotopic aspects, the loss of the fixed point property under the product with the interval and under the product of two manifolds with this property, other important examples, smooth actions without fixed points of cyclic groups on Euclidean spaces, the Atiyah-Bott index theory in the category of smooth manifolds, fixed points of fibering maps, Smith theory, Lefschetz numbers for iterations of maps and many other applications of the fixed point theory). A deep result concerning the Lefschetz number of the iterations of f was obtained by Dold, and in a more complete form (together with a connection with the properties of f itself) in Babenko and Bogatyj (1985, 1986, 1987). The connections between the Lefschetz number, Nielsen number and the true number of fixed points are completely understood (see, for example, Zhang (1986)).

Under similar restrictions, the *Lefschetz number of a pair* of maps p, q : $X \to Y$ is defined, provided that one of them, say p, has acyclic inverse images of points. The nonvanishing of this number implies the existence of *points of coincidence* (that is, the points $x \in X$, such that $p(x) = q(x)$). Typically, in this situation one considers the set-valued map $f = p^{-1}q$. The first result of this type was obtained by Eilenberg and Montgomery. The most complete results (together with a brief survey of earlier results) can be found in Górniewicz and Skordev (1978) and Skordev (1981b).

Many authors have generalized also the well known *Lefschetz theorem on the coincidence points* of two maps of piecewise linear manifolds. The final version was obtained in Davidyan (1980, 1983). The *Lefschetz number of two continuous maps* is defined for arbitrary n-dimensional manifolds (possibly with a boundary), one of which is proper and the other has a compact closure of its image. As always, nonvanishing of this number implies the existence of the points of coincidence. Schlagbauer (1972) contains results about the coincidence points for maps of closed, piecewise linear manifolds of different dimensions.

There exist many generalizations of the *Borsuk-Ulam theorem*. These include (and have already appeared in some monographs) theorems of Bourgin-Jiang, Conner-Floyd, and also the results of Nakaoka, Munkholm, G.S. Skordev and the others, concerning maps of $\mathbb{Z}_p$-*spaces* into Euclidean spaces and manifolds. In these results, under certain essential requirements of acyclicity, it is possible to estimate the dimensions of sets of orbits which are mapped to a point, or the number of the coincidence points for pairs of maps as well as the dimension of some special sets of points of coincidence of such maps. The most advanced results in this direction (with application of homology and cohomology, characteristic classes, the concept of the index) were obtained in Volovikov (1980, 1982), where other results, among them some obtained by the author, are mentioned. See also Hemmi, Kobayashi and Yosida (1987), Kobayashi (1986) and (Skordev (1975).

Problems of the following type have the same origin. Under specific conditions of acyclicity of elements of two finite systems $\{A_i\}$ and $\{B_i\}$ of mutually disjoint closed subsets of the sphere S^n whose union C is linked with some r dimensional cycle in $S^n \setminus C$, this cycle is linked already with some $A_p \cup B_q$ (Bing, Wilder and the others). Analogously, if such systems cover a closed orientable manifold, then under certain conditions it must be covered by some $A_p \cup B_q$. Far-reaching generalizations of results of this type to systems of sets (including infinite ones) of arbitrary finite order whose elements have intersections that are acyclic in a suitable sense, which are contained in arbitrary topological spaces or generalized manifolds, were obtained in Bykov (1973a). As corollaries one can obtain theorems concerning the minimal number of sets that cover, separate points or link cycles, concerning their multiplicity, etc.

Integer cohomology (and homology) of compacta that are *cohomology (homology) spheres* (that is, that have the same groups as spheres) over fields of coefficients were described in Bykov (1973b, 1973c). In each case, a classification of the corresponding "spheres" is provided, which in the case of cohomology turns out to be equivalent to the classification of Abelian, torsion-free groups of rank one. In the case of homology, they are in a one-to-one correspondence with infinite regular 2-adic fractions. In addition a description of $\mathrm{Ext}(H, \mathbb{Z})$ was obtained for torsion-free groups H of rank one.

In terms of the local behavior of homology and cohomology, the points of a space X can be divided into "inner" and "peripheral" (see Lawson and Madison (1970)). The analysis of different definitions of such points, of the connection between the concept of homological "stability" of points and local homology and cohomology, of different properties of sets of points of a corresponding type (dimension, density, etc.) is conducted in Kintishev (1973b) and Rothe (1983a) (the papers contain a discussion of the origins of these concepts).

6.2. Development of the Theory. The standard interval I in the homotopy property for cohomology with coefficients in G can be replaced by an arbitrary connected space, if a space X is compact (see Deo (1975), Sigmon (1972)), or by any connected compactum T if G is a compact group or a field and X is paracompact (see (Sigmon (1972)) (the *generalized homotopy property*). The general case (for both cohomology and homology) was studied in Nguyen Le Anh (1983). For the validity of the generalized homotopy property a crucial criterion for the points $t_0, t_1 \in T$ playing the same role as the end points of I is that the exact sequence $0 \to G \to H^1(S; G) \to H^1(T; G) \to 0$ be split, where S is the compactum obtained from T by attaching to t_0, t_1 the endpoints of I. It was established that for a given group G, this property depends on its validity for the group $\mathbb{Z}$ of integers; and in the case of homology this property depends also on the specific situation when X consists of a single point.

Nguyen Le Anh (1984) contains the *Vietoris-Begle theorem* for homology with coefficients in arbitrary countable groups G in the category of metrizable compacta (see Sect. 6.3, Chap. 3). A less complete result was obtained earlier in Kintishev (1977a). Nguyen Le Anh (1984) contains references to certain other results, in particular, to analogous theorems in the case when G is a compact group or a field.

Since the chains $C_*^c(A; G)$ defining the homology of $A \subset X$ are the same as the chains with supports in A, they form a subcomplex in $C_*^c(X, B; G)$, where $B = X \setminus A$, and therefore we obtain the following *exact sequences*:

$$\ldots \to H_k^c(A; G) \oplus H_k^c(B; G) \to H_k^c(X; G) \to \mathcal{J}_k^{(A,B)} \to$$

$$\to H_{k-1}^c(A; G) \oplus H_{k-1}^c(B; G) \to \ldots$$

$$\ldots \to H_k^c(A; G) \to H_k^c(X, X \setminus A; G) \to \mathcal{J}_k^{(A,B)} \to H_{k-1}(A; G) \to \ldots$$

The second of these sequences becomes the sequence in Sect. 2.4 if A is an open subset of X. The groups $\mathcal{J}_k^{(A,B)} = \mathcal{J}_k^{(B,A)}$ coincide with $\varinjlim_U H_k^c(X, X \setminus U, G)$, where U is an open subset of X, such that $X \setminus U$ is the union of two closed subsets contained in A and B respectively. It is natural to call these groups the *junction homology between complementary sets A and $B = X \setminus A$.* If A is closed, then $\mathcal{J}_k^{(A,B)} = \varinjlim_V H_k^c(X, A \cup (X \setminus V); G)$, where V is a neighborhood of A in X. In this case $\mathcal{J}_k^{(A,B)}$ frequently turns out to be naturally isomorphic to the $(k-1)$-dimensional *hyperhomology* (in the sense of Cartan and Eilenberg (1966)) of the inverse system of chain complexes $\{C_*^c(V \setminus A; G)\}$ for the inverse limit functor $\varprojlim$ and therefore it is also natural (in the indicated situations) to call $\mathcal{J}_k^{(A,B)}$ the $(k-1)$-dimensional *homology of the surrounding of a closed subset A.* We note that when $A = \{x\}$ is a point, then $\mathcal{J}_k^{(A,B)} = H_k^x$ is the group of local homology (in particular, H_k^x coincides with the $(k-1)$-dimensional hyperhomology of the system of chain complexes $\{C_*^c(V \setminus x; G)\}$ that defines the groups I_{k-1}^x, see Sect. 2.2). Similar results are also true for cohomology. Instead of the groups $H_k^c(A), H_k^c(X), H_k^c(X, X \setminus A), \mathcal{J}_k^{(A,B)}, H_k^c(X, X \setminus U), H_k^c(X, A \cup (X \setminus V)), C_*^c(V \setminus A), H_k^x, I_{k-1}^x$ we consider respectively $H^k(X, X \setminus A), H^k(X), H^k(A), \mathcal{J}_{(A,B)}^k, H^k(U), H^k(V \setminus A), C^*(X, A \cup (X \setminus V)), I_x^k$ and H_x^{k+1}.

A.V. Zarelua has described a general approach to the description of several types of *characteristic classes*. Characteristic classes are defined for certain types of resolutions of sheaves (see Zarelua (1986, 1987)), in particular, corresponding to the maps $f : X \to Y$, compare Sect. 6.1, Chap. 3, which are covering spaces or fibrations). Zarelua also considered their connection with the classical characteristic classes.

6.3. Some Generalizations. The most developed construction of homology with nonconstant coefficients (in *copresheaves*) is that proposed in Benyaminov (1970) together with the discussion of the properties of universality of homology and an attempt at their interpretation as derived functors. Compare also with Sect. 1.2, Chap. 5.

In the eighties some other constructions of homology for non-compact spaces were proposed (and discussed) in the literature. In addition to the fact that they were inspired by the shape theory of such spaces, no information about their purpose or advantages, and about their connection with homology H_*^c with constant coefficients (which are simply not mentioned) was given. Frequently the new groups are also called the Steenrod-Sitnikov homology, see for example Cordier (1987), Koyama (1984), Lisitsa and Mardešić (1983, 1985), even though, beginning with Milnor (1960), which is quoted in the above-mentioned papers, this term was traditionally used in connection with H_*^c (compare with Sect. 1.5, Chap. 2,; see also Sect. 3.4, Chap. 5).

The new homology theories, called also strong or coherent, represent a version of hyperhomology (compare with the previous section) for the $\varprojlim$-functor of inverse systems of chain complexes $\underline{C}_* = \{C_*'^{\lambda}\}$ appearing in definitions, that is, the chains complexes $\underline{C}_*$ in the Abelian category of the inverse systems spectra of Abelian groups. Accordingly, they can be obtained as homology (with respect to total gradation) of the double complex $\varprojlim L^*(\underline{C}_*)$, where L^* is an arbitrary lim-acyclic resolution (for examples of such resolutions see Jensen (1972). The chain complex considered in Lisitsa and Mardešić (1983, 1985) coincides with $\varprojlim L^*(\underline{C}_*)$ if we take in place of L^* the well known *Roos resolution* described, for example, in §4 of Jensen (1972) (this is how this complex was constructed in Miminoshvili (1984)). In contrast to the ordinary situation (the difference exists only in the infinite dimensional case) the total gradation is defined not by direct sums but by direct products of homogeneous parts of the double complex $\varprojlim L^*(\underline{C}_*)$.

In the category of cell complexes the homology under consideration (as follows immediately from the very nature of the inverse system $\underline{C}_*$ and chain complexes that form it) coincides with ordinary homology, that is, with H_*^c and in the category of compact spaces with the homology H_* (see Miminoshvili (1984)). Consequently, there always exists a natural transformation of H_*^c into the theory under consideration. In the general case, very likely the considered homology can be different from H_*^c (Yu. T. Lisitsa, and also Mardešić and Prasolov (1987)). It is not likely, however, that they have any advantage over H_*^c; rather they have deficiencies when compared to H_*^c (such as being nontrivial in negative dimensions, lacking the additivity properties, compare with Mardešić and Prasolov (1987). For lim-acyclic $\underline{C}_*$ (in any case, according to the argument above, for finite dimensional spaces) they should coincide with the homology of the complex $\varprojlim \underline{C}_*$ (this property was used, for example in Sect. 1.5). In general situations it is necessary to consider two spectral sequences typical for hyperhomology, (in Sect. 1.5 above, one of them is trivial). In the papers indicated above such issues are not considered.

References*

Algebra, Topology, Geometry, Foundations and Logic, History of Mathematics, Ordena Trud. Krasnogo Znameni Tbilis. Math. Inst. im. M.A. Razmadze, Tbilisi (1985), 84–106

Aleksandrov, P.S. (1947): Combinatorial Topology. M.-L. GITTL, 1947, 660 pp. (Zbl.70,179 (Vol. 1); Zbl.97,159 (Vol. 2)); Zbl.37,97; English translation: Vol. 1 (1956), Vol. 2 (The Betti groups) (1957), Graylock Press, Rochester

Aleksandrov, P.S. (1955): Topological duality theorems, 1. Closed sets. Tr. Mat. Inst. Steklova **48**, 110 pp. Zbl.65,161; English translation: Transl., II. Ser., Am. Math. Soc. 30, 1–102 (1963)

Aleksandrov, P.S. (1959): Topological duality theorems, 2. Non-closed sets. Tr. Mat. Inst. Steklova **54**, 136 pp. Zbl.92,154; English translation: Transl., II. Ser., Am. Math. Soc. 30, 103–233 (1963)

Alexander. J.W. (1935): On the ring of a compact metric space. Proc. Natl. Acad. Sci. USA **21**, 511–512. Zbl.12,231

Alexander, J.W. (1938): A theory of connectivity in terms of gratings. Ann. Math., II. Ser. **39**, 883–912. Zbl. 19,374

Altshuler, L.D. (1985): On functorial dependence between homology and cohomology groups. Mat. Zametki **38**(4), 599–607. Zbl.602.55004; English translation: Math. Notes 38, 839–843 (1985)

Babenko, I.K., Bogatyj, S.A. (1986): Algebraic multiplicity and periodic points of maps. Geometry, Differential Equations and Mechanics, the 7th Topical Conference, February 1985, Moscow, 35-48 (Russian)

Babenko, I.K., Bogatyj, S.A. (1986): Lefschetz numbers, local indices and periodic points. Dokl. Akad. Nauk SSSR **291**(3), 521–524 (Russian). Zbl.637.58022

Babenko, I.K., Bogatyj, S.A. (1987): Arithmetic properties of Lefschetz numbers of iterations. Baku Intern. Topol. Conf. Proceedings, part 2, Baku, 26 (Russian)

Bacon, Ph. (1974): Axioms for Čech cohomology of paracompacta. Pac. J. Math. **52**(1), 7–9. Zbl.291.55007

Barratt, M.G. (1962): A note on anomalous singular torsion groups. Q. J. Math. **13**(49), 75–79. Zbl.108,177

Barratt, M.G., Milnor, J. (1962): An example of anomalous singular homology. Proc. Am. Math. Soc. **13**(2), 293–297. Zbl.111,354

Bartik, B. (1968): Aleksandrov-Čech cohomology and maps into Eilenberg-MacLane polyhedra. Mat. Sb. **76**(2), 231–238. Zbl.172,483; English translation: Math. USSR, Sb. 5, 221–228 (1968)

Benyaminov, E.M. (1970): On Deheuvels homology of metric compacta. Dokl. Akad. Nauk SSSR **195**(3), 523–525. Zbl.217,487; English translation: Sov. Math., Dokl. 11, 1487–1489 (1971)

Benyaminov, E.M., Sklyarenko, E.G. (1967): On local cohomology groups. Dokl. Akad. Nauk SSSR **176**(5), 987–990. Zbl.167,516; English translation: Sov. Math., Dokl. 8, 1221–1225 (1967)

Berikashvili, N.A. (1980): Steenrod-Sitnikov homology theory on the category of compact spaces. Dokl. Akad. Nauk SSSR **254**(6), 1289–1291. Zbl.482.55007; English translation: Sov. Math., Dokl. 22, 544–547 (1980)

---

*For the convenience of the reader, references to reviews in Zentralblatt für Mathematik (Zbl.), compiled using the MATH database, and Jahrbuch über die Fortschritte der Mathematik (Jrb.) have, as far as possible, been included in this bibliography.

Berikashvili, N.A. (1986): On the uniqueness of homology theory on the category of compact spaces. Tr. Tbilis. Mat. Inst. Razmadze, Akad. Nauk Gruzin. SSR **83**, 19–25 (Russian). Zbl.616.55003

Borel, A. (1960): Seminar on Transformation Groups. Ann. Math. Stud. **46**, Princeton, New Jersey, Princeton Univ. Press 1960, 245 pp. Zbl.91,372

Borel, A., Moore, J.C. (1960): Homology theory for locally compact spaces. Mich. Math. J. **7**, 137–159. Zbl.116,403

Botvinnik, B.I., Kuz' minov, B.I. (1979): Equivalence conditions for homology theories on the category of bicompacta. Sib. Mat. Zh. **20**(6), 1233–1240. Zbl.438.55003; English translation: Sib. Math. J. 20, 873–879 (1980)

Bredon, G.E. (1967): Sheaf Theory. New York: Mc Graw-Hill, 272 pp. Zbl.158,205

Bredon, G.E. (1968): On the Künneth formulas and functorial dependence in algebraic topology. Am. J. Math. **90**(2), 522–527. Zbl.159,246

Bredon, G.E. (1969): Wilder manifolds are locally orientable. Proc. Natl. Acad. Sci. USA **63**(4), 1079–1081. Zbl.186,270

Bykov, V.M. (1973a): On closed covers of manifolds and linking of cycles. Dokl. Akad. Nauk SSSR **211**(5), 1027–1030. Zbl.288.55008; English translation: Sov. Math., Dokl. 14, 1140–1143 (1974)

Bykov, V.M. (1973b): On homology and cohomology spheres. Usp. Mat. Nauk **28**(6), 197–198 (Russian). Zbl.289.55006

Bykov, V.M. (1973c): On certain properties of generalized manifolds and cohomology spheres. Thesis of Dissertation for the Candidate of Phys. Math. Sciences, Moscow Univ., 8 pp (Russian)

Cartan, H. (1948-49): Séminaire de topologie algébrique. Paris: Ecole Norm. Supér. 1948–1949. Zbl.35,246

Cartan, H. (1950-51): Cohomologie des groupes, suite spectral, faisceaux. Séminaire. Paris: Ecole Norm. Super. 1950–1951 (2nd éd 1955. Zbl.67,154)

Cartan, H., Eilenberg, S. (1956): Homological Algebra. Princeton: Princeton Univ. Press, 488 pp. Zbl.75,243

Chogoshvili, G.S. (1940): On homology theory of topological spaces. Soobshch. Akad. Nauk Gruzin. SSR **1**, 337–340 (Russian). Zbl.25,235

Chogoshvili, G.S. (1946): Duality principle for retracts. Dokl. Akad. Nauk SSSR **51**(2), 87–90 (Russian). Zbl.60,408

Chogoshvili, G.S. (1951): On equivalence of functional and spectral homology theory. Izv. Akad. Nauk SSSR, Ser. Mat. **15**(5), 421–438 (Russian). Zbl.45,259

Chogoshvili, G. (1962): On homology theory of non-closed sets. General Topology and its Relations to Modern Analysis and Algebra, Proc. Symp. Prague. Sept 1961, Prague: Publ. House Czechosl. Acad. Sci., 123–132. Zbl.113,168

Chogoshvili, G.S. (1966): Generalized products and limits and their applications to homology theory. Usp. Mat. Nauk **21**(4), 23–34. Zbl.171,438; English translation: Russ. Math. Surv. 21(4), 13–23 (1966)

Cordier, J.M. (1987): Homologie de Steenrod-Sitnikov et limite homotopique algébrique. Manuscr. Math **59**(1), 35–52. Zbl.643.55003

Davidyan, V.R. (1980): On coincidence points of two maps. Mat. Sb. **112**(2), 220–225. Zbl.438.55011; English translation: Math. USSR, Sb. 40, 205–210 (1981)

Davidyan, V.R. (1983): On coincidence points of two maps of manifolds with a boundary. Usp. Mat. Nauk 38(1), 149–150. Zbl.527.55007; English translation: Russ. Math. Surv. 38(1), 176 (1983)

Deo, S. (1975): On the tautness property of Alexander-Spanier cohomology. Proc. Am. Math. Soc. **52**, 441–444. Zbl.304.55005

Deo, S., Muttepawar, S. (1983): Inverse limit theorem and expansion theorem for cohomological dimension. Bull. Aust. Math. Soc. **28**(1), 67–75. Zbl.524.55004

Dold, A. (1972): Lectures on Algebraic Topology. Berlin-Heidelberg-New York: Springer, 377 pp. Zbl.234.55001 (2nd ed. 1980)

Dranishnikov, A.N. (1987): On homology dimension mod p. Mat. Sb. **132**(3), 420–433. Zbl.622.55002; English translation: Math. USSR, Sb. 60(2), 413–425 (1988)

Dranishnikov, A.N., Shchepin, E.V. (1986): Cell-like maps. The dimension raising problem. Usp. Mat. Nauk **41**(6), 49–90. Zbl.627.57017; English translation: Russ. Math. Surv. 41(6), 59–111 (1986)

Dugundji, J. (1981): The Lefschetz theorem for self-maps of compacta. General Topol. and Modern Analysis. Proc. Conf., Riverside, Calif., May 28–31, 1980, New York, 177-184. Zbl.479.55002

Dydak, J. (1986): Steenrod homology and local connectedness. Proc. Am. Math. Soc. **98**(1), 153–157. Zbl.603.55006

Dyer, E. (1981): Some remarks on fixed point theory. General Topol. and Modern Analysis. Proc. Conf., Riverside, Calif., May 28-31, 1980, New York, 185–208. Zbl.456.55002

Edwards, D.A., Hastings, H.M. (1976): Čech and Steenrod Homotopy Theories with Applications to Geometric Topology. Lect. Notes Math. **542**, 296 pp. Zbl.334.55001

Eilenberg, S., MacLane, S. (1942): Group extensions and homology. Ann. Math., II. Ser. **43**(4), 757–831. Zbl.61,406

Eilenberg, S., Steenrod, N. (1945): Axiomatic approach to homology theory. Proc. Natl. Acad. Sci. USA **31**, 117–120. Zbl.61,405

Eilenberg, S., Steenrod, N. (1952): Foundation of Algebraic Topology. Princeton: Princeton Univ. Press, 328 pp. Zbl.47,414

Eisenack, G., Fenske, C. (1978): Fixpunkttheorie. Bibliographisches Inst. Mannheim, Wien, Zürich, 258 pp. Zbl.369.47001

Godbillon, C. (1967): Cohomologie à l'infini des espaces localement compacts. C.R. Acad. Sci., Paris, Ser. A **264**(9), 394–396. Zbl.147,233

Godement, R. (1958): Topologie Algébrique and théorie des Faisceaux. Paris: Hermann, 283 pp. Zbl.80,162

Górniewicz, L., Skordev, G. (1978): On coincidence of continuous maps. Fundam. Math. **101**, 171–180 (Russian). Zbl.426.55003

Grothendieck, A. (1957): Sur quelques points d'algèbre homologique. Tohoku Math. J. II. Ser. **9**, 119–211. Zbl.118,261

Hemmi, Yu., Kobayashi, T., Yosida, T. (1987): The Borsuk-Ulam theorem for $\mathbb{Z}_p$-maps from S^{2m+1} to a $\mathbb{Z}_p$-complex. Mem. Fac. Sci. Kochi. Univ. A8, 27–30. Zbl.621.55002

Hörmander, L. (1973): An Introduction to Complex Analysis in Several Variables. 2nd ed., Amsterdam-London: North Holland Publ. Co., 224 pp. Zbl.271.32001

Husemöller, D. (1966): Fibre Bundles. New York: Mc Graw-Hill, 300 pp. Zbl.144,448

Hyman, D.M. (1968): A category slightly larger than the metric and CW-categories. Mich. Math. J. **15**(2), 193–214. Zbl.157,539

Inasaridze, Kh.N. (1986): On a characterization of exact bifunctorial homology theories. Soobshch. Akad. Nauk Gruzin. SSR **121**(2), 241–243 (Russian). Zbl.606.55007

Inasaridze, Kh.N., Mdzinarishvili, L.D. (1980): On the mutual relationship between continuity and exactness in homology theory. Soobshch. Akad. Nauk Gruzin. SSR **99**(2), 317–320 (Russian). Zbl.448.55002

James, I.M., Whitehead, J.H.C. (1958): Homology with zero coefficients. Q. J. Math., Oxf. II. Ser. **9**(36), 317–320. Zbl.86,160

Jensen, C.U. (1972): Les Foncteurs Dérivés de $\underleftarrow{\lim}$ et leurs Applications en Theorie des Modules. Lect. Notes Math. **254**, 103 pp. Zbl.238.18007

Kahn, D., Kaminker, J., Schochet, C. (1977): Generalized homology theories on compact metric spaces. Mich. Math. J. **24**(2), 203–224. Zbl.384.55001

Kharlap, A.E. (1975): Local homology and cohomology, homology dimension and generalized manifolds. Mat. Sb. **96**(3), 347–373. Zbl.325.57002; English translation: Math. USSR, Sb. 25, 323–349 (1976)

Kintishev, Ya.Ts. (1977a): On the Vietoris-Begle theorem for exact homology. C.R. Acad. Bulg. **30**(1), 7–8 (Russian). Zbl.418.55004

Kintishev, Ya.Ts. (1977b): Homologically unstable points. Vestn. Mosk. Univ., Mat. Mekh. No. 3, 63–69. Zbl.357.55001; English translation: Mosc. Univ. Math. Bull. 32(3), 57–62 (1977)

Kobayashi, T. (1986): The Borsuk-Ulam theorem for Z_p-map from a $\mathbb{Z}_p$-space to S^{2n+1}. Proc. Am. Math. Soc. **97**(4), 714–716. Zbl.598.55002

Kolmogorov, A.N. (1936a): Dualität im Aufbau der kombinatorischen Topologie. Mat. Sb. **1**, 97–102. Zbl.14,38

Kolmogorov, A.N. (1936b): Homologierung des Komplexes und des lokal bikompakten Raumes. Mat. Sb. **1**, 701–706. Zbl.16,139

Kolmogorov, A. (1936c): Les groupes de Betti des espaces localement bicompacts. C.R. Acad. Sci., Paris, **202**, 1144–1147. Zbl.13,422

Kolmogorov, A. (1936d): Propriétés des groupes de Betti des espaces localement bicompacts. C.R. Acad. Sci., Paris, **202**, 1325–1327. Zbl.13,423

Kolmogorov, A. (1936e): Les groupes de Betti des espaces métriques. C.R. Acad. Sci., Paris, **202**, 1558–1560. Zbl.14,38

Kolmogorov, A. (1936f): Cycles relatifs. Théorème de dualité de Alexander. C.R. Acad. Sci., Paris, **202**, 1641–1643. Zbl.14,39

Korobov, A.A. (1969): On cohomology of algebras with an identity. Dokl. Akad. Nauk SSSR **184**(1), 24–27. Zbl.203,347; English translation: Sov. Math., Dokl. 10, 17–20 (1969)

Koyama, A. (1984): Coherent singular complexes in strong shape theory. Tsukuba J. Math. **8**(2), 261–295. Zbl.564.55007

Kuz'minov, V.I. (1967): On derived functors of the inverse limit functor. Sib. Mat. Zh. **8**(2), 333–345 (Russian). Zbl.162,549

Kuz'minov, V.I. (1968): Homological dimension theory. Usp. Mat. Nauk **23**(5), 3–49. Zbl.179,279; English translation: Russ. Math. Surv. 23(5), 1–45 (1968)

Kuz'minov, V.I. (1980): On equivalence of homology theories on the category of bicompacta. Sib. Mat. Zh. **21**(1), 125–129. Zbl.481.55005; English translation: Sib. Math. J. 21, 93–97 (1980)

Kuz'minov, V.I., Shvedov, I.A. (1975): Hyperhomology of the limit of a direct spectrum of complexes and homology groups of topological spaces. Sib. Mat. Zh. **16**(1), 62–74. Zbl.306.55004; English translation: Sib. Math. J. 16, 49–58 (1975)

Kuz'minov, V.I., Shvedov, I.A. (1976): On Bredon's cohomological dimension of hereditarily paracompact spaces. Dokl. Akad. Nauk SSSR **231**(1), 24–27. Zbl.357. 55005; English translation: Sov. Math., Dokl. 17, 1506–15009 (1977)

Lawson, J., Madison, B. (1970): Peripheral and inner points. Fundam. Math. **69**(3), 253–266. Zbl.203,561

Lefschetz, S. (1935): Chain-deformations in topology. Duke Math. J. **1**, 1–18. Zbl.11,180

Lefschetz, S. (1942): Algebraic Topology. Am. Math. Soc. Coll. Publ. **27**, 389 pp. Zbl.61,393

Lisitsa, Yu. T., Mardešić, S. (1983): Steenrod-Sitnikov homology for arbitrary spaces. Bull. Am. Math. Soc., New Ser. **9**(2), 207–210. Zbl.532.55003

Lisitsa, Yu. T., Mardešić, S. (1985): Strong homology of inverse systems of spaces. II, Topology Appl. **19**(1), 45–64. Zbl.559.55008

Mardešić, S., Prasolov, A.V. (1987): Non-triviality of $\underleftarrow{\lim}^1$ and strong homology. Baku Int. Topol. Conf., Baku, 90

Massey, W.S. (1978a): How to give an exposition of the Čech-Alexander-Spanier type homology theory. Am. Math. Mon. **85**(2), 75–83. Zbl.377.55005

Massey, W.S. (1978b): Homology and Cohomology Theory. New York, Basel: M. Dekker Inc., 412 pp. Zbl.377.55004

Mdzinarishvili, L.D. (1980): On the Künneth formula and the $\varprojlim$ functor. Soobshch. Akad. Nauk Gruzin. SSR **99**(3), 561–564 (Russian). Zbl.457.55015

Mdzinarishvili, L.D. (1982): A generalization of Künneth formula. Soobshch. Akad. Nauk Gruzin. SSR **106**(2), 257–260 (Russian). Zbl.534.18008

Mdzinarishvili, L. (1984): Universelle Koeffizientenfolgen für den $\varprojlim$-Funktor und Anwendungen. Manuscr. Math. **48**, 255–273. Zbl.572.18006

Mdzinarishvili, L.D. (1987): On homological dimensionality. Proc. Baku International Top. Conf., part 2, Baku, 187

Milnor, J. (1960): On the Steenrod homology theory. Berkley, (mimeographed notes)

Milnor, J. (1962): On axiomatic homology theory. Pac. J. Math. **12**, 337–341. Zbl.114,396

Miminoshvili, Z.P. (1984): On exact and semi-exact homology sequences of arbitrary spaces. Soobshch. Akad. Nauk Gruzin. SSR **113**(1), 41–44 (Russian). Zbl.546.55006

Miryuk, V.G. (1973): Homology and cohomology of sets and their neighborhoods. Mat. Sb. **92**(3), 306–318. Zbl.285.55002; English translation: Math. USSR, Sb. 21, 303–315 (1974)

Mitchell, W.J.R. (1977): A problem of Bredon concerning homology manifolds. Transformation groups, Proc. Conf. Newcastle 1976, London Math. Soc. Lect. Notes Ser., No. **26**, 305–306. Zbl.386.57006

Mitchell, W.J.R. (1979): Homology manifolds, inverse systems and cohomological local connectedness. J. Lond. Math. Soc., II Ser. **19**(2), 348–358. Zbl.405.57009

Nguyen Le Anh (1981): On homological dimension of compacta. Vest. Mosk. Univ., Mat. Mekh. No. 2, 29–31. Zbl.462.55001; English translation: Mosc. Univ. Math. Bull. 36(2), 35–37 (1981)

Nguyen Le Anh (1983): Generalized homotopy axiom. Mat. Zametki **33**(1), 117–130. Zbl.515.55004; English translation: Math. Notes 33, 58–64 (1983)

Nguyen Le Anh (1984): On a Vietoris-Begle theorem. Mat. Zametki **35**(6), 847–854. Zbl.571.55006; English translation: MN 35, 444–447, (1984)

Petkova, S.V. (1972): On the axioms of homology theory. Dokl. Akad. Nauk SSSR **204**(3), 557–560. Zbl.272.55014; English translation: Sov. Math., Dokl. 13, 716–719 (1972)

Petkova, S.V. (1973a): On the axioms of homology theory. Mat. Sb. **90**(4), 607–624. Zbl.256.55008; English translation: Math. USSR, Sb. 19, 597–614 (1974)

Petkova, S.V. (1973b): Some uniqueness theorems in homology and cohomology theories. Usp. Mat. Nauk **28**(3), 195–196 (Russian). Zbl.285.55007

Petkova, S.V. (1977): Spaces with finitely generated homology and cohomology. Serdica Bulg. Math. Publ. **3**, 349–358 (Bulgarian). Zbl.442.55005

Petkova, S.V. (1982): On coincidence of homology for homologically locally connected spaces. C. R. Acad. Bulg. **35**(4), 427–430 (Russian). Zbl.511.55005

Pontryagin, L. (1934): A general topological theorem for duality for closed sets. Ann. Math., II Ser. **35**, 904–914. Zbl.9,156

Raymond, F. (1961): Local cohomology groups with closed supports. Math. Z. **76**(1), 31–41. Zbl.100,191

Rothe, D. (1983a): On peripheral points. Math. Nachr. **114**, 255–261 (Russian). Zbl.573.55004

Rothe, D. (1983b): Peripheral cohomological local connectedness. Fundam. Math. **116**(1), 53–66. Zbl.527.55011

Rothe, D. (1983c): On characterization of homological dimension by local homology and cohomology. Math. Nachr. **113**, 53–57. Zbl.538.55002

Saneblidze, S.A. (1986): Extraordinary homology theories for compact spaces. Tr. Tbilis. Mat. Inst. Razmadze **83**, 19–25 (Russian). Zbl.616.55001

Schapira, P. (1970): Théorie des Hyperfonctions. Lect. Notes in Math. 126, 157 pp. Zbl.192,473

Schlagbauer, H. (1972): On the dimension of coincidence sets. Abh. Math. Semin. Univ. Hamburg **37**(3-4), 236–245. Zbl.237,55006

Sigmon, K. (1972): A strong homotopy axiom for Alexander cohomology. Proc. Am. Math. Soc. **31**(1), 271–275. Zbl.227.55008

Sitnikov, K.A. (1951): The duality principle for non-closed sets. Dokl. Akad. Nauk SSSR **81**(3), 359–362 (Russian). Zbl.43,383

Sitnikov, K.A. (1954): Combinatorial topology of nonclosed sets I. The first duality principle; spectral duality. Mat. Sb. **34**, 3–54 (Russian). Zbl.55,163

Sklyarenko, E.G. (1964): On some applications of sheaf theory in general topology. Usp. Mat. Nauk **19**(6), 47–70. Zbl.141,208; English translation: Russ. Math. Surv. 19(6), 41–62 (1964)

Sklyarenko, E.G. (1969): Homology theory and the exactness axiom. Usp. Mat. Nauk **24**(5), 87–140. Zbl.204,560; English translation: Russ. Math. Surv. 24(5), 91–142 (1969)

Sklyarenko, E.G. (1971a): Uniqueness theorems in homology theory. Mat. Sb. **85**, 201–223. Zbl.214,217; English translation: Math. USSR, Sb. 14, 199–218 (1972)

Sklyarenko, E.G. (1971b): On the theory of generalized manifolds. Izv. Akad. Nauk SSSR, Ser. Mat. **35**, 831–843. Zbl.269.57007; English translation: Math. USSR, Izv. 5, 845–857 (1972)

Sklyarenko, E.G. (1979): On homology theories associated with the Alexander–Čech cohomology. Usp. Mat. Nauk **34**(6), 90–118. Zbl.426.55005; English translation: Russ. Math. Surv. 34(6), 103–137 (1979)

Sklyarenko, E.G. (1980a): On homologically locally connected spaces. Izv. Akad. Nauk SSSR, Ser. Math. **44**(6), 1417–1433. Zbl.456.55006; English translation: Math. USSR, Izv. 17, 601–614 (1981)

Sklyarenko, E.G. (1980b): Poincaré duality and the relationship between the Tor and Ext functors. Mat. Zametki **28**(5), 769–776. Zbl.457.18012; English translation: Math. Notes 28, 841–845 (1981)

Sklyarenko, E.G. (1984): Some applications of the $\underleftarrow{\lim}{}^1$-functor. Mat. Sb. **123**(3), 369–390. Zbl.573.55005; English translation: Math. USSR, Sb. 51, 367–387 (1985)

Skordev, G.S. (1970a): On resolutions of a continuous map. Mat. Sb. **82**(4), 532–550. Zbl.205,269; English translation: Math. USSR, Sb. 11, 491–506 (1971)

Skordev, G.S. (1970b): On dimension raising maps. Mat. Zametki **7**(6), 697–705. Zbl.205,268; English translation: Math. Notes 7, 421–426 (1970)

Skordev, G.S. (1971): On resolutions corresponding to a closed map. Mat. Sb. **86**(2), 234–247. Zbl.224.55008; English translation: Math. USSR, Sb. 15, 227–240 (1972)

Skordev, G.S. (1973): On A-ANR spaces. Bull. Acad. Pol. Sci. Ser. Sci. Math. Astron. Phys. **21**(12), 1123–1130 (Russian). Zbl.271.54012

Skordev, D.S. (1975): On the theorem of Bourgin-Jiang. Serdica Bulgar. Math. Publ. **1**, 183–198. Zbl.362.55007

Skordev, G.S. (1976a): On the fixed point theorem of Schauder. Serdica Bulgar. Math. Publ. **2**, 122–125. Zbl.359.47034

Skordev, G.S. (1976b): Fixed points of acyclic maps. God. Sofij. Univ. Fak. Mat. Mekh. **67**, (1972-1973), 7–18 (Russian). Zbl.361.54031

Skordev, G.S. (1981a): Fixed-point index for ANR's. Forschungsschwerpunkt Dynamische Systeme, Fachbereich Math. (Informatik, Elektro.) Physik Univ. Bremen, Rep. 46

Skordev, G.S. (1981b): Coincidence of maps in Q-simplicial spaces. Fundam. Math. **113**, 67–79. Zbl.483.55002

Skordev, G.S. (1982): Resolutions of a closed map. God. Sofij. Univ. Fak. Mat. Mekh **71**(1), (1976-1977), 87–118 (Russian). Zbl.523.18008

Skordev, G.S. (1983): The spectral sequence of Zarelua. Pliska Bulg. Math. Stud. **6**, 121–149 (Russian). Zbl.558.55001

Spanier, E.H. (1966): Algebraic Topology. New York: Mc Graw-Hill, 528 pp. Zbl.145, 433

Steenrod, N.E. (1940): Regular cycles of compact metric spaces. Ann. Math., II Ser. **41**, 833–851. Zbl25,234

Swan, R.G. (1958): The Theory of Sheaves. Math. Inst. Univ. of Oxford, 174 pp.

Volovikov, A.Yu. (1980): On a theorem of Bourgin-Jiang. Usp. Mat. Nauk **35**(3), 159–161. Zbl.436.55004; English translation: Russ. Math. Surv. 35 (3), 196–200 (1980)

Volovikov, A.Yu. (1982): Maps of free $\mathbb{Z}_p$-spaces into manifolds. Izv. Akad. Nauk SSSR, Ser. Mat. **46**(1), 36–55. Zbl.503.55001; English translation: Math. USSR, Izv. 20, 35–53 (1983)

Wells, R.O. (1973): Differential Analysis on Complex Manifolds. Englewood Cliffs, N.J.: Prentice-Hall Inc., 252 pp. Zbl.262.32005

Zarelua, A.V. (1969): Finite-to-one maps of topological spaces and cohomology manifolds. Sib. Mat. Zh. **10**(1), 64–92. Zbl.194,238; English translation: Sib. Math. J. 10, 45–63 (1960)

Zarelua, A.V (1972):. Cohomological structure of finite-to-one maps. Tr. Tbilis. Mat. Inst. Razmadze **41**, 100–127 (Russian). Zbl.321.55010

Zarelua, A.V. (1977): On a resolution of a continuous map and a spectral sequence associated with it. Tr. Tbilis. Mat. Inst. Razmadze **56**, 99–117 (Russian). Zbl.399.55018

Zarelua, A. (1979): Sheaf theory and zero-dimensional mappings. Applications of sheaves. Proc. Durham 1977, Lect. Notes Math. **753**, 768–779. Zbl.443.55016

Zarelua, A.V. (1983): Limits of local systems of sheaves and zero–dimensional maps. Tr. Mat. Inst. Steklova **54**, 98–112. Zbl.547.54008; English translation: Proc. Steklov Inst. Math. 154, 105–120 (1985)

Zarelua, A.V. (1986): Locally homotopically contractible differential sheaves. Usp. Mat. Nauk **41**(6), 91–108. Zbl.623.55003; English translation: Russ. Math. Surv. 41(6), 113–134 (1986)

Zarelua, A. (1987): Characteristic classes of sheaf resolutions and classical characteristic classes. Baku Intern. Topol. Conf., Part 2, Baku, 115

Zhang Hingguo (1986): The least number of fixed points can be arbitrarily larger than the Nielsen number. Acta Sci. Natur. Univ. Pekin. 1986, No. **3**, 15–25. Zbl.615.55005

Subject Index

Encyclopaedia of Mathematical Sciences
Editor-in-Chief: R. V. Gamkrelidze

Analysis

Volume 13: **R.V. Gamkrelidze** (Ed.)
Analysis I
Integral Representation and Asymptotic Methods
1989. VII, 238 pp. 3 figs. ISBN 3-540-17008-1

Volume 14: **R.V. Gamkrelidze** (Ed.)
Analysis II
Convex Analysis and Approximation Theory
1990. VII, 255 pp. 21 figs. ISBN 3-540-18179-2

Volume 26: **S.M. Nikol'skii** (Ed.)
Analysis III
Spaces of Differentiable Functions
1991. VII, 221 pp. 22 figs. ISBN 3-540-51866

Volume 27: **V.G. Maz'ya, S.M. Nikol'skii** (Eds.)
Analysis IV
Linear and Boundary Integral Equations
1991. VII, 233 pp. 4 figs. ISBN 3-540-51997-1

Volume 19: **N. K. Nikol'skij** (Ed.)
Functional Analysis I
Linear Functional Analysis
1992. V, 283 pp. ISBN 3-540-50584-9

Volume 20: **A. L. Onishchik** (Ed.)
Lie Groups and Lie Algebras I
Foundations of Lie Theory.
Lie Transformation Groups
1993. VII, 235 pp. 4 tabs. ISBN 3-540-18697-2

Volume 41: **A. L. Onishchik, E.B. Vinberg** (Eds.)
Lie Groups and Lie Algebras III
Structure of Lie Groups and Lie Algebras
1994. V, 248 pp. 1 fig., 7 tabs. ISBN 3-540-54683-9

Volume 15: **V.P. Havin, N.K. Nikol'skij** (Eds.)
Commutative Harmonic Analysis I
General Survey. Classical Aspects
1991. IX, 268 pp. 1 fig. ISBN 3-540-18180-6

Volume 72: **V.P. Havin, N.K. Nikol'skij** (Eds.)
Commutative Harmonic Analysis III
Generalized Functions. Applications
1995. Approx. 260 pp. 34 figs. ISBN 3-540-57034-9

Volume 42: **V.P. Khavin, N.K. Nikol'skii** (Eds.)
Commutative Harmonic Analysis IV
Harmonic Analysis in R^n
1992. IX, 228 pp. ISBN 3-540-53379-6

Algebra

Volume 57: **A.I. Kostrikin, I.R. Shafarevich** (Eds.)
Algebra VI
Combinatorial and Asymptotic Methods of
Algebra. Nonassociative Structures
1995. VII, 287 pp. 4 figs. ISBN 3-540-54699-5

Volume 58: **A.N. Parshin, I.R. Shafarevich** (Eds.)
Algebra VII
Combinatorial Group Theory.
Applications to Geometry
1993. VII, 240 pp. 39 figs. ISBN 3-540-54700-2

Volume 73: **A.I. Kostrikin, I.R. Shafarevich** (Eds.)
Algebra VIII
Representations of Finite-Dimensional Algebras
1992. VI, 177 pp. 98 figs. ISBN 3-540-53732-5

Volume 77: **A.I. Kostrikin, I.R. Shafarevich** (Eds.)
Algebra IX
Finite Groups of Lie Type. Finite-Dimensional
Division Rings
1995. VII,287 pp. ISBN 3-540-57038-1

Topology

Volume 17: **A.V. Arkhangel'skii, L.S. Pontryagin** (Eds.)
General Topology I
Basic Concepts and Constructions.
Dimension Theory
1990. VII, 204 pp. 15 figs. ISBN 3-540-18178-4

Volume 51: **A.V. Arhangel'skii** (Ed.)
General Topology III
Paracompactness, Function Spaces,
Descriptive Theory
Due 1995. VII, 229 pp. ISBN 3-540-54698-7

Volume 12: **S. P. Novikov** (Ed.)
Topology I
1995. V, 319 pp. 78 figs. ISBN 3-540-17007-3

Springer

Tm.BA95.09.19

Springer-Verlag and the Environment

We at Springer-Verlag firmly believe that an international science publisher has a special obligation to the environment, and our corporate policies consistently reflect this conviction.

We also expect our business partners – paper mills, printers, packaging manufacturers, etc. – to commit themselves to using environmentally friendly materials and production processes.

The paper in this book is made from low- or no-chlorine pulp and is acid free, in conformance with international standards for paper permanency.

Printing: Saladruck, Berlin
Binding: Buchbinderei Lüderitz & Bauer, Berlin